GHANA IN SEARCH OF DEVELOPMENT

Dedicated to Enyonam, Elorm and Eyram

The Making of Modern Africa

Series Editors: Abebe Zegeye and John Higginson

Doctors and the State
The struggle for professional control in Zimbabwe
Dorothy Mutizwa-Mangiza

Class Formation and Civil Society: The Politics of Education in Africa
Patrick M. Boyle

The Social Services Crisis of the 1990's
Edited by Anna Kajumulo Tibaijuka

Oil and Fiscal Federalism in Nigeria
Augustine A. Ikein and Comfort Briggs-Anigboh

Environment, Health and Population Displacement
Andrew E. Collins

International Banking and Rural Development
Pade Badru

Structural Adjustment and Mass Poverty in Ghana
Kwabena Donkor

Contemporary Issues in Socio-economic Reform in Zambia
Edited by Herrick C. Mpuku and Ivan Zyuulu

The State and Organised Labour in Botswana
Monageng Mogalakwe

Ghana in Search of Development

The challenge of governance, economic management and institution building

DAN-BRIGHT S. DZORGBO

Routledge
Taylor & Francis Group
LONDON AND NEW YORK

First published 2001 by Ashgate Publishing

Reissued 2018 by Routledge
2 Park Square, Milton Park, Abingdon, Oxon OX14 4RN
711 Third Avenue, New York, NY 10017, USA

Routledge is an imprint of the Taylor & Francis Group, an informa business

A Library of Congress record exists under LC control number: 99075543

ISBN 13: 978-1-138-63745-0 (hbk)
ISBN 13: 978-1-315-20514-4 (ebk)

Contents

Preface

Since they attained political independence mainly in the late 1950s and the 1960s, many African countries have embarked on the project of promoting socioeconomic development and welfare but with mediocre results. In the last two or so decades, many countries in Sub-Saharan Africa have even stagnated or retrogressed. As a result of this, many of these countries since the 1980s, have come to accept the intervention of IMF and World Bank's economic policy packages, popularly known as structural adjustment programmes, as a means for promoting economic growth and development.

This book deals with this African development predicament. In particular, it focuses on the development experience of Ghana. Ghana achieved its independence in 1957 to become the first country in Sub-Saharan Africa to do away with colonialism. Ghana was favourably endowed in its economic and social life, and judged to have considerable head start over many countries in Sub-Saharan Africa at this time and consequently all predictions pointed to a bright future. But strangely, by the end of the 1970s, Ghana also became the first country in twentieth century Africa to have experienced a classic socioeconomic decline and is now struggling to survive. How and why did this happen?

This book tries to answer this question. It focuses mainly on the interplay between state politics and socioeconomic development. Drawing largely on the experiences of early European development and late development of the Newly Industrializing Countries in recent memory, the book develops a comparative-analytic theoretical framework that weaves together sociological, economic and political, as well as institutional perspectives, which suggests that many African countries have weak or lack the bureaucratic governing structures, macroeconomic management systems and institutions that conduce to economic growth and development. This, the book argues, derives mainly from the political conditions operating in contemporary Africa.

At the empirical level, the book provides a model, which suggests that Ghana's postcolonial development has suffered in several respects mainly as a result of the failure or inability of governing elites to develop "consensual" politics and a clearly specified long-term development objective that could be widely understood, accepted and have relevance for policy-making. The book argues that the inter-elite rivalry and ideological conflict that characterized the independence struggle continued to be played out in the postcolonial era, such that the political system was unable to ensure a smooth transition from one model of development to another. It further argues that military praetorianism and neopatrimonial domination of the state, with their attendant arbitrary rule and lack of public accountability and transparency in the ruling process have undermined the capacity of the state to develop and rationalize itself not only for the developmental tasks it assumed, but also, to provide a predictable and calculable socioeconomic and political environment that could sustain development.

This book maintains that a detailed historic-sociological analysis of Ghana's socioeconomic and political environment, and the political conditions that give rise or shape to such an environment, provides a promising point of departure from which we could investigate what failed Ghana. It thus critiques certain prevailing orthodoxies: modernization, dependency and neoliberal economic theories in development studies, for their inadequate attention to the state, institutional frameworks and the specific contexts within which development struggles are waged. It develops the thesis that, without some stability in the governing process, a rational-bureaucratic state, respect for property institutions, and predictability in the macroeconomic management systems, as well as calculability in the legal and administrative framework, long-term development cannot be sustained. Finally, the book contends that, *governance*, *economic management* and *institution building* are basic challenges facing Ghana in the search for development.

Acknowledgements

The scholarship enterprise is never the work of one person. Sir Isaac Newton is reputed to have said that the reason why he managed to see further was because he had stood on the shoulders of giants. At the final stages of this study, it is not easy to recall, let alone, repay one's intellectual debts. The authors on whose shoulders I have stood can easily be discerned from the footnotes and references.

Besides these giants, there are others who have helped me along the way. I am particularly grateful to Associate Professor Pablo Suarez who supervised this work, which was originally my doctoral dissertation. His intellectual support and interest in the study as well as his penetrating criticisms, comments and suggestions greatly helped me to sharpen my own thoughts and arguments. He was actively engaged in the study and always brought to my attention relevant literature that I had missed. But his contribution goes beyond academic support. He also supported me in diverse ways during the period this study was undertaken.

I am also very thankful to Professor Björn Eriksson who also critically read this work and offered valuable suggestions. Professor Tom R. Burns did not only read and gave useful advice, but also drew my attention to some relevant research findings of his own. Associate Professor Mekuria Bulcha, was very supportive of me, both academically and socially. Associate Professor Tekeste Negash and Dr. Mohammadrafi Mahmoodian brought some useful comments and suggestions, particularly, at the last stages of the study. I am very grateful to them all for their help.

My thanks also goes to the many colleagues and friends who created an encouraging and stimulating environment at the Sociology Department of Uppsala University where I studied. Similarly, I will like to express my gratitude to the Department for providing me financial and material support during my doctoral studies. I also thank the Scandinavian Institute for Africa Studies for providing a travel grant that enabled me to visit Ghana to collect data for the study. Last but not least, I want to thank my most loved ones, Enyonam and Elorm for their love and understanding as well as endurance in holding on to our dream. Without their patience and sacrifice, I could not have had the peace of mind to undertake this study.

The acknowledgement of help from others, however, does not confer a scholarly consensus on the views expressed in this study. Any shortcoming remains exclusively mine.

I hope that I managed to see further and not live in the shadows of giants. But the story of what failed Ghana and how Ghana can reclaim its development is not finished yet. It is often said that there are no dogmas in academic research, and that the only dogma is that no research finding is dogmatic. There would be fewer, greater pleasures, than to have others engage in the debate on Ghana's development problems. But as someone who has directly experienced much of the turbulent and sad postcolonial development history analyzed here, I hope I have managed, with all the objectivity I could summon, to not only illuminate in a small way some of the internal contradictions that militate against Ghana's socioeconomic progress, but also to outline, a feasible strategy for achieving a "better" order for Ghanaians.

Dan-Bright S. Dzorgbo
Uppsala University, Summer 2000

List of Abbreviations

7YDP	Seven Year Development Plan
31stDWM	31st December Women's Movement
ADC	Agricultural Development Corporation
AFRC	Armed Forces Revolutionary Council
APRP	All People's Republican Party
CDOs	Civil Defence Organisations
CDRs	Committees for the Defence of the Revolution
CECEC	Committee for Economic Cooperation with Eastern Countries
CECSC	Committee for Economic Cooperation with Socialist Countries
CEPS	Customs, Excise and Prevention Service
CFAO	Compagnie Francais D'Afrique Occidentale
CIA	Central Intelligence Agency
CIBA	Council of Independent Business Association
CMB	Cocoa Marketing Board
CPP	Convention People's Party
CVC	Citizens Vetting Committee
ECA	United Nations Economic Commission for Africa
ECLA	United Nations Economic Commission for Latin America
GCP	Ghana Congress Party
GDP	Gross Domestic Product
GIHOC	Ghana Industrial Holding Corporation
GIPC	Ghana Investment Promotion Centre
GNP	Gross National Product
GNPC	Ghana National Petroleum Company
GNTC	Ghana National Trading Company
GPRTU	Ghana Private Road and Transport Union
GTMC	Ghana Textile Manufacturing Company
IDC	Industrial Development Corporation
IFIs	International Financial Institutions
IMCs	Interim Management Committees
IMF	International Monetary Fund
INCC	Interim National Coordinating Committee

ITG	International Tobacco Ghana Limited
JCC	Joint Consultative Committee
MAP	Moslem Association Party
MNCs	Multi-National Corporations
NAC	National Agricultural Committee
NAL	National Alliance of Liberals
NATO	North Atlantic Treaty Organisation
NDC	National Democratic Congress
NEC	National Economic Committee
NGOs	Non-Governmental Organisations
NIB	National Investment Board
NICs	Newly Industrialising Countries
NIEO	New International Economic Order
NLC	National Liberation Council
NLM	National Liberation Movement
NPP	Northern People's Party
NRC	National Redemption Council
NUGS	National Union of Ghanaian Students
OAU	Organisation of African Unity
OFY	Operation Feed Yourself
OPEC	Organisation of Petroleum Producing Countries
PAMSCAD	Programme of Action to Mitigate the Social Cost of Adjustment
PAP	People's Action Party
PDA	Prevention Detention Act
PDCs	People's Defence Committees
PNDC	Provisional National Defence Council
PP	Progress Party
RWAFF	Royal West African Frontier Force
SAPs	Structural Adjustment Programmes
SCOA	Societe Commerciale Ouest Africaine
SFC	State Farms Corporation
SMC	Supreme Military Council
SOEs	State Owned Enterprises
SSNIT	State Security and National Insurance Trust
TC	Togoland Congress
TUC	Trade Union Congress
UAC	United African Company
UGCC	United Gold Coast Convention
UGFC	United Ghana Farmers Council
UNDP	United Nations Development Programme
UNICEF	United Nations Children's Fund

UNP	United Nationalist Party
UP	United Party
USAID	United States Agency for International Development
UTC	United Trading Company
VALCO	Volta Aluminum Company
WACB	West African Currency Board
WB	Workers' Brigade
WDCs	Workers' Defence Committees
YFL	Young Farmers' League

Map of Ghana

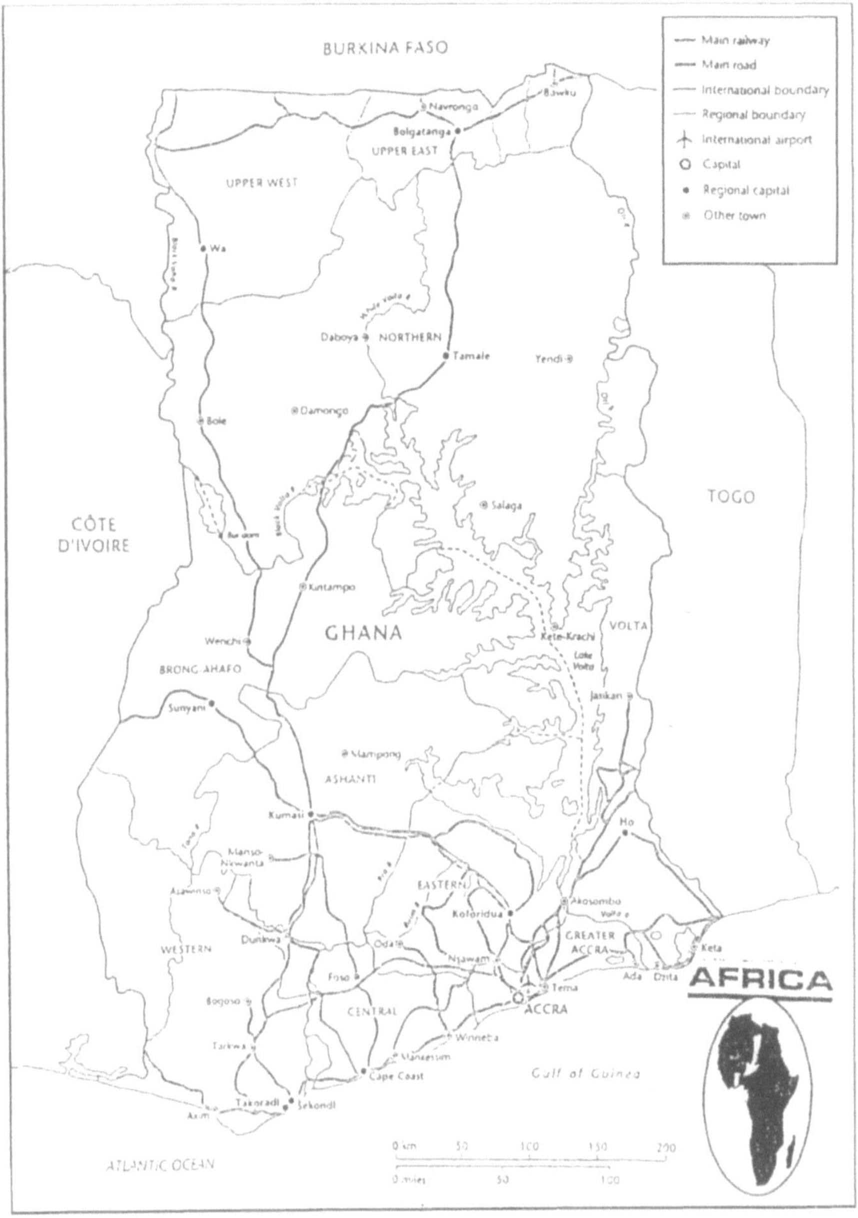

BURKINA FASO
Main railway
Main road
International boundary
Regional boundary
International airport
Capital
Regional capital
Other town
Navrongo
Bawku
Bolgatanga
UPPER EAST
UPPER WEST
Wa
Daboya
NORTHERN
Tamale
Yendi
Damongo
Bole
Salaga
TOGO
CÔTE D'IVOIRE
Kintampo
GHANA
Kete-Krachi
VOLTA
Wenchi
BRONG-AHAFO
Jasikan
Sunyani
Mampong
ASHANTI
Kumasi
Ho
EASTERN
Koforidua
Akosombo
GREATER ACCRA
Keta
WESTERN
Dunkwa
Oda
Nsawam
Foso
Ada
Tema
ACCRA
AFRICA
Bogoso
CENTRAL
Winneba
Tarkwa
Cape Coast
Gulf of Guinea
Takoradi
Sekondi
Axim
ATLANTIC OCEAN

1 The Paradox of Ghana's Postcolonial Development

Only governments can raise
or lower the levels of nations
— Voltaire

Introduction

This study inquires into how and why Ghana failed to promote and let alone sustain socioeconomic development in the postcolonial period. The decade from the mid-1970s to the mid-1980s represents a critical epoch in the annals of modern African socioeconomic development. During this era, especially the latter half, many countries in Sub-Saharan Africa,[1] experienced severe socioeconomic crisis that threatened to halt their viability as modern nation-states (World Bank, 1981, 1989a). This was particularly so in Ghana. In response to this "African crisis", a marked reorientation has emerged in Africa's development discourse. Since the early 1980s, neoliberal economic development ideology, mediated by "structural adjustment programmes" (hereafter SAPs) implemented in liberal democratic political frameworks, has been "imposed" on African countries by Western aid agencies as a means for rejuvenating their dilapidated economies. In Ghana, where these programmes have been consistently followed for the last 17 years or so, the results have been only modest and some analysts have warned that Ghana's so-called success is fragile and aid-driven. The current uncertainty about Ghana's development, even in the wake of SAP, brings to the fore the urgent need to determine what factors are militating against the promotion of socioeconomic development. This concern delimits the scope of the study.

In 1957, when Ghana, the then Gold Coast and once Britain's "model colony" (Austin, 1964:3), attained political independence (the first Black African country to do so), the general mood was that an irreversible process of socioeconomic and political development was about to unfold. The Ghanaian economy at this time no doubt bore the major hallmarks of the so-called developing countries: an agricultural economy with the production of a single primary commodity whose world price was subject to fluctuations, the extraction of few minerals for exports, limited manufacturing activities, high illiteracy and poverty, etc. Being a poor and relatively small country, Ghana also depended on international trade for the importation of much of the commercial and technological goods critical for socioeconomic development. It is in this respect that the early development economists who assessed Ghana's economy characterized it as "fragile" (cf. Kay, 1972).

Yet, if Ghana shared similar characteristics with other developing countries, it did possess some limited material and human resources, which according to Szereszewski (1966:56) placed it in a privileged position among such countries. In the 1950s, Ghana was the world's leading exporter of cocoa with an average output of 370,000 tonnes per annum. Even though cocoa production was the main export-oriented activity, accounting for as much as 50 percent of the Gross Domestic Product (GDP), it provided much national wealth. Indeed, the cocoa economy, mainly a peasant agriculture, brought improvement in living standards to the extent that before World War II, Ghanaian farmers in the cocoa-growing areas were judged to be better-off monetarily and also "immensely better-off" in welfare than the majority of peasants in South-Eastern Europe (Rimmer, 1992:2). By the criterion of per capita income, Ghana was a middle-income country, judged to be the richest or at least one of the richest countries in Sub-Saharan Africa (Killick, 1978:4-8; Ahene, 1992:43; Bequele, 1985). In the 1960s, Ghana's per capita income of $490 was approximately the same as that of Mexico and South Korea (Herbst, 1993:17; Shaw, 1993:62), although this does not mean that Ghana and these countries are similar in other respects. For Cote d'Ivoire, a neighbouring country comparable to Ghana in terms of natural resources, climate and physical environment as well as social structure (cf. Foster and Zolberg, 1971; Berg, 1971; Young, 1982:1; Rimmer, 1992), per capita income was only US$283 (cf. Morawetz, 1977). Ghana also inherited about £200 million as foreign reserves mainly from previous cocoa exports (Seidman, 1978:4; Shillington, 1992:8).

Furthermore, Ghana was reputed to have had more stock of educated and skilled manpower than any other country in Sub-Saharan Africa. According to Kirk-Greene (1986:30), Ghana had more schools and health services per capita and a better road system than any other British territory in Africa. In addition, Ghana boasted a prosperous emerging middle class and rural capitalists (Hill, 1963, 1970), a distinguished history of higher education, administrative elites and widespread respect for representative institutions (Austin, 1964:1-2; Sandbrook, 1986:2-3). The country produced and still produces about 10 percent of the world's gold, and substantial amounts of diamonds, bauxite and manganese, as well as tropical hardwoods. Ghana also is endowed with an abundant supply of arable land and a relatively favourable climate, which provide potential for further agricultural expansion. It is little wonder that Ghana's relative economic prosperity in this era attracted economic migrants from other African countries particularly Nigeria, Burkina Faso, Mali, and Togo (Amin, 1973:41-48).

On the basis of these material and human advantages, both the academics and politicians of the day shared the view that Ghana would develop into a coherent national unit and a "successful" modern economy and serve as a model for other African countries (Austin, 1964; Huq, 1989:1; Hyden, 1983:xi). But this was not to be. By the time Ghana could celebrate the silver jubilee of its independence, the most promising economy in Sub-Saharan Africa had atrophied, with all the conventional socioeconomic indicators of well-being pointing down. Three or four decades of postcolonial development have witnessed a brief spurt of socioeconomic progress followed by protracted economic and social stagnation and then decline. Between the mid-1970s and the mid-1980s, Ghanaians experienced what can be described as an extraordinary and self-reinforcing economic decline, characterized by negative growth, hyper-inflation, acute food shortage, massive unemployment, and other deplorable circumstances. In addition, much of the country's impressive features of modernity (e.g. transportation and communication networks, the education and health systems) attained in the first decade after independence deteriorated and finally collapsed.

Development experts and keen observers of the Ghanaian scene were quick to note that Ghana had "squandered" its assets. A wave of nostalgia for the colonial and immediate postcolonial period was discernible among the older generation of people while the younger ones were overwhelmed by a sense of hopelessness and despair about the future. Some warned about an impending social revolution. Ghanaians, especially the intellectuals who became fed up with economic hardship, opted for exit avenues and left for neighbouring countries such as Togo, Cote d'Ivoire and Nigeria in search of jobs and basic necessities of life, which were lacking in Ghana's shortage economy. Europe and North America became the destination for others. The irony is that Ghana, which was once the haven for Africans, was now "exporting" economic migrants.

Ghana's economic crisis seems to have no parallel in Africa. Ghanaians may be the ones to have led the way for Africa's march to independence, but, as Frimpong-Ansah (1991:3) contends, they were also the first to have suffered a classic case of economic stagnation and decline in modern Africa. Few scholars dispute this conclusion, even though what failed Ghana remains a bone of contention in both academic and political circles. In broader Third World comparative terms, only the decline of Argentina[2] in Latin America can approximate the Ghanaian experience. Ghana currently has a per capita annual income of $390, which is just about half that of Cote d'Ivoire, while South Korea has $9700 (*World Development Report* of 1997). At the time of this writing, South Korea has become one of the fastest growing countries in East Asia (White and Wade, 1985; Brian and Rachel Griffiths, 1987; Harris, 1987). By the mid-1980s, Cote d'Ivoire was considered by some analysts as one of the relatively "successful" countries in Africa (Crook, 1990; Marcussen and Torp, 1982; Lubeck, 1987), while Ghana became a low-income country struggling to survive (Hettne, 1990:24). The economic decline was so critical that some analysts wrote Ghana off as a hopeless case (cf. Herbst, 1993). In one World Bank assessment, the average Ghanaian was poorer in the 1980s than at the time of independence (World Bank, 1989a:1).

Given the paradox of postcolonial development presented above, it seems Ghana must begin afresh to create the conditions that will ensure socioeconomic development and improvement in the welfare of its people (Ninsin, 1989:1). While it is certainly an exaggeration to argue that Ghana has lost three or four decades of development, the contrast between the inherited socioeconomic promise and the languor of contemporary times, indicates that Ghana's socioeconomic performance has been unhealthy not only in relation to the broader Third World development experience, but also within the context of its historical experience (Bequele, 1985).

Since the early 1980s, a new development thinking has emerged in Ghana reflecting a consensus among Ghanaian policy-makers as well as international and bilateral aid agencies, the International Monetary Fund (IMF) and its sister institution, the World Bank,[3] that capitalist "economic reform" is a prerequisite for sustained economic growth and development. The IMF and the World Bank are now in the forefront "helping" Ghana to reform its economy with economic policy packages known as SAPs.

Basically, these programmes entail retrenchment in public sector employment; desubsidization of food and social services; the restoration of flexible foreign exchange regimes or the devaluation of local currencies; liberalization of external trade; increases in taxes to decrease the overall deficit and hold down consumption (demand management); increases in producer prices for export crops; higher interest rates, which dampen borrowing, deflates the economy and constrains imports; privatization of state-owned enterprises (Mosley and Toye, 1988; Nelson, 1990). In recent years, legal and administrative reforms to encourage private sector activity and foreign investment have been included. Underlying these policies is the assumption that economic growth and development can only be resumed if state-sector activity is limited while the private sector is promoted and the market liberalized (cf. Gibbon and Bangura, 1992).

Beginning in the early part of the 1990s, these international financial and aid agencies, as well as the governments of the dominant Western nations (e.g. the US, Britain, France, Canada, Germany, etc.) supporting economic reform programmes in Ghana and Africa overall, renewed their interest in the political dimension of development by adding yet another requirement to what needed to be done about the socioeconomic débâcle: "good governance" (World Bank, 1992). Accordingly, the causes of the crisis are not only related to the statist management of the economy, but also to the character of the political environment: the "failure of governance". As the World Bank cogently puts it, "underlying the litany of Africa's development problems is a crisis of governance" (World Bank, 1989a:60). This political concern, Lancaster (1993) argues, arose mainly from the failure of a decade-old SAPs to produce definite success in Africa. African governments are thus being encouraged to institute "good governance", mainly in the form of multiparty democracy, in addition to "economic reforms", as a precondition for further financial and development aid.[4] This political conditionality is premised on the belief that democratic governance and economic reform will reinforce each other and facilitate a stable process of development (Kohli, 1993; Shaw, 1993).

The efficacy of this development ideology in turning around the economies of Ghana and other African countries is, however, a subject of great controversy. The IMF and World Bank see the impact of their SAPs as positive, while the United Nations Economic Commission for Africa (ECA) sees them as negative and detrimental to Africa's long-term development.[5] Some scholars have also criticized the IMF/World Bank policies in Africa, especially for the social cost they entail for the urban unemployed, women and children, prompting UNICEF (1987) to call for "adjustment with a human face". Some critics also see SAPs as a violation of national sovereignty and the "imposition" of the neoliberal economic strategy and pressures for democratization as an attempt by the "triumphant" West in a post-Cold War era to disseminate a development paradigm that facilitates the internationalization of capitalism and western-style democracy. Even Africa's most ambitious and comprehensive "adjusting" country, Ghana, which has been internationally acclaimed as the most "successful" case of the IMF/World Bank's "solution" has been denounced by analysts who argue that the Ghanaian "success story" is nothing but a zealous attempt by the IMF/World Bank to claim that their free-market economic policies work. It is argued that Ghana relies too much on foreign aid; that the huge external debt Ghana is accumulating in recent years could be an obstacle to further development (cf. *West Africa*, 17-23 May, 1993); that many Ghanaians have not benefited from whatever improvements there are in the economy (Sandbrook, 1993:11); and that unemployment and poverty are out of control and Ghana is facing deindustrialization as a result of the elimination of protective measures (Fatton, 1992:123).

A more recent assessment (which revisits the World Bank-ECA debate) by Mosley and Weeks (1993) indicates that there is little evidence that Sub-Saharan Africa is in the throes of economic recovery. This suggests that the African development problem is more complicated than previously acknowledged and that there is something fundamentally wrong with policy prescription. Obviously, it is a reflection of the fact that existing development literature does not adequately enable us to understand the factors that are inhibiting development in African countries and consequently, the conditions or factors that actually drive development forward (McCarthy, 1994: xviii).

One major characteristic feature of development sociology or studies is its strong policy involvement. It is more explicitly normative than social science, in general. Its main implicit assumption is that the conditions of the so-called developing countries are unsatisfactory and ought to be changed. Development problems are usually defined as *national* problems, and development practice involves mapping out strategies that can bring about structural transformation in the cultural, political, social and economic spheres to solve collectively problems of human existence. Thus, Hettne contends that, in development studies, the distinction between development theory (the pursuit of academic or scientific knowledge) and development strategy or ideology (plans based on sets of ideas for action) often gets blurred. Devising strategies to overcome problems of underdevelopment necessarily implies an agent or institution, and usually it is the state (cf. Hettne, 1990:3). Consequently, the state becomes a crucial condition and factor for development. Yet, as one analyst argues, "too often, the role of the state [is] either taken for granted (in its planning and developmental responsibilities) or it [is] dismissed (as decadent or a creature of other class forces). Not uncommonly, the prevailing tone of much writing seemed concerned with allocating blame (internal and external) for developmental failure rather than exploring analytically the mechanisms and interplay of politics, policy and the state which brought about the generally dismal (although occasionally impressive) performances that were recorded" (Leftwich, 1983:92).

Many factors, both external and internal, have been offered to explain the apparent failure of development in Africa. On the external side, one can mention the colonial legacy, unequal exchange, unfavourable terms of trade and low world market prices for primary products, trade barriers and protectionists policies in the advanced countries, aid dependency, etc. Among the internal factors are hostile climatic conditions, low levels of savings and investment, limited markets, inappropriate policies, just to name a few. Alone, and in combination, these factors are serious obstacles to development, but in explaining why postcolonial development has so far failed in Ghana, this study focuses mainly on the interplay between state politics and socioeconomic development. It seeks to reveal the nature of the postcolonial state, its mode of interaction with society, and how politics puts severe limitations on the promotion of development. The study conceives the state as a crucial component in developmental process. Thus in somehow broader terms, this is a study of how states promote or fail to promote development.

The Problem: The State in the Context of Underdevelopment

In the social sciences, there is a certain pervasive but not easily acknowledged tension between theories and evidence about the world around us. This is most striking in development studies, where there is some abiding desire to formulate universally valid principles and formal models (based on stylized versions of the development history of the West) and a need to understand the great variety of actual development experiences and particular conditions that sustain economic growth and long-term development (Brohman, 1995; Carter, 1995).

In general, social scientists dealing, in particular, with Africa's development or the lack of it, have concentrated on economic issues and policies, trade regimes, and the negative impact of the external economic environment, and have increasingly paid less attention to the state's *role* in laying the institutional conditions that can sustain long-term development and also to its activating role in the development process. Recently, Hyden has drawn attention to what can be considered a "neglect" of the state from development studies, noting that, although the state is conceived as indispensable to development, analysts have often worked on the premise that the state is a developmental, or an effective public instrument for controlling and promoting development (Hyden, 1996).

Development theories, whether informed by bourgeois social science (i.e. modernization theory or neoliberal economic theory) or (neo)Marxism, (i.e. dependency theory) have treated the state respectively as a capable agency for development and as inconsequential component in the development process (cf. Hettne, 1988:2-3; Evans, 1992). Hence, how the state affects the African development process is not well understood and illuminated. Often, development is conceived of as an autonomous process, independent of state politics and institutional framework. In a recent review of development theories, Carter (1995:603) asserts that "the problem of 'the state', as it presents itself in the context of 'underdevelopment' has been undertheorized and little researched". Writing about the failure of development in Africa, Ergas also maintains that the African state "still remains something of a *terra incognita* for ... [many] analysts" (1987:2). Indeed, the conceptualization of development in

Africa has strayed away from a focus on the African state as an agent of development. The African state has thus become more or less like a "sacred cow" whose nature, anatomy and functioning, as well as its relation to developmental process, have not been explicitly explored in many scholarly works. Generally too, less attention has been devoted to internal problems that hamper the state from developing over time, the instrumental rationalities, capabilities and institutional framework critical for promoting development within particular national contexts.

This lacuna in development research means that many important issues that need to be addressed in many African countries to pave the way for development have remained obscured. Consequently, Africa is episodically subjected to all kinds of fashionable or hegemonic foreign ideologies and fads with the result that nothing seems to be working, often provoking doom-laden journalism and intellectual pessimism that there are no "quick fix", "easy answers" and "models of development" which can be applied to Africa (cf. Timbalake, 1988:174). If the African development problem has proved so intractable, it is being suggested here that existing development theories have not adequately explored analytically the interplay between African state politics and socioeconomic development.

Modernization Theory and the Problem of Political Order

At least since the end of World War II, developmentally-oriented social scientists have devoted much attention to studying how to promote development in the so-called Third World. The first major postwar theory about development in sociology is modernization theory, which tends to see development as an all-encompassing process involving not only economic but also cultural, psychological, social and political dimensions. Modernization theorists (predominantly American social scientists) were concerned with the explication of the problems of underdevelopment and the generation of prescriptions that could guide social action. Drawing largely on the experience of the "great transformation" that unfolded in nineteenth century Europe, which transformed predominantly agrarian and traditional societies into predominantly industrial or complex ones, modernization theorists provided a typological distinction of societies: "traditional" and "modern", and sought to demonstrate that there was a unilinear path of "development" that all societies were destined to traverse. Thus the dominant themes of modernization theory arose from established sociological traditions and involved the reinterpretation, often conscious, of the concerns (evolutionism, increasing differentiation, diffusionism, structural functionalism, etc.) of classical sociologists, viz: Comte and Spencer. To modernization theorists, modernity and development were often cognate terms.[6]

The dominance of modernization theory in development studies reflected the hegemony of "structural functionalism" in sociology in the 1960s and the Parsonian pattern-variables analytical framework of affectivity-affective neutrality, particularism-universalism, ascription-achievement, diffuseness-specificity and collectivity-self, which were held to describe the behavioural orientations and organizations that typify "traditional" and "modern" societies. From an evolutionary sociological perspective, modernization theorists advocated economic advancement to bring about the gradual replacement of affectivity by affective-neutrality, ascription as a standard by achievement, functional diffuseness by functional specificity, particularism by universalism, and collectivity by self (see Hoselitz, 1960).

At a more mundane level, modernization in economic terms meant industrialization and urbanization, as well as the technological transformation of traditional modes of production. Socially, it involved the expansion of education, the weakening of traditional ties and the ascendancy of achievement over ascription criteria as the basis of individual progress. In political terms, modernization was identified as the shifting of loyalties and orientations from the local peripheries to the national centres mediated through bureaucratic-legal domination, liberal democratic institutions and values. Culturally, it meant the increasing secularization of societies, predicated upon the spread of scientific knowledge and technology. Finally, in psychological terms, modernization was seen primarily as the acquisition of a *state of mind* by people, which included expectations of progress, the need for achievement and a readiness to adapt to change. Modernization was thus held as an irreversible and non-violent process through which the so-called Third World would, in Parsonian terms, upgrade their "adaptive capacities", and once this process began to unfold, it was supposed to bring liberal democracy in its wake (cf. Rostow, 1960; Almond and Coleman, 1960; McClelland, 1961; Lerner, 1964).

The main assumption of modernization theorists was that Africa, and indeed the Third World, was poised for a European-type social transformation. African societies were held as "traditional" and at the lowest stage of the evolutionary continuum but were to undergo an imminent, all-embracing social change. In other words, African societies, economies and polities would follow a relatively well-defined unilinear path of economic, social and political development similar to those of industrialized Euro-American societies, albeit with little outside help in the form of finance capital, technology and expertise combined with "social engineering" entailing the drawing of plans by development economists and experts to be implemented by developmentally-oriented elites. As Eisenstadt notes, "historically, modernization is a process of change towards those types of social economic and political systems that have developed in Western Europe and North America from the seventeenth to the nineteenth centuries" (Eisenstadt, 1966:1). Lerner (1964:iix) underlines its universalizing dimension by arguing that the "Western Model" is only historically Western; sociologically, it is global. Such an all-embracing conceptualization of development meant that modernization theorists had to draw on diverse disciplines in constructing the tenets of their theory. As such, the theory became an amalgam of propositions and hypotheses drawn from disciplines such as economics, sociology, political science and psychology. Given its multidisciplinary background, modernization theory emphasized different methods and concerns, provoking Harrison (1988) to suggest that it should be regarded as a *perspective* rather than a *theory*.

In the immediate postcolonial period, modernization theory provided the intellectual framework within which African governments sought to realize the goals of socioeconomic development and nation-building (Lancaster, 1988; Apter and Rosberg, 1994:5). But events in the Third World in the 1960s (the period in which most African countries got independence) shattered the optimism of easy economic and political development. Political instability, authoritarianism and revolutionary struggles have come to characterize many of the Third World countries. These events were seen as indicators that economic development had failed or was failing. The vision of economic development fostering liberal democracy was supposedly confounded by what was being seen as political instability and decay. As military coups, political and ethnic violence, as well as political irredentist movements became endemic, a view soon emerged that the process of economic and political development was not inexorable or easily understood. Nor, for that matter, was the direction of political change in developing countries predictable or necessarily productive of democracy or stability (O'Brien, 1979). These economic and political "disjunctures" made a mockery of the general assumptions of modernization theory and standard development economics.

Few modernization theorists, if any, had anticipated such events; hence the need to hurriedly re-examine earlier hypotheses and assumptions about economic development and political democracy in order to accommodate these new experiences. A new revision of modernization theory was offered, actually a self-criticism, suggesting that the early optimism was unwarranted and that authoritarian politics and elites, rather than liberal democratic politics, might be necessary for promoting development and managing the conflicts that accompanied socioeconomic transformation. In an often quoted book *Political Order in Changing Societies*,[7] which deals with the relationship between rapid social change and political development, Huntington maintains that just as there is an "economic gap" between the developing and the developed world so is there a "political gap": a situation, which results from the rapid mobilization of new groups into politics in an environment where political institutions were slowly developing. Quoting de Tocqueville, Huntington notes that "if men are to remain civilized or become so, the art of association together must grow and improve in the same ratio in which the equality of conditions is increased. The political instability in Asia, Africa, and Latin America derives from the failure to meet this condition: equality of political participation was growing much more rapidly than the art of association together" (Huntington, 1968:4-5).

Huntington's analysis points to the destabilizing effect of the transition from traditional to modern societies. Traditional and modern societies, he maintains, are stable, but transitional societies are unstable. As he argues, "it is not the absence of modernity but efforts to achieve it which produce political disorder" (1968:41). Violence and instability are, in large part, the product of rapid social change and the rapid mobilization of new groups into politics coupled with low development of political institutions (1968:4).

In Huntington's view, developmental objectives require substantial resources, and, since resources are scarce, governments must limit the kinds of demands made upon them so as to avoid the risk of losing legitimacy by failing to meet too many demands. The more governments lost their legitimacy and authority, the more likely it is that their power would be challenged by other rival elite groups who might not necessarily abide by rules that conduce to the orderly "circulation of elites" in a political system. The tone of Huntington's analysis suggests that the Hobbesian nightmare of "war of all against all" was haunting developing societies (O'Brien, 1979; Markovitz, 1977:7), and there was the need for political authoritarianism to create political order.[8]

This "incompatibility thesis" about the relationship between economic development and democracy was corroborated by Apter (1965) in his book *The Politics of Modernization*, which appeared three years before Huntington's. Whereas Huntington sees unfulfilled economic expectations and "political gap" as determinants of political decay, Apter sees the social and cultural environments as determinants of the character of politics. For Apter, the heterogeneous character of African societies forces their political systems to operate in fragile environments. Thus, strong states are necessary to manage the systemic strains inherent in the modernizing process. He distinguishes between a "reconciliation system", as characterizing western societies, and "mobilization system" as characterizing developing societies. He argues that mobilizing people and resources for development is crucial in the transition towards modernity, thus, initially political authoritarianism (which he pleads should be regarded as "predemocratic" as against "undemocratic") is the most appropriate means for overcoming societal heterogeneity, conflict and discontent, as well as for fostering national unity and consensus (see also Almond and Powell, 1966). In the late 1960s, modernization theorists no longer shared the optimism that the development process would be easy and conducive to democracy. Their concern was how political elites could establish "political order" or "effective" political institutions, which, according to O'Brien (1979) may or may not be formally democratic in character, for the purpose of accomplishing development.

Huntington and Apter's political "realism" was an attempt to rescue a foundering modernization theory. The new revision was elitist in nature, and, by implication it meant "farewell" to democracy. But the analysis leaves the impression that authoritarian elites would gradually liberalize politics as they develop capitalist economies, which would in turn facilitate the development of pluralistic polities (Carter, 1995:596).

In spite of this "internal" revision, modernization theory was still perceived as inadequate for addressing Third World development problems, mainly because it was ahistorical, Eurocentric and teleological (Chabal, 1992). It universalized Western institutions and values and linked them to progress, while denigrating those of the so-called Third World societies and tying them to stagnation and underdevelopment. The typological distinctions of society offered by modernization theory are largely fictional, as many of the supposedly beneficial traits of modern society are only rarely found in real world industrial societies, while the detrimental traits of traditional societies are equally difficult to find in developing countries. Most societies contain some combination of traditional and modern traits, a fact that is obscured by the mutual exclusivity of the ideal types in the modernization model (Brohman, 1995:125). It also lacked an adequate analysis of social relations, class structures, ethnicity and gender.

Furthermore, as Leftwich (1983:88) maintains, modernization theory "underplayed the role of politics and the state in promoting socioeconomic development", by assuming that all authoritarian regimes are "developmental states", which is actually not always the case. The developmental capability of the state was taken for granted and its actual functioning, as well as how its actors use state power in society has remained more or less a "black box".[9] Critical development issues remained untheorized. Since modernization theory failed to problematize the state in these terms, it also failed to recognize that some forms of authoritarian domination may not conduce to development. A corollary of this weakness is the failure to consider the possibility that social forces such as military rebellions and recurrent coups attending the state might undermine its ability to grow steadily to acquire legal-rational and bureaucratic capabilities for promoting development. Its "bourgeois" revision was a recipe for military take-overs and dictatorship.

Yet at this time, it was the neglect of history and global inequality among nations in the international capitalist system that would bring modernization theory down. For as a theory that is supposed to deal with development problems of countries that had suffered from colonialism, it astonishingly ignored analysis of historical antecedents and structural relations that exists among nations and global inequality that characterizes the international capitalist system. Hoogvelt's scathing sarcasm of Parsons's sociology, which informed the modernization theory, cannot be overemphasized here. As she notes:

> in Parsons' approach one gets the impression that the history of mankind has been one happy, relaxed and peaceful exchange of ideas, stimulating progress here, there and everywhere where contact between societies was made. Cultural diffusion appears as a friendly merchant traveller, a timeless Marco Polo, innocently roaming the world, gently picking up a few ideas in one place and harmlessly depositing them in another. Incredulously, the 'domination', 'exploitation', 'imperialism', and 'colonialism' are not discussed in any of Parsons' work on evolution (Hoogvelt, 1976:18).

In the end, modernization theory could not establish its relevance to the developing world because it was seen as a celebration of the achievements of the advanced industrial countries (Broham, 1995:125). Given the Cold-War context in which it was dominant, combined with its Parsonian neo-evolutionary twist, it was perceived as an attempt by social scientists, largely American, to establish a non-Marxist social science to contain Communism and advance American and Western hegemony. In reaction, came the radical critique from the Left, namely neo-Marxist theory of underdevelopment and dependency theory,[10] which takes greater cognizance of international linkages and emphasizes systemic inequality in the global economy focusing on sustained exploitation of the Third World areas by the more developed capitalist nations.

Dependency Theory and the Problem of Global Exploitation

Originally posited to address Latin American economic problems, *dependency theory,* was soon elaborated to account for underdevelopment throughout the Third World. In one of the well-known books on dependency theory, *Latin America: Underdevelopment Or Revolution,* Frank (1969, see also Frank, 1967) vehemently assails modernization theory as an intellectual smokescreen behind which lies the exploitation of the Third World by the First World. Frank dismisses the idea that underdevelopment was a stage or simply a lack of development that *could* be overcome by the injection of capital/technology and diffusion of Western values/attitudes, as some ECLA (United Nations Economic Commission for Latin America) scholars and modernization theorists have argued. On the contrary, Frank posits a "see-saw" theory, which argues that *under*development is a process through which economic development of the advanced countries (the centre/metropole) *causes* underdevelopment of the Third World (periphery/satellite). Development and underdevelopment, Frank maintains, are opposite faces of the same coin, and both development and underdevelopment are related products of a single but dialectically contradictory economic structure of capitalist development.

For Frank, as for others who work within this research tradition, the same historical process of expansion and development of capitalism throughout the world simultaneously generated and continues to generate economic affluence in the North and structural underdevelopment in the South (Frank, 1969:9). As such, a development theory and policy for the Third World that is not based on their past socioeconomic history that gave rise to their present underdevelopment is inadequate (Frank 1969:3; see also Amin, 1972). Frank's contribution was thus innovative, though some of the propositions prefigure in the neo-Marxist works of Baran ([1957]/1964), and, somehow, in the earlier works of theorists of imperialism (Hilferding, Lenin, Luxemburg, etc.), and also later, with modifications, in neo-Marxist analyses of Furtado (1964); Cardoso and Faletto (1979); dos Santos (1970,1973); Emmanuel (1972); Galtung (1971); Rodney (1972) and Amin (1974, 1976). It was, however, Wallerstein's (1974, 1979) world system theory, which distinguishes "core", "semi-periphery" and "periphery" nations, which marks the apogee of the development of the dependency perspective.

Since different scholars work in the dependency theory tradition and have, of course, assigned different weights to the several constituents properties of "dependent" or "peripheral status", emphasizing respectively trade, finance, ownership of productive assets, technology or ideology, and cultural imperialism, there have also been divergent views as to what, precisely, dependency theory is supposed to explain. The theory is thus not a monolithic whole, giving rise to Palma's (1978) proposition that dependency explanations should be regarded as a *perspective* rather than *theory*, as the case of modernization has been.

This caveat notwithstanding, three particular tenets are often held as constituting the corpus of dependency theory. These are: (i) the strategic causal factors for understanding the structure and dynamics of Third World social formations are related to their specific mode of insertion into the international economic system; (ii) the peripheral economies are not only deliberately *underdeveloped* from outside, but have also become *dependent* on markets, capital and technology of the metropole that a large proportion of their natural resources and annual incomes are siphoned off, which retards the growth of capital stock in these poor countries; and (iii) dominant elites in the peripheries are imperialism's Trojan horse or *comprador* class[11] (in Frank's analysis "lumpenbourgeoisie"), who connive with transnational corporations and International Financial Institutions (IFIs), such as the IMF and World Bank, to exploit their own countries because they benefit from such an association (Frank, 1969:388; Amin, 1976). What these propositions imply, as Booth (1985:52) notes, is that the development problems, the social structures and politics of the so-called developing countries, are to be understood primarily in terms of the particular nature of their insertion into the international capitalist system, rather than in terms of largely domestic considerations.

Thus, whereas modernization theory denies the possibility of conflicting rationality between the national and international spheres by arguing that integration of all national economies whether developed or developing into the international capitalist system is always beneficial for all countries (Bienefeld, 1982:2), dependency theorists see this "interdependence" as purely a zero-sum game which leads to the aggrandizement of the "core" and the impoverishment of the "periphery". The theory is sensitive to global inequalities and poverty that exist and continue to exist among nations. It emphasized asymmetrical relations and *unequal exchange* (see Emmanuel, 1972) among nations, and argues that peripheral Third World states are weak in the international states system; as such, they cannot develop within it. Dependency theorists thus concede that capitalist development is impossible in the Third World (cf. Palma, 1978:889) and have looked up to "progressive" Third World regimes to "delink" their countries from the world capitalist system and pursue socialist and *autocentric* development (Frank, 1967; Amin, 1976); a view which Frank (1991) has recently modified by noting that delinking from the capitalist world economy which he advocated in the 1960s is no longer a realistic policy and that socialism in Third World countries offers scarce alternatives.

Like modernization theory before it, dependency theory has been criticized on logical, theoretical and empirical grounds (see Smith, 1980; Booth, 1985; Roemer, 1981; Lall, 1975; Palma, 1981, Warren, 1980). Indeed, in the wake of the phenomenal economic development or economic differentiation of some countries of East and South Asia (Hong Kong, South Korea, Taiwan, Singapore, Indonesia) and of Latin America (Brazil) or the so-called Newly Industrializing Countries (NICs) of the Third World in the last two or three decades, some scholars have questioned the standard assumption of dependency theory that Third World countries cannot develop within the existing international capitalist system without first delinking from it. The theory has emphasized the uniform negative or stagnant effects of international capitalism and has not anticipated its recent dynamizing impulses in the Third World.

Consequently, auxiliary concepts such as *dependent development* have been introduced particularly in the Latin American debate to accommodate these new realities. Some dependency scholars now maintain that an alliance between the peripheral state, the local productive bourgeoisie and transnational corporations can promote development in the periphery. For example, Cardoso (1972) writes that *dependency, monopoly capitalism and development* are not contradictory terms: there occurs a kind of *dependent capitalist development* in the sectors of the Third World integrated into the new forms of monopolistic expansions. Cardoso thus acknowledges some possible capitalist development, which means that capitalist development is possible in the periphery via an alliance between the local bourgeoisie and foreign capital, though he adds a disclaimer that such development might benefit the international capital and governing elites more than it would the general population. Peter Evans (1979) also makes similar arguments about the alliance between multinationals, the state and local capital in the case of Brazil, thus in the Latin American debate an innovation has emerged in dependency theory.

In the 1970s, dependency theory was the orthodoxy of the Left in many African countries, and it was and is still considered in some circles as offering the "correct" and the most "progressive" intellectual framework for conceptualizing development problems. In Ghana, dependency theory has attracted many intellectuals and most political leaders have equally subscribed to its explanations of the country's economic collapse. Many of Nkrumah's neo-Marxist publications in the late 1950s and early 1960s have something to say about the negative effects of capitalism and imperialism, which share some affinity with dependency explanations. In addition, his political efforts in Ghana and on the African continent to keep imperialism at bay, his coinage of the term *neocolonialism* to denote a situation where the Third World state is formally independent but its economy is tied to that of the metropole, and thus "teleguided" from outside, all helped to provide fertile ground for dependency theory and "neocolonialism" explanations (Africa's version of dependency theory) to thrive in Ghana.

Neo-Marxist and dependency explanations have also had considerable influence over policies in Ghana. For example, the National Redemption Council/Supreme Military Council (NRC-SMC) regime of Col. Acheampong, which toppled the liberal democratic regime of Busia in 1972, sought to capture the "commanding heights" of the Ghanaian economy because of what it perceived as "neocolonial" domination. President Rawlings, at least in the earlier stage of his political career, maintained that his 31 December 1981 revolution was meant to cut off the stronghold that multinationals exerted on the Ghanaian economy (see *West Africa,* 28 June 1982). Consequently, between 1981 and 1982, his Provisional National Defence Council (PNDC) confiscated a considerable share of private assets in the form of bank accounts, houses, cars, and manufacturing industries as a means to dislodge society from the control and power of foreign capital and their local allies (defined as professionals, merchants, businesspeople, traders, etc.) and as a prelude to the socialist transformation that was supposed to end the exploitation of the poor by the rich (see Ahiakpor, 1985). The mainstream political discourse then sees Ghana's postcolonial decline as fundamentally caused by the workings of the global capitalist system, the machinations of Western imperialism and its key institutions, such as the IMF and the World Bank, and multinational corporations in a symbiotic relationship with the domestic ruling comprador class (cf. Price, 1984).

In the academic circles too, dependency theory and neocolonial explanations have also enjoyed some prestige. Indeed many of the political economy studies by Ghanaian scholars are Marxist or dependency-inspired and have emphasized the external factors and the injuries inflicted on the Ghanaian economy by foreign capital in collaboration with local comprador agents. In one recent study, Ninsin writes that the ruling groups or petty-bourgeoisie "have depended heavily on external economic interests... have achieved considerable success in [their] class project with the active connivance of external economic and political interests, the bureaucrats in the leading institutions of state and their counterparts in commerce" (1989:2). Similarly, Hutchful (1979:41) argues that, military coups and repeated changes in regime in Ghana represent a situation where different fractions—or alliances of fractions—of the ruling petty bourgeoisie have seized power in order to negotiate directly with imperialism over the place of the neo-colony and its internal classes in the international capitalist system of production, and to attempt to "mediate" the continuing crisis in the Ghanaian political economy.

In his conceptualization of the "petty-bourgeoisie", Hutchful notes that it is a "class" composed of the commercial stratum of merchants and petty-traders, owners of small enterprises in the service and extractive industries (transport, construction, motor repairs, etc.), small-scale manufacturers, an intelligentsia of professional men (lawyers, doctors, teachers, etc.), and the rural sector of capitalist cocoa farmers and rentier chiefs. This class, he further maintains, also includes those in the public sector: high civil servants, judges, magistrates, military and police officers, managers of state enterprises and specialists—in other words, the state bureaucracy as well as strategic political actors who mobilize and mediate between the interests inherent in the system (Hutchful, 1979:41-5).

When concepts subsume so many phenomena, they are likely to lose their explanatory power. By classifying all these social groups as one, the analysis portrays almost everybody as belonging to an exploiting class, which as this study will show, has not always been the case. To anticipate some objections, it is difficult to categorize cocoa farmers and the state as belonging to an "exploiting class" when the state's policies worked to the detriment of the cocoa economy, or to see the bureaucrats and those who dominated the state as sharing a common interest when state's policies brought the living standards of bureaucrats to a point where they were forced to seek greener pastures outside Ghana. The grip that neo-Marxist, dependency and neocolonial explanations have on politics and national intellectual discourse has compelled Price (1984:163-4) to remark that there exists "a near consensus about the essential causes of... Ghana's economic problems, thus rendering debate beside the point". This consensus, he contends, "focuses on a set of loosely-knit and widely-held 'neo-Marxist' notions of neo-colonialism". Such then was the state of the Ghanaian debate until the IMF and World Bank came to offer their own diagnosis of and solution to the problem.

Dependency theory has brought out a crucial factor— global exploitation— that modernization theory glossed over. In this respect, the theory has made a significant contribution, or what Hout (1993) calls progressive "problem shift", in the development discourse by problematizing external factors that impose limitations on the so-called developing countries in their struggle to overcome underdevelopment. But this study contends that dependency explanations also have some limitations. To begin with, excessive concern to understand how these "externalities" impinge negatively on the domestic arena has unjustifiably diverted attention away from consideration of internal factors that also affect development. It can also be argued that, however detrimental and distorting the effects of the metropoles, they also present opportunities for ideas, capital and technologies that can be utilized by the developing countries.

Furthermore, there is no reason to assume that indigenous dominant elites in business or the bourgeoisie would always be willing to connive in a subordinate position with foreign capital to exploit their own countries. Nor, for that matter, should one assume that governing elites of liberal orientation, will not work to lessen foreign economic influence. In addition, given the inability of socialist giants such as the former Soviet Union and China to disengage from the asymmetrically structured world system, it is doubtful that dependency theory's solution is a realistic option for smaller countries in Africa lacking technology, skills, and other resources at this early stage of their development. Furthermore, dependency theory has generally treated the state and domestic institutions as if they do not matter in development or simply as precipitates of international capitalist system. This is rather strange given the ubiquitousness and centrality of the state in the peripheral economies and the deliberateness with which socioeconomic development is supposed to be pursued in national contexts.

At the same time, the dependency explanations can be taken as implying that only global redistribution of wealth could bring about effective development in the Third World. While this would certainly be beneficial, to wait for such an action is to engage in utopianism in a world that as Gilpin (1987:25) argues, is governed more by "economic nationalism" than any ideology. This can also underrate the capacity of political agency to effect positive change in the social environment and to articulate a relevant strategy not only against the overpowering effects of dominant nations, but more importantly, to carry out successful socioeconomic development.

In Ghana where dependency theory has gained prominence in the social scientific and political discourse for sometime, there seems to be a certain tendency to virtually ignore "internal" factors or to "trim" reality to "fit" the theory. This tendency in research, Moore has long warned, "always carries the danger that research may overemphasize facts that fit theory beyond their importance in the history of countries" (Moore, 1968:xiii). Ghana's development discourse has become something like a "social philosophy" dealing with what "ought to be" rather than "what is" the reality of the development process. While the effects of predatory accumulation of global capitalism can be debilitating for developing countries, an undue "externalization" of the development problem can become a convenient apology for domestic policy failures and even serve as a smokescreen that camouflages political practices and governing systems that are inimical to development. Dependency explanations can create political and intellectual inertia that devalues the efforts and activities of the people, which could contribute to their own development. On the other hand, one can argue that it is precisely because the impact of international forces and development is uniform, homogenizing and less variable worldwide, that the state and other domestic factors also make a difference in development. We can thus not fully comprehend recent development crisis in most African countries without understanding how African states empirically function and manage the development process.

Economic Crises in the Three Worlds and the Resurgence of Neoliberalism

In recent years, the IMF and the World Bank, now providing loans to African governments, have provided neoliberal political economy analysis of the African development crisis, focusing the debates on the internal conditions in Africa. Yet still, the analyses seem unsatisfactory, mainly because they have remained largely general, Eurocentric, and not focused on understanding the nature of the state and how, specifically, it interacts with the wider society to promote socioeconomic development. This is mainly because the "African crisis" has occurred in an era of two major world crises. The response of dominant political leaders to these crises has had a backlash on Africa's development discourse in recent years. These crises are: (i) the collapse of state socialism in the former Soviet Union and Eastern Europe and the consequent demands for economic *perestroika* and political *glasnost*; and, (ii) the crisis of the Welfare State and Keynesianism in the developed western countries, which has propelled neo-conservative political elites into power (spawning "Thatcherism" in Britain and "Reaganomics" in the US) seeking policy means to dismantle the Welfare State and promote more free-market economies and liberal democracy (see Hettne, 1990).

This resurgence of the Right in political economy in the West has led to what Toye (1987) calls a "counter-revolution" in development theory and policy. The neoliberal "counter-revolutionaries" disparage all previous paradigms à la Frank, Amin and Wallerstein, and development economics that dominated the postcolonial development practice and favoured a strong role for the state in the national economy as "bad" sciences that got their assumptions and prescriptions wrong. In this regard, Lal (1983) maintains that the demise of *development economics* would be conducive to the health of both economics and developing economies. Development economists are now accused of polluting economics (Kay, 1993). Some analysts argue, though for other reasons, that we live in an historic epoch which is being conditioned by and oriented towards the operation of the capitalist economic order and that there is only one history (Fukuyama, 1992) for the evolution of societies: economic and political liberalism.[12] From this understanding, a wholesale attack has come on the African state in recent years and African policy-makers are being urged to concede a historical determinism that accepts free-market and liberal democracy as inevitable.

The IMF, the World Bank, and neoliberal economists maintain that Africa's economic crisis was basically self-inflicted through past *etatist* policies, the "distortion" of the market, artificial price fixing, subsidies, price controls and similar factors (World Bank, 1981). It is argued that an overextended state-sector activity (or *dirigisme* in Lal's terminology) has stifled entrepreneurial activities and the emergence of the bourgeoisie, and has also generated patron-client relationships through which political elites buy off political challenge from vociferous urban groups. Furthermore, governing elites are said to transfer rural resources to subsidize food and social services for themselves and urban dwellers as a strategy to moderate their rule. The "urban coalition" (as this unspoken pact is often called), the argument goes, was maintained at the expense of rural people whose surplus continued to underwrite the upkeep of the "parasitic urban dwellers" (see Bates, 1981, 1987). This "politics", it is argued, has undermined the productive capacities of the agricultural sector upon which African countries depend for foreign exchange necessary for supporting their national economies. The policy prescription from these analysts which underlies the structural adjustment programmes (SAPs) in Africa is that governments should roll back the arm of the state and let the market operate freely and within a democratic framework. The imperfections of the market are said to be better than those of the state. SAPs and western style political democracy have now attained the status of development ideology in many African countries.

The liberal political economy analysis, provides some new insights into the dynamics of African economies. Yet, if one adopts the "urban coalition" thesis of the liberal economists that the state extracts resources from the rural areas to spend on itself and those in the urban areas, the analysis fails to account for why the Ghanaian state pursued policies that severely hurt nearly all socioeconomic groups in the urban centres in the 1970s and early 1980s. Furthermore, the preoccupation by the neoliberals to roll back the state in Africa's socioeconomic processes seems to be sidestepping the crucial role that states play, particularly in the early or intermediate stages of development. While some students of African economic and political development assert that Africa's crisis has a political dimension (see for example, Callaghy, 1988:68; Ayittey, 1991), the neoliberal attack on the state is seen by some as tantamount to "throwing the baby out with the bath-water". Andrian Leftwich (1993, 1994), for example, has criticized the current IMF/World Bank's development ideology as naive because it seems to be based on a misreading of the actual development history of nineteenth and twentieth centuries, which rather reveals that the state has been main actor of development.

Although some scholars now perceive the neoliberal agenda in Africa as a "reaction that has gone too far" (see Killick, 1989) the general impression left by professional economists, as Skocpol (1985:41) observes, is that the state does not matter in economic transformation in the Third World. Indeed, the general perception today is that the state is no longer the principal agent of socioeconomic change but an obstacle to progress. Consequently, as Healey and Robinson (1992) note, the enthusiasm for its interventionist role in the economy has waned in recent years. But, as this study will show, the state is still an important agent in socioeconomic change, and the problem is with understanding how its mode of operations and intervention limit socioeconomic progress rather than doctrinaire subjection of the development process to purely market forces.

Understanding how the state limits socioeconomic progress provides a better position from which to offer useful prescriptions that could make its intervention in the social environment more positive and effective. Given that capitalism is the only viable—or rather fashionable—development ideology currently available, the state is still important for laying the basic rules, institutional context, and regulations for its development. This point has been recently reiterated by Leys and Berman (1994:8), when they observed that the "...development of capitalism... requires... a capable enabling state, and the difficulties of capitalist development [and this could be logically extended to mean broader socioeconomic development] in Africa are inseparable from the acute crisis of the Africa state". The task then is to revitalize the development discourse through a rigorous analysis of the specific ways through which the state does inhibit and can promote development.

The main development problem conceived in this study is not, as neoliberal economists would have it, that the state interferes badly in markets (although this counts to some extent), but rather, that the nature of state politics and the state's mode of interaction with the larger society in many African countries such as Ghana, has not only created an unpredictable and incalculable socioeconomic and political environment hostile to long-term development, but has also reduced the state's capability and effectiveness in managing socioeconomic processes. Given this, there is need to go beyond the debates of democracy, markets and authoritarianism to understand what forces and dynamics weaken the state's effectiveness and capability in socioeconomic processes.

This study proposes, following Evans, Rueschemeyer and Skocpol (1985) to "bring the state back in" the Ghanaian development discourse. While the nature of society, to a large extent, determines the character of state politics, on the one hand, and the state's politics also shape the nature of society on the other, this study examines how the state acts upon society through its governing structures, macroeconomic management systems and institutions, which provide the basic framework within which development takes place. This approach is particularly justified since in fairly new states, there are often political struggles over rules, institutions and development agendas, and also over who has the right and ability to make countless rules and decisions which provide a framework that guides the actions of human actors, be they individuals, groups, investors, corporations, organizations, business communities and others whose activities in a large measure (in addition to that of the state) determine the socioeconomic progress of society.

Understanding Ghana's Development Problem: An Alternate Approach

The theoretical problem being broached in this study is that both modernization and dependency theories as well as neoliberal development theory, which is being touted as an innovation created to address the key contradictions of previous development theories, have proved inadequate in addressing the root causes of the development problems that face many African countries, such as Ghana. This is mainly because theories of development have for too long been de-contextualized and de-historicized as a result of the quest for universal models and explanations. Rather than focusing on how people in different countries respond to a common body of universal problems, the tendency, as Hermassi (1980:13) maintains, has been "to fasten on an exclusive definition of reality—for some, the spread of modernity, for others, the emergence of a world system—so that the chance to grasp the special conditions under which relatively unique configurations of institutional arrangements will affect and give shape to these transformations has been forfeited". In searching for general laws and models or how the forest looks from a distance, analysts have sacrificed the variations in local contexts or have ignored how the grasses in the forest look. Variations in state effectiveness and domestic institutional arrangements in different national development contexts are ignored in the development studies. Although these are crucial issues in the promotion of development, they are often treated as "noises" that need to be filtered out to increase the parsimony of "grand" development theories (Brohman, 1995). Yet it is precisely these variations that account for the differential development performance of nations. Writing recently on this problem, Ake notes that:

> because the development paradigm largely ignored the specificity and historicity of African countries, it put them in a position in which everything was relevant to them and nothing was uniquely significant for understanding them. Hence the mounting anarchy of development studies and paradigms constructed for other purposes and other kinds of experience, meaningless for being incomplete and out of context, were applied in ways and for purposes that are not always clear and to reality that defy comparability (Ake, 1996:13).

Indeed, in recent years, analysts have bemoaned the institutional success of modernization, dependency and neoliberal economic theories and have pleaded for a re-conceptualization of underdevelopment problems to renounce the generality and big themes of development theory so as to bring analyses closer to reality or real life of people and societies (cf. Brohman, 1995; Hyden 1983; Leys and Berman, 1994:8-9; Leys, 1996).

They have urged in various ways that, in order to understand and explain processes of development, scholars need to become more sensitive to the complexities of various development experiences and to forge more appropriate analytical tools within development study for interpreting such experiences. Brohman (1995:123), for example, in his assessment of these theories, asserts that the development field currently faces three broad challenges: (i) to identify untenable concepts, categories, and theoretical propositions about Third World development; (ii) to trace the historical and ideological origins of these concepts and theoretical propositions; and (iii) to "unlearn" preconceived notions and prevailing discourses that prevent the understanding of the complexities of development in the various Third World areas.

Equally, in his recent book which "takes stock" of the contending development theories at the end of the twentieth century, Leys (1996), who has been involved in development theory debates over three decades, notes that the assumptions that underlined development theory since the 1950s are no longer valid, and, as such, this calls for more historical and more explicitly political theoretical effort. Similarly, Hume and Turner (1990:216-217) suggest that development theories ought to abandon their "conceptual and theoretical straight-jackets" in favour of approaches that reflect the diverse realities of Third World societies and that are focused on the practical problems of development. Other analysts emphasize the disillusionment and tensions created by Eurocentric theories and "top-down" orientation of development strategies in many so-called Third World countries and plead for the deconstruction or reconstruction of development paradigms to enable the incorporation of indigenous development thinking (ethnodevelopment) into development discourse and practice (cf. Brohman, 1995; Carter, 1995). What these disclaimers suggest is that the existing paradigms have reached a certain impasse that calls for innovations in the conceptualization of development problems.

Thus, at the theoretical level, this study de-emphasizes these foregoing approaches and seeks to develop historically-grounded conceptual tools and apparatuses that can shed light on how the state presents itself in the African, and particularly, Ghanaian development context. It is the assumption of this study that the state affects in no insignificant way the socioeconomic transformation in a given society. The assumption is not unjustified, because economies are politically embedded and this embeddedness is critical for the way they perform over time (see Jones, 1987; North, 1990). Development can be essentially conceived as the outcome of political decisions (Jackson and Rosberg, 1994:291). Generally, the modern state has taken on a variety of tasks ranging from the provision of social peace, rules and laws, to social welfare. It makes collective decisions and policies to organize and develop society; it creates, modifies and changes—purposefully or unintentionally—the institutional framework within which citizens and economic actors operate. It is mainly from the political realm that the basic ordering of society comes. The state, its policies, and the institutions it establishes are thus critical for domestic organization. To a large extent then, the state sets the parameters for society's development and also can determine its dependence on the outside world. Thus, to ignore the state in the development debate is to overlook a significant variable in the developmental process.

In general, state and political leaders are confined in one way or another by external circumstances. These circumstances can be very restrictive in the African context. But it would be wrong to suggest that state policies do not matter in development or governments do not have latitude to shape their social surroundings in a manner that facilitates development. To believe otherwise, as Jackson and Rosberg (1994:308) argue, is to subscribe to a determinism that not only eliminates political choice and possibilities that have been crucial to modern development everywhere, but also to deny that governments can turn difficult socioeconomic conditions around.

The emphasis of this study is on the domestic arena, and only secondarily is it concerned with external factors. Four main reasons account for this bias:

(i) as already noted, many of the scholarly works on Ghana's development originate from neo-Marxist tradition and turn to emphasize externally-induced problems thus a significant residual from the internal realm is unaccounted for;

(ii) the undue focus on "externalities" can lead to the glossing over of the limitations on development that originate from internal contradictions;

(iii) although external factors do limit socioeconomic development in Ghana, any solution to this dilemma will also require the mobilization of indigenous resources and social forces; and

(iv) if there is a favourable change in the international economic arrangements, Ghana might be unable to take full advantage of such positive changes if the internal contradictions to be pointed out in this study remain firmly in place.

In inquiring into what failed Ghana, the study provides fresh insight into the multifaceted processes through which the state has affected and still affects the development process. More specifically, it is hypothesized that inter-elite rivalry and its attendant political instability, the personalization or patrimonialization of the state, and arbitrary rule, have combined to undermined the capacity of the state to rationalize itself not only for the developmental task it assumed, but has also hindered the state from evolving a reasonable predictable and calculable socioeconomic and political environment, as well as administrative and legal institutions that could facilitate economic growth and development. The study develops the thesis that, without some degree of stability in the governing process, supported by a rational-legal state, trust and reciprocity in state-society relationship, respect for property and means of production, as well as calculability in the institutional framework of society, development cannot be sustained. Simply put, there can be no sustainable long-term development in a context that institutionalizes uncertainty, unpredictability and anarchy. In a nutshell, the study will argue that without a developmentally effective state and minimum "enabling conditions", conceptualized as *governance*, *economic management* and *institutions* undergirding a clearly specified long-term development objective that is widely understood and accepted and has relevance for policy-making, difficulties are likely to be encountered in the efforts to achieve development.

Definition of Key Concepts and Terms

As most social science concepts and terms are nebulous, evasive and sometimes controversial, it will be useful at this juncture to specify the ways in which the main concepts and terms: *development*, *state*, *governance*, *economic management* and *institutions* are used in the study and how they relate to the promotion of development.

The concept of *development,* like "progress", is value-laden, suffering from both ideological and normative freights. This is because the criteria for measuring development is not laid down by sociological theory but ideology, which in itself is contentious. For example, both Marxist and liberal scholars may diverge on what constitutes "development". The multi-disciplinary nature of development studies makes matters worse, since different disciplines have stressed different indicators for assessing development. Yet, over the last three or four decades, three main issues are discernible in the conceptualization of development. There has been a movement away from the narrow definitions stressing: (i) economic growth and increased per capita income (as was the case in earlier development economics) to (ii) a much broader social, equity and welfare dimensions of development, for example, education, health, (as development sociologists and the United Nations Development Programme (UNDP) stress) and (iii) in contemporary times, to a concern to protect the earth's finite resources so as not to jeopardize inter-generational equity. This last concern has come to be known as "sustainability of development" (see e.g. *Our*

Common Future: The Brundtland Commission Report of 1988) and has generated a number of "green movements" in contemporary times.

Though development can be a controversial concept, it is used here to mean economic and social development upon which many scholars, irrespective of ideological persuasions, are likely to agree. Economic development means the quantitative expansion of goods and services or wealth of society, which is often measurable by Gross National Product, per capita indices or expressed as economic growth. Social development is much broader concept which means that economic growth or wealth is converted into social facilities and services (e.g. health, education and other social and welfare facilities) that could enable societal members to enjoy a better quality of life. This means that a society can be experiencing economic development or growth, but this might not be necessarily expressed in the quality of life of the people (see Seers, 1979). Thus, whether social development will be achieved or not will depend on mainly state politics. *Development* and *socioeconomic development* are used in the study interchangeably. Economic growth is not the same as development, but it is *sine qua non* for the attainment of broader socioeconomic development.

However, development does not simply imply only conditions or goals that may be reached, as sometimes implied in early modernization theory (Rostow, 1960; Eisenstadt, 1966). Rather, it also denotes a strategy and a process, which also involves causal conditions and relationships that encourage sustained economic growth over years and upon which broader developmental goals depend. Many analysts will agree that in several African countries (where the livelihood of the majority hovers around the satisfaction of basic needs), development involves, as Leftwich (1983:83-84) notes, "all the methods and processes whereby the level and efficiency of the forces and means of production, their organization and control, can be raised or improved, and their products so distributed as to liberate human beings from vulnerability, hunger, poverty, disease and ignorance". In a more Marxian philosophical perspective, development is concerned with how to replace the domination of circumstances and chance over individuals by the domination of individuals over chance and circumstances.

Development therefore entails the organization of people, resources and technical innovation to address material problems of human living. This societal organization requires state politics, consequently, state and development are intimate. The importance of this process or conditions of development cannot be overemphasized, for, as Ingham (1993) argues, the failure of nations to understand development as a *process* and not only as a *goal* is one important reason for many difficulties they encounter in designing consistent and effective policies. This study sees development as increasing socioeconomic advance, but also stresses both the process and conditions that could make it happen.

The term *state*, according to Scruton (1996:528), derives its modern usage from Machiavelli, but it has no commonly acceptable definition. Surely, this is not the occasion to enter the complexity of definitions and theories of the state. Following Chazan, Mortimer, Ravenhill and Rothchild (1992:39), the term *state* is seen in this study as the organized aggregate of relatively permanent institutions of governance. According to these authors, the state can be seen as a set of associations and agencies claiming control over a defined territory and its population. The main components of the state are, consequently, decision-making structures, (executive, parties and parliaments) decision-enforcing institutions (bureaucrats, parastatal organizations and security forces) and decision-mediating bodies (primarily courts, tribunals and investigatory commissions). States can be said to have come into existence since human beings began to live together in society and had to follow some "commonsensical" laws to make group life possible. States have come in different shades (centralized and decentralized) depending on the cultural and environmental conditions of particular societies at specific points in time. Yet, in the modern sense the state is conceived of as a complex set of an institutional legal order with coercive faculties overarching all other organizations in a given society (cf. Poggi, 1987).

The definition of the state as an institutional legal order makes it possible to separate notions such as *government* commonly associated with it. The term *government* is often used differently in different contexts. It can also be used to denote the process of governing, i.e. the exercise of political power. It can also refer to the existence of that process, a condition of "ordered rule", or it can mean the manner, method, or system of governing in a society, or the structure and arrangement of offices and how they relate to those governed. Finally, it can be used in reference to the specific occupants of public office who are in position to make binding decisions and even change or modify the state at any given time.

While it is useful to keep these distinctions analytically separate, it needs to be noted that the terms *state* and *government* are sometimes used interchangeably. Also important to note is that, the state, as Post (1991:36) reminds us, is subject to a peculiar anthropomorphism, which causes it to be treated in intellectual and popular discourses as though it were an actual person directly acting upon society and individuals. This personification of the state suggests that the state as a constellation of institutions (apparatuses) regulates, directs and influences its surrounding environment and people. But the state is *not* an entity that does things as it often appears in discussion. Only *people* can do things. Obviously, it is the people who gain access to and acquire a monopoly over the state or the institutions of power that do things.

Implicit in liberal development discourse is the view that the state is a neutral institution and a benevolent arbiter of the various competing social forces in society and that it exists to promote the "common good": economic growth, efficiency and social welfare and for that matter the general development of society. This "consensus" and "functionalist" understanding of the state: that it serves common interest and it exists because it performs some function for society is, however, often contested.

In *The Communist Manifesto of 1848*, Marx and Engels (1973:110-111) proclaim that the executive of the modern state is "a committee for managing the common affairs of the whole bourgeoisie". In much of the Marxist-inspired analyses the state is often perceived as an agglomeration of public institutions that serve the interest of dominant social forces rooted in the structure of production. Such an "instrumentalist" view of the state often sees economic power as determining and conditioning political power in a capitalist society (Shaw, 1985).

Since the state is an instrument of domination and is associated primarily with the exercise of power (which implies access to material resources, privileges and prestige) there are unceasing debates about whose interest state power serves; how it serves it and how autonomous it is from society and vested interests, classes, etc. (see Poulantzas, 1973, 1975; Miliband, 1974; Offe and Ronge, 1988; Habermas, 1975). If the state is dominated by individuals, then the state as some discrete entity can never be autonomous or even have a qualified semi-autonomy (Post, 1991). The state is also seen as an instrument of domination by the bourgeois class, a domination that is mystified by what Gramsci (1971) conceptualizes as *hegemony*: the cultural leadership exercised by the ruling class, that is, their practice and defence of bourgeois ideas, symbols and values as the mainstream ideas of society. These are not idle debates because the state dominates all other myriad organizations in society and has the coercive faculty such as the military, as Weber (see Gerth and Wright Mills, 1977) recognizes, to back its rule and implement its decisions. This class-based model appears more realistic than the liberal functionalist view, because people who control the state are just as human with interests of their own.

But while the model provides a framework for analyzing the capitalist mode of production and its power structures, it might not always be useful in the analysis of the African state, which presides over societies that have not yet undergone a complete capitalist transformation whereby class forces could shape the nature and activities of the state (see Leys, 1996:166-7). For example, in many African countries, the state does not reflect the activities of the bourgeoisie or is not used necessarily to defend the economically dominant interests in society. Marenin's (1987) observation that the nature of the aggregation of the people with access to and control of the state is complex, especially in terms of their relations with productive forces in society might suggest a more cautionary approach to discussions on the Africa state situation than the doctrinaire acceptance of the thesis formulated by Marx and some of his followers.

Indeed, in many Africa contexts, where classes are weak and where the state's reproductive sustenance is assured by the control it exercises over the exports of primary commodities, minerals, petroleum oil, and other resources, as well as foreign aid, those who dominate the state often display some autonomy from otherwise important social classes such as the bourgeoisie, farmers and labour and turn to serve interests other than those of these groups and even that of the general society. This argument comes close to Marx's (1973) view formulated in the *Eighteenth Brumaire of Louis Bonaparte* that the weakness of class saturation favours an independent executive, not identified with any sectional interest and capable of developing an independent project of its own for the advancement of society. The nature of the state then can also be found, as Marenin (1987:61) notes, in the organizational processes by which contending groups manage to institutionalize and reproduce their wishes and interests and clothe them in the distinctive normative claim of the state, namely, that the state is the only agency in society that has an interest in serving the public good.

The debates on these issues are rather complex and beyond the scope of this study. And though scholars disagree on the necessary functions and on how much the state acts as an agent and to what extent it is autonomous of hegemonic groups, they nevertheless see the state as standing separate from society in order to do what needs to be done, those things that no social agency other than the state can do: to represent and defend the nation-state vis-à-vis other states; to establish the minimum of order and stability necessary for general security and routines of life; to act as arbiter of vested interests; to provide a relatively stable governing process, rational and reliable rule systems; to organize national resources and the people and to manage the public realm effectively (cf. Marenin, 1987), so as to ensure socioeconomic progress, be it capitalist, socialist or mix of these. The state must be able to manage the public realm effectively to raise the wealth of society, otherwise, it is of no use. But before the state can do this effectively, it needs to control its agencies and be rationalized into an effective bureaucratic agency.

Governance, or, more precisely, *good governance,* has recently appeared in the development literature expressing the concern about the nature of the political environment within which development takes place. Much has been said in recent years about the "failures" of the African states: that they are authoritarian; that they control society; that ruling elites concentrate their power; that the decision-making processes within them are opaque; that they extract resources from society but spend much of it on themselves and lack the capacity to spur development, just to list a few (cf. Bratton and Rothchild, 1992). In response to these criticisms has come the call for *good governance*. The concept of *good governance* as is currently used has two meanings.

The first, which is the World Bank's (1992) meaning, seeks to enhance the administrative and managerial capacity of the state to enable it to play a more effective role in development. The World Bank focuses on four areas in the administration of the public realm: (i) *accountability*, by which public officials are to be responsible for their actions; (ii) *a legal framework for development*, which ensures a system of rules and laws that provides clarity, predictability and stability for private economic actors, which are also impartially and fairly applied to all and which provide the basis for conflict resolution through an independent judicial system; (iii) *information*, by which is meant information about policies, budgets and government's intentions, are reliable, made public and accessible to all, since they are crucial for private sector calculation; and (iv) *transparency*, the call for open governments to limit corruption and enhance accountability and the relationship between the state and the private sector. In a nutshell, the World Bank is calling for efficient, independent, accountable and open public service. This is an apolitical governance and is, in several respects, Weberian (see Chapter 2).

The second meaning, associated more with western governments but endorsed by the IMF/World Bank because both share the same views on development policies, is political i.e. political restructuring: mainly the adoption of multiparty politics, the existence of freedom of speech and the rule of law as well as free press with the hope that some of the "debilitating" features of authoritarian African states can be minimized. Though democracy is consistent with material development and the world's most developed nations are "democratic", it is not clear whether development needs democracy.

There are some theoretical arguments in favour of democracy and development but the historical evidence suggests otherwise. A recent review of the evidence from successful "late development" since the mid-nineteenth century countries such as Germany, Japan, Taiwan, South Korea, etc. suggests that these countries made their greatest progress under conditions that cannot be characterized as democratic in the western sense (cf. Leftwich, 1994). This suggests that democracy is not a necessary condition for development, although it seems that once a certain threshold of development is reached, the pressure from below for democratic rule becomes irresistible. A temporary loss of democratic freedom may be a justified price for development, since effective development has often required that some quite radical steps are taken early in the developmental cycle, or, as Rostow (1971) argues, economic growth is a prerequisite for the establishment of participatory political systems. This might entail an effective bureaucratic state, or what O'Donnell (1979) calls "bureaucratic authoritarianism", which will not be necessarily democratic.

In this regard, some scholars see the pressure for democracy as not directed to the remodelling of a "developmental state",[13] but an attempt to curtail the state's activity and power in order to release the market from the restraints of statism, and this is driven mainly by the current hegemony of global neoliberalism (cf. Maclean, 1997:132). Given the possibility that democracy might not always go hand-in-hand with development, the issue of democracy and development in the good governance debate is muted in this study.

The World Bank position is not without merit given the so-called problems (see Chapter 2) of the African state in recent times. Surely, few would contest the proposition that good governance is an important ingredient of development. But *governance* can also mean stability in the governing process, cooperation in state-society relationship and the use of state politics in a publicly responsible manner to facilitate productive investment and activities. A climate of political stability is one way of ensuring predictability and social peace in the governing process and providing social actors with a stable environment and opportunities to enjoy the benefits that the resources of their society and their own efforts can bestow on them. But stability is not enough if it is not accompanied by the development of a relationship of trust and reciprocity between the state and society. At a more practical level, governance entails a system of rule through which governing elites, or to use Sklar's (1994:122) term, "managerial bourgeoisie", predominate and respond to the early stage of their country's development by using public institutions to organize the people and other economic actors for the purpose of increasing national productive capacity, and ultimately, promoting development. In other words, a system of governance that facilitates development is linked with the existence of a "developmental state".

Economic Management is a term that is more likely to occur in economics or business administration than in development studies. Here, it refers to the state's system of managing the national economy, i.e. the system of administration that defines how resources are extracted, transferred, distributed and used in a manner that continuously ensures the state's own reproduction and the mechanism for resource reallocation to various groups dependent on the state without endangering the productive infrastructure or system of the society. Economic management systems are closely tied to economic development policies and, invariably, to state development policies (Clapham, 1985). It also involves the existence of some system of public accountability that ensures the use of national resources, e.g. export earnings, foreign loans and grants, in a manner that is beneficial to the larger society.

Generally, economic management entails political control of resources and its manipulation to run the state and pursue collective goals of society. This is usually a *technical matter* requiring proper national bookkeeping, financial accounting and auditing systems, supervision and vigilance to avoid unnecessary waste and privatization of national resources. This technical and management issue is usually not raised in development studies. If raised at all, it is rather presented as merely strategies for economic development between which governments can choose, especially in so far as these can be related to "socialist" or "capitalist" development paths, instead of also reflecting choices that must be pragmatically made within the peculiar economic production context which governments are operating. The most important concern in national economic management is to ensure the reproduction capacity of the state and the pursuit of national economic viability. Thus, the alternatives of capitalist or socialist development strategies may be better regarded as different routes for attaining this common objective of government action, rather than as serving, in their own right, as the principal objective of government (Clapham, 1985:90 see also Hettne, 1990:45). Economic management in this sense, is raised in this study, to draw attention to the effective management of the public realm as a problem that confronts many African governments.

Institution is a central concept in sociology. At least since sociology was programmatically freed from social philosophy and since Durkheim's exhortation that sociologists study "social facts as things", as well as his claim (a view that he later modified) that sociology is pre-eminently "the science of institutions, their genesis and functioning" (Durkheim, [1895]/1982:45), the study of institutions and how they shape human action has been central to sociological analyses. This centrality of institutions to sociology has compelled one analyst to remark that "the theory of institutions is the sociological counterpart of the theory of competition in economics" (cf. Jepperson, 1991:143). Yet, for some time, social scientists particularly those in development studies, have overlooked the importance of *institutions* and *institutionalization* in the development process. This has led to the development process being conceived as independent of institutions and institutional context. The focus on institutions in this study is intended to bring their significance for human action, economic activities and development into the development debate. The key question then is what is an institution?

Within sociology, an *institution* is identified any time human activity, role, or practice attains a state of permanence, property, or becomes a norm; and *institutionalization* refers to the process of such attainment (see Berger and Luckman, [1966]/1991; 70-73). Thus, from the sociological perspective, social phenomena as diverse as cultural practices such as the handshake, voting, marriage and formal organizations such as the university, the presidency, etc.) are institutions. Sociologists group all these phenomena together as institutions because they are all "production systems", "enabling structures" and "performance scripts" (Jepperson, 1991:144-145). If, indeed, sociologists' *institutional* baggage contains simple things like handshakes and the university, then it is desirable to tighten the concept of *institutions* to render it useful for the study.

Institutions, in this study, are seen as societal rules that define and organize the complex actions and behaviours of social actors. Traditionally, such institutions arise from the cultural domain, which makes *institutions* sometimes coterminous with *norms*. In modern times, however, institutions are mainly the products of political decisions to regulate and order complex social life. These formal rules are the institutions addressed in this study. Modern rules are not given, but are human-created mechanisms to guide, facilitate and strengthen people's ability to work together in all types of organization. But to argue that rules order human action is not to accord them properties of permanence, fixity, or "consensus" as the claim is often made with regard to the Parsonian *institutional* approach that seeks to integrate the polity, the economy and society (Parsons, 1951). Societal rules are relatively enduring, but they are also nearly always being created and evolved by the state.

These institutions can also be regarded as rule systems (see Burns and Flam, 1987), in so far as they structure human action and provide framework for human actors (individuals, groups, organizations, businesses, communities and other collectivities) to engage in various activities that underlie economic development. Institutions also function to order society, to reduce the uncertainties and "transaction costs" (i.e. the cost of production and exchange) involved in human interaction and in the economy (North, 1990:3). They ensure predictability, continuity, and make future obligations possible. Thus, institutions evoke the imagery of the social production system. They may range from rules formulated in national constitutions, to decrees, economic policies, business and investment codes, property right rules, and written laws that regulate public offices and roles. Embedded in institutions is a double-edged property: institutions are facilitating and also constraining structures, now empowering social actors and now controlling them; rules impose cost or even punishment. Seen in this way institutions are the "rules of the game" of society (Wunsch and Oluwu, 1990:10) or the "nuts and bolts" that hold the societal machinery. Without them, social life cannot be possible.

Given the importance of institutions in a given social system, it can be conversely argued that a failure of the state (the most important institutional author in modern societies) to formulate and nurture rational rule systems (as a result of political upheavals and arbitrariness or internal wars) can only lead to uncertainty and unpredictability of the social life, and, consequently, undermine socioeconomic progress. *Deinstitutionalization,* i.e. the reverse of institutionalization, is detrimental to progress. Its effects can be manifold, for example, peasant-producers will withdraw from participating in the formal market, savings will dwindle in credit markets, wealth will be transferred abroad, high transaction cost will be introduced into the economy, the economy will attract little or no investment (both local and foreign), there will be an increase in rent-seeking and in speculative rather than long-term, productive activities. In such a context, stagnation and even decline must be expected (Wunsch and Owulu, 1990).

The nature of a people's *institutions* thus affects social actors, be they peasants/farmers, traders, foreign or local investors, businesses, or corporations and ultimately, the promotion of development. Because modern societies and economies are extraordinarily complex, producing an array of goods and services at different levels and places and under varying and changing circumstances, rules that encourage and facilitate a broad repertoire of interaction and organizational forms are essential if economic production is to be expanded and sustained. Thus in the quest for sustainable long-term development, institutions or rule systems that act as "commonsensical" framework for purposive social action (Cotterrell, 1992) cannot be overemphasized. Without the ability to make effective collective choices and to provide relatively enduring rational rule-systems, public goods and productive activities within a given social system cannot be elicited. When rules are predatory or changed arbitrarily, they constrain or undermine cooperation, exchange and long-term productive activities critical to sustained economic growth and development. Ideally therefore, institutions should be set up or reformed so that actors are led by their predominant interest to act in ways that further socially desirable ends.

The forgoing conceptualizations seek to draw attention to the importance of creating an enabling, nondistortionary, institutional environment for development. The state as seen in this study is not only a referee content in making and enforcing the rules for society; but also a creator of institutions, indeed many times the dominant creator of the economic rules and, consequently, the main determinant of the economic game. Each moment, the state and its agencies are creating and enforcing the rules of society, reshaping their contours and determining the institutional framework of society which invariably affect transaction costs, economic productions and outcomes. As the World Bank (1997) notes, when played well, the state's activities in this respect can catalyze development, and when played badly, can produce stagnation and, in the extreme, economic and social disintegration.

The state, then, is in a unique position: not only must it establish, through a social and political process, the formal rules by which all other organizations must abide; but as an organization itself, it, too, must abide by those rules. The state is thus intimately connected to development through its system of governance, economic management, and institutions. The foregoing terminological discussions can be put into more philosophical terms. Following Rawls's (1972) argument about the original state for constructing social justice, it is clear that, for the purpose of socioeconomic progress, governments need to devise governing and economic management systems as well as institutions for their societies, as if, one is embarking on a project to construct a city without knowing where one will be located after its construction.

Sources of Data and Research Method

Data for this study comes mainly from primary and secondary sources. The primary materials include government official publications, government statements and reports in newspapers and periodicals. Primary information was also gathered through interviews and observation during a three-month fieldwork in 1995. The fieldwork made possible a first hand assessment of the economic and political reform Ghana embarked upon during the last 16 years or so. It enabled the assessment of the impact of SAP on the cocoa economy and on the people. The secondary sources include previous scholarly works (books and articles) done on Ghana's past and contemporary development. Other secondary sources include academic and World Bank publications on Africa and Ghana that are deemed relevant to the study.

Socioeconomic statistics can be useful indicators of a country's development trajectory, but they can also be sterile if not buttressed with qualitative explanations as to how and why things went they way they did. This study is essentially a qualitative case-study that falls within the domain of historical and development sociology. The empirical material is presented chronologically rather than thematically, portraying the process of postcolonial development as it unfolded.

Talking about developmental failure necessarily implies some understanding of what constitutes developmental success. Gunder Frank has contended that economic development and underdevelopment are opposite sides of the same coin. Indirectly, Frank is implying that Western development was a "success" while that of the so-called developing countries was not. Indeed no civilization, culture, or society can establish its identity or progress without comparing itself to other civilizations, cultures, or societies, and the so-called advanced countries are no exception. Thus, lessons from "successful" developers can be crucial for understanding the situations of societies such as Ghana struggling to overcome underdevelopment.

Also, to speak of development is usually to speak of events and historically grounded processes and structures within specific nation-states. The nation-state is the main unit for promoting socioeconomic development and for scientific analysis pertaining to the well-being of people. In case-studies such as this, one can dwell upon socioeconomic and political processes internal to particular nation-states; conclusions reached may be tested for internal consistency or validity and indeed they may fit the empirical evidence assembled, but, as Jones (1987) contends, since there may be no generally accepted over-arching theory of long-term development or economic change, there would be no proper check on the chosen variables that the researcher holds as crucial for his or her task.

A common problem that the case-oriented researcher faces is how to develop procedures that can lead to a balanced assessment and validation of explanations and conclusions. Comparisons and contrasts with other civilizations are thus important in historically-oriented studies for assessment of the development experience of other societies. Otherwise, the conclusions would remain uncontrolled. Critical observation of *actual* development process of other societies can also yield insights into how to promote development in others.

Furthermore, such a method can illuminate how certain conditions may combine in different and sometimes contradictory ways in other environments to produce certain outcomes. At a methodological level, therefore, the theoretical framework for the study is historical and comparative analytic[14] identifying parallels, patterns of consistencies, contrasts and deviations that can illuminate the Ghanaian case. The development experiences of other societies are specified to check this study's conjectures, arguments and hypotheses and thus enhance its validity.

The Organization of the Study

The study is organized into eight chapters. Chapter 2 develops the theoretical framework for the study. Since it is suggested in this study that existing theories have proved inadequate for understanding developmental problems, a more relevant analytical framework is proposed. The attempt here is not to inaugurate a new theory as such, but to develop a network of concepts, propositions and understandings analytically derived from historical and contemporary development experiences of "successful" developed or developing countries, which can be used to address the issues that lie at the heart of this research. It is hoped that this approach will shed light on some of the development problems that have remained obscure from the vantage points of modernization, neo-Marxist and/or dependency and neoliberal perspectives.

Chapter 3 provides a historical backdrop against which the study assesses the postcolonial development experience. This chapter describes precolonial societies in terms of economic and political institutions, along with the economic and political transformation provoked by colonialism and the elite struggle for political independence. Chapter 4 begins the postcolonial development history and focuses on Nkrumah's rule (1951-1966), during which (particularly the later part) a radical attempt was made to transform Ghana into a modern socialist industrialized nation. This effort accounts for the economic problems and political tensions in the society which precipitated the military overthrow of the Nkrumah regime in 1966. Chapter 5 deals with the period from 1966 to 1969, when the military regime of the National Liberation Council (NLC) made attempts to reverse and even dismantle the so-called socialist order, and also tried to impose a free market economy and liberal democracy.

Chapter 6 examines the rule of the second civilian regime of Busia and his Progress Party (1969-1972), which followed the NLC. This regime sought to carry out a liberal economic programme but, was also ousted by the military in 1972. This chapter also deals with the rule of the new military regime: the National Redemption Council (NRC) and Supreme Military Council (SMC) of Colonel Acheampong, which attempted to reverse Busia's economic development policies by reintroducing a Nkrumah-type economic and political order. This regime was also overthrown by the military in 1979.

Chapter 7 addresses the rule of Flt. Lt. Rawlings: the Armed Forces Revolutionary Council (AFRC), a 112-day rule in 1979 that ended with a surrender of power to a civilian regime headed by Limann. The chapter also depicts the brief rule of Limann's People's National Party (PNP). This civilian regime was removed from power by Fl. Lt. Rawlings through yet another military coup in 1981. Chapter 7 also examines Rawlings' Provisional National Defence Council's (PNDC) rule from 1981 to 1983, when the country was supposed to have undergone a socialist revolutionary transformation; a transformation that was, however, abandoned in favour of capitalist economic reform via structural adjustment programmes. The final section of Chapter 7 updates the empirical evidence on the recent economic and political reforms. Finally, Chapter 8 provides a summary of the study and draws some theoretical and policy implications.

Notes

1 "Sub-Saharan Africa" refers to the African region south of the Sahara minus South Africa. Unless otherwise stated, Africa is used in this study to refer to this region, even though the United Nation Economic Commission for Africa, (ECA) refers to the whole continent as Africa.

2 See Tom R. Burns (1996), 'The Crooked Ways of Socio-economic Development: What Argentina Can Teach Us About Sustainable Dynamic Systems of Social Production', paper presented at Uppsala Theory Circle Seminar, Uppsala University, Sweden.

3 The World Bank and the IMF have become perhaps the greatest source of borrowing for African countries in recent times. Before the 1980s, these two institutions worked rather separately. The IMF dealt with balance of payment problems and stabilization policies, while the World Bank with development projects. In recent years, however, these two institutions are working hand-in-hand, with IMF programmes designed to work within the World Bank's framework of structural adjustment or economic reform programmes. It is upon this basis that the structural adjustment programmes are usually connected to the IMF and the World Bank. See Eriksen (1990).

4 The West's current aid imperative was articulated by Douglas Hurd, a former British Secretary of State and Commonwealth Affairs when he noted that governments which persist with repression, corrupt management, or with wasteful and discredited policies, should not expect the West to support their folly with scarce resources which could be used better elsewhere (see *West Africa*, 25 June-1 July, 1990:1077). In 1990, for instance, the Scandinavian countries joined the US in a crusade to cut off aid to Arap Moi's government of Kenya, and at the November 1991 meeting in Paris of the Consultative Group for Kenya, the country's aid donors established explicit political conditions for assistance, thus Kenya becoming a precedent for the rest of Africa. See Diamond (1994).

5 The World Bank and the ECA offer two major antagonistic diagnoses of the crisis and assessments of SAPs. The Bank (1981, 1989a) sees the causes of the African crisis as mainly self-inflicted, and the impact of the SAPs as more positive for countries that have implemented economic reform policies consistently than those which have not (see World Bank, 1989b). The ECA (1989a, 1989b), however, sees the causes mainly as externally-induced and the impact of the SAPs as negative. For a lively discussion of this controversy, see Partiff (1990).

6 While modernization and development may coincide, there is, however, an analytical difference between the two concepts. Modernization connotes a state of being current or up-to-date. Development, however, is always a value state which may or may not have been achieved in some other social context, and which may not even be achievable. Inevitably then, there would be disagreement on whether or not development is occurring (cf. Harrison, 1988:xiii).

7 Carter (1995) notes that, a survey of US professors in 1974 found that Huntington's (1968) *Political Order* was the book most often cited as among the "most important in the field". As Africa's problem is being seen as also related to the "failure of governance", political institutionalization and lack of political accountability, *Political Order* is once again is a basic reference.

8 Interestingly, on the issue of effective political order, Huntington invited policy-makers in Third World countries to take a look at the political stability of Leftist regimes and pointed to communist totalitarian governments in Moscow and Peking (and not Washington) where the lessons of establishing order could be learnt.

9 One exception to this general criticism is Riggs (1964), whose study focuses on the state and institutions of developing countries. For him, developing countries do have Western-type rational bureaucratic institutions but they are mainly forms lacking in substance. Borrowing the concept *prismatic* from physics, Riggs describes institutions in "traditional" societies as fused and undifferentiated, but those in the "modern societies" as diffracted and differentiated. According to him, in between these traditional and modern societies lie "transitional" or developing countries whose institutions are neither completely fused nor diffracted. Like a prism receiving white light from one source and diffracting it in the opposite direction, the institutions of developing countries, including the state, exhibit features of formal rationality, but these are a veneer covering deep-seated features of substantive rationality. Riggs is trying to ground a phenomenology of the internal logic of state politics in most developing countries that is inimical to development. But as O'Brien (1979:59) notes, his views were not followed because his writing was turgid and his analysis "heretical", not following the dominant orthodoxy of functionalism in that era.

10 Neo-Marxist *underdevelopment theory* argues that peripheral economies are deliberately *underdeveloped* from the outside, while *dependency theory* accepts this argument but adds that the peripheries have become *dependent* on markets, capital and technology of the metropoles and this condition limits capital accumulation in the poor countries and consequently their autonomous development. In this study, the focus is on dependency theory which is more comprehensive and more popular neo-Marxist development theory to date.

11 According to Paul Kennedy (1990:7), the term "comprador" was originally a Portuguese word used to describe the native representatives of Portuguese trading companies operating in various parts of Asia. By earning their profits through the supply of auxiliary services for foreign interests, comprador entrepreneurs facilitate their country's continued subordination to outside influences thereby perpetuating national economic dependency and underdevelopment.

12 This is the core of Fukuyama's "the end of history" thesis. The seemingly victory of liberalism and markets over socialism and central planning according to Fukuyama means that human history is moving in one direction: economic and political liberalism. This, according to the thesis, may constitute the end point of mankind's ideological evolution and also the inevitable end of the whole story of social development.

13 According to Leftwich (1994), the idea of the "developmental state" is associated with Friedrich List's classic critique of Quesnay and Smith, where he argued that the less advanced nations required "artificial means" to catch up with the advanced nations. In the development literature, the "undemocratic" states of Korea, Indonesia, Taiwan, China, etc. in the last three two or so decades are often conceived of as "developmental states" because of their ability to achieve socioeconomic development, at least measured by the Human Development Index (HDI): e.g. per capita, life expectancy and educational attainment. Some of these countries are now coming under pressure for democratization. Equally, it seems that the African conditions require "developmental states", because although minimum states championed by the World Bank will not do much harm, it is doubtful if such less activating states can lift African countries from their present economic predicament.

14 According to Ragin (1987:1), virtually all empirical social science researches involve comparison of some sort. But as he maintains, sociologists traditionally reserve the "comparative method" to the branch of social science that deals with cross-societal differences and similarities. The "comparative method" alluded to in this study, is not to be understood in this strict sense, but in the sense that, the theoretical framework of the study draws upon certain development experiences of other societies that can be analytically useful to understanding the Ghanaian case. In other words, Ghana's

development experience is not compared with any other country as implied by the comparative method.

2 Bringing the State Back In: State-Development Nexus

> In the birth of societies it is the leaders of the states who create the institutions; afterwards it is the institutions that shape the leaders— Montesquieu

Introduction

This chapter provides theoretical insight into the relationship between the state and development based mainly on the experiences of "early developers" of Western Europe, but also the "late developers" of some non-Western societies. Against this backdrop, the chapter examines the origin, nature and functioning of the African state and draws some parallels and contrasts that illuminate some of the problems militating against socioeconomic development in Africa, particularly Ghana, which provides the context for this study.

This chapter is divided into four major sections. The first focuses on Weber's historical and institutional theory of capitalist development in Western European societies. Weber's contribution is important for three reasons: (i) there is not yet any actual successful development that has not already implied "capitalism"; (ii) in general, in terms of development, after Europe, all other countries are followers; and (iii) capitalism is currently being "accepted" by many African leaders as a viable development alternative, hence, the early European capitalist development experience can provide a useful analogy for understanding some of the problems "late developers" such as Ghana presently encounter.

The second section discusses patrimonial political domination, which Weber considers as inimical to capitalist development. This section also examines the forces or rather conditions that historically shaped the emergence of the modern bureaucratic state, its internal development, and the consequent institutional complexes it initiated that facilitated modern capitalist development. The third section leaves the context of Europe and focuses on the role of the state in the socialist transformation of Europe and the development of the Newly Industrializing Countries (NICs). The final section brings the discussion into the African context. There, it is argued that, at this early stage of their development, most African countries experience a new form of patrimonial political domination or what is rather described as neopatrimonialism often reinforced by praetorianism (i.e. excessive military intervention in politics), which nurtures certain pathologies that undermine the evolution of the state into a rational-legal institution, and, consequently, its capacity and effectiveness to shape positively its social environment in a way that can be conducive to socioeconomic development.

Max Weber on the Genesis of Modern Capitalist Development

On the origin of modern capitalist development in the western world, Max Weber' view that Protestant ideas, particularly the Calvinist doctrine of predestination, provided the attitudinal and motivational forces for rationalized entrepreneurial capitalism is now legend in sociological discourse. In *The Protestant Ethic and the Spirit of Capitalism* (1930),[1] Weber asserts that the rationalization of economic attitudes—or modern capitalism—arose from the idea of predestination propagated by Protestants, particularly Calvinists. The idea of predestination in Calvinism holds that the salvation of an individual has already been determined, that is, God has predestined the fate of all human souls. Some are saved and some are damned; God has decided which is which. But Weber discovered a psychological contradiction in Calvinism. If God has already made his decision about who are his "favourites", why did Calvinists bother to fulfil his commandments? This doctrine should lead to religious nonchalance and its psychological contradiction should be unbearable for its adherents. But such was not necessarily the case, because according to Weber, Calvinism also teaches that one could know his or her spiritual position by observing signs of salvation in the real world, mainly by one's success in life. Simply stated, success in work or business is a sign of being selected by God. Conversely, a failure in life, was a sign of damnation.

This doctrine necessitated the most rigorous control of the believers' behaviour in all facets of life. Hard work which brings success, then turned into a vocation and a calling. One main consequence of this ethic is that it created a new positive attitude towards work and innovators who were calculating, daring and "above all temperate and reliable, shrewd and completely devoted to their business, with strictly bourgeois opinions and principles" (Weber, 1930:69). In this study, Weber demonstrates how Protestantism has changed previously negative views about capitalism into a positive one. The overall logic of Weber's thesis, is that, Calvinism unintentionally, or, even paradoxically, created the "spirit of capitalism". The "spirit of capitalism" is thus the secularization of the "Protestant ethic".

The "ethic thesis" is, however, a subject of great debate (see Coleman 1990:6-10; Robertson, 1959; Tawney, 1959). In spite of this, some analysts have considered it Weber's distinctive contribution to the origin of capitalist development and have extended it into development studies (see McClelland 1961; Lerner 1964 on modernization theory).

If Weber's contribution at the level of religious ideas is accepted, then it leads to the conclusion that societies without a Calvinist-Protestant ethic are incapable of realizing capitalist development. This would certainly fly in the face of the actual development experience of Japan and late twentieth century development of other non-Protestant countries such as South Korea, Taiwan, Singapore, Indonesia, and now China. It will equally fly in the face of some ethnic groups such as the Ibos in Nigeria who have exhibited extraordinary pro-business attitudes suggesting that a "functional equivalence" to, or something much more than the Protestant ethic might be at work and that the "ethic thesis" does not explain everything about the origin of modern capitalist development.

Weber, however, to his credit, does not concede capitalist development impossible in non-Western societies, as dependency theorists have sometimes led us to believe. No doubt Weber argues that modern capitalism was first fully-fledged in the Occident mainly due to the psychological impetus provided by Protestant ideas, but he also maintains in his later works that capitalist development was made possible mainly because the bureaucratic-legal state supplies a particular institutional framework that conduces to it; a significant shift of position, but relatively unknown, probably because of the popularity of the "ethic thesis" and its use to reinforce the West's sense of cultural supremacy.

This later position which Collins (1980) refers to as Weber's "last theory of capitalism", provides the basis for the development of the conceptual argument in this study. This theory is found mainly in *General Economic History* ([1927]/1981), but also somehow discernible in *Economy and Society* ([1968]/1978). These two books are so voluminous and encyclopaedically organized that to grasp the "theory" one needs to systematically piece together the disparate ideas. This "mature" theory is both historical and institutional in perspective. In it, the "spirit" elements that constitute the core of the *Protestant Ethic* are accorded a small role in the overall picture of the rise of modern capitalism.

It is important to note that Weber is not the only one who deals with the question of capitalism; Marx also studies it. Marx's main pre-occupation is to unravel the structure and mode of functioning of capitalist societies and to determine the key elements that shape their dynamism. Marx draws attention to the negative consequences of capitalism and argues that it deserves to be supplanted by socialism. Marx, however, focuses more on the *dynamics* of capitalist production system and its effects than on its *origin*. In this respect, instead of conceiving these sociological giants as in disagreement, we should see them as complementing each other.[2] That is, Weber leaves off where Marx begins. Indeed, in *General Economic History,* which forms the main foundation of his last theory on capitalist development, Weber avoids the types of statements in *The Protestant Thesis* that are often seen by critics as anti-Marxian, and, rather, devotes attention to many of the issues that are dear to Marx and Engels.

General Economic History begins with a discussion of primitive agrarian communism. Later in the book, there is mention of private appropriation of the means of production and the proletarianization of workers so that they are compelled to sell their labour to the capitalists. In the closing chapter, there is the reference to the threat of working class revolution once capitalism matures and work discipline loses its religious legitimation. Weber's attitude to Marx's ideas, as Collins (1980:937) maintains, is critically respectful. Despite the points of convergence, class conflict is central in Marx's conceptualization of the *development* of capitalism, while, for Weber, other factors, such as the state and the institutions the state creates are, much more important and class conflicts are not necessarily or exclusively the most important conflicts in society (see also Callaghy, 1988:71).

The key issue Weber addresses in his "institutional theory" is how the state aided, although not always deliberately, the emergence of capitalism in the western world. Like most Weberian categories, Weber uses *capitalism* as an analytic concept and makes it abundantly clear that, in one form or the other, it has existed in a variety of historical time and space, for example, as far back as ancient Babylon (Collins, 1980). Yet Weber maintains that, for an economy to be considered as capitalist, the provision of human needs must be in the hands of private businesses seeking profit, and this provision must be organized on a sufficiently large scale that "if we imagine [it] taken away the whole system must collapse" (Weber, 1981:275-8).

Weber also sees capitalism as present wherever the industrial provision for the needs of a human group is carried out by the continuous *rational* activity (or based on rational calculation) of private entrepreneurial organization (1981:275). In other words, productive activity needs to be geared mainly towards exchange and profit, which, in Marx's analysis, would mean that commodities must acquire *exchange value* as distinct from *use value*. The main idea stressed by Weber's definition of capitalism is *rationality* in production and exchange undergirded by proper and methodical accounting and book-keeping with the main purpose of making profit or money. Weber does not restrict *rationality* to economic activity alone, but regards the whole evolution of western societies as a rationalizing process through which the state itself becomes rationalized and is in turn dominant.[3]

The concept of *rationality* is, therefore, applied in the description of the organization of societies. Weber distinguishes two main types of *rationality*: *formal* and *substantive*. *Formal rationality* refers to the adoption of means-end calculus (based on abstract principles, regulations, laws and rules applied universally) to social action and organization (Weber, [1968]/1978). In modern societies, where social actions and institutional orders are dominated by formal rationality, decisions are made without regard to personal qualities of individuals or any arbitrarily applied standard of criterion or reference. The bureaucratic organization of modern societies appears in Weber's work as the extraordinary manifestation of formal rationality. *Substantive rationality*, on the other hand, refers to considerations to ethical values and social justice, and, as Cohen (1981:xxx) notes, *substantive rationality* directly orders patterns of action in relation to constellation of past, present, or future "value postulates". The scope of these patterns of action is indeterminate. *Substantive rationality* does not order action on purely means-ends calculus. As such, from the perspective of *substantive rationality,* the calculations of the capitalist or bureaucrat are conceived as "irrational" (see Cohen, 1981). At the risk of generalization, one can say that modern societies are dominated by *formal rationality* while traditional societies by *substantive rationality*.

From the definition of capitalism Weber offers, early modern Europe and contemporary Africa are not capitalist. In many African countries today labour is free, but it is not compelled to enter the labour-market, market exchange and rational monetary systems are burdened by political restraints, private appropriation by entrepreneurs of the means of production and the use of rational machine technologies in the productive process are only embryonic. Indeed, one can imagine the "capitalist" enclaves that exist in Africa today as being taken away without the whole system collapsing.

In this respect, modern capitalist development in Africa remains largely potential. A question that readily comes to mind is what can be done to pave the way for capitalist development? This question is particularly pertinent in present times, where capitalism is being accepted as the most viable route to development. Writing about the wave of economic and political liberalism apparently sweeping across Africa in recent times, Sklar (1994:123), tells us that "the spirit of capitalism in Africa has now emerged from a long shadow of ideological distrust, attributable mainly to its historic association with colonial rule. Its growth surely will invigorate the African bourgeoisie and shape its nature as African societies begin to free themselves from oppressive statism". It is here that the state's role in capitalist development to which Weber gives considerable sociological attention, becomes relevant.

Since, for Weber, capitalism has existed in different forms in contradistinction to the modern capitalism based on rational permanent enterprise, he uses different terms to label these earlier forms: "commercial capitalism", "traditional capitalism", "political capitalism", "booty capitalism", "adventurer's capitalism" and "patrimonial capitalism". Weber, however, does not consider these forms of capitalism as "progressive" because they are parasitic and their reproduction is not based on principles of formal rationality and methodical productive activity. Weber recognizes that the above-mentioned obstacles to capitalism and its various "unprogressive" manifestations were very much part of early European civilization. But what interests Weber is how, in the history of the West, these obstacles and forms of capitalism came to be sublimated by modern rationalized capitalism. In other words, he seeks to understand how formal or instrumental rationality underlying modern capitalist development came to usurp the motivating cultural force of nearly all substantive ethical values in society.

In *General Economic History*, Weber no longer seeks explanations for capitalism from the ethos of ideas, but from the institutional framework that the modern state provides. The result of this endeavour is the "historical and institutional" theory of capitalist development; a broader and much sophisticated framework than its "younger" idealist thesis. This theory presents a sequence of causal conditions entailing six elements conducive to the emergence of large scale capitalism (Weber, 1981:275-78). They are re-arranged more logically here:

(i) formally free labour, but which is compelled to sell its labour;
(ii) market exchange not burdened by irrational restraints or political arbitrariness;
(iii) private appropriation by entrepreneurs of all the means of production;
(iv) rational monetary system and the commercialization of economic life;
(v) rational machine technologies for production; and
(vi) calculable law, both in adjudication and in public administration.

In Weber's scheme, the last factor of a calculable legal system is regarded as an intermediate condition supported by an economic ethic combining universal commercialisation with moderate pursuit of repetitive gain. This condition, which allows the other elements to predominate, is fostered mainly by the rational-legal bureaucratic state and by legal citizenship (a factor that Weber seems to have exaggerated because capitalism can develop without it), and, more remotely, by a complex of administrative, military and religious factors.[4]

Weber grounds a conceptual argument that on the surface, seems to fall within liberal political economy presented by Adam Smith, David Ricardo, John Stuart Mill, and others. But it is inaccurate to see Weber's contribution just as "old wine in a new bottle". Whereas (neo)liberal economists stress the unbridled operation of the market as the main agent propelling capitalist development, Weber sees the rise of the modern bureaucratic state as a crucial precondition for the institutional underpinnings of modern capitalism. In his commentary on Weber's theory, Collins (1980:931-2) contends that the state is the factor to which Weber gave the greatest attention, but it is the most overlooked in his theory of capitalism.

For Weber, it is the state that provides the rational administrative and political framework, or what Weber himself calls the adjudicative and calculable law (Weber, 1981:277), within which these elements came to dominate and which subsequently made possible long-term modern capitalist development. The key term Weber employs to conceptually capture these conditions is *calculability*. It is the state, as Collins (1980) painstakingly argues in his systematization of Weber's "last theory of capitalism", which provides the historical key to all the institutional factors conducive to modern capitalist development. The main aspect of the state that is germane to modern capitalism is the development of a governing structure in the form of the provision of rational political order underpinned by calculable and adjudicative rule-systems or institutions that accord citizens personal security, respect for their private property and predictable political framework for social action. Weber also mentions geographical conditions as more favourable to capitalism in Europe than in China or India but gives them little significance in his scheme.

On the other hand, Weber maintains that modern capitalism does not necessarily originate from extensive trade and accumulation. This position challenges some Marxist-inspired literature, particularly Wallerstein's (1979) view that the generalized world system of exchange and production is the main factor responsible for modern capitalism. If this position is right, then it means that, for African countries trade with the outside world alone would never be sufficient to spur capitalist development. There are additional factors that, from Weber's perspective, must originate from the political realm. For Weber, it is mainly the institutions that the bureaucratic state provides that affect the outcomes of capitalist development. In this regard, Collins (1980:932) maintains that "one may even argue that the bureaucratic state was the proximate cause of the impulse to rationalization, generally—above all, via the late seventeenth—and eighteenth-century spirit of enlightened absolutism, which set the state for the industrial revolution" (see also Ira Cohen, 1981:lxxii).

These "contextual variables" that the state provides are the key mediating factors that enable the application of "instrumental rationality" to every aspect of modern capitalist production, which also brings in its wake further subjection of other spheres of life, including what Habermas (1987) terms, "the lifeworld" to instrumental rationality or at least to make them compatible with it. In this regard, one can regard modern capitalist development and indeed development in general as a rationalization process in which the state assumes a crucial role in shaping through public policies its surrounding social environment in a manner that makes things calculable and predictable for economic actors. On the other hand, the very process of societal development entails, among other things, finding the most optimal and effective ways of connecting given ends to means, i.e. progressive rationalization of all aspects of human endeavours.

Weber's explanations of the origin of modern capitalism in the West, which Collins calls a theory, is not without problems. First, it neglects external factors, such as colonial plunder and exploitation of resources from abroad, which dependency theorists argue, have also helped to fuel Europe's capitalist and industrial development. Furthermore, it is an unrealistic view of any economy that has ever existed. But following Collins, this account should be treated as an ideal type, which should help distil the salient features of the state-development nexus in national contexts that is the orientation of this study.

Max Weber on the Patrimonial State as Obstacle to Capitalist Development

The discussion so far has focused on mainly factors that Weber identifies as crucial to capitalist development. But Weber is also concerned with the identification of another factor, namely patrimonial political domination, which can militate against such a propensity. In order to understand how patrimonial politics inhibits capitalist development, it will be useful to provide Weber's categorization of legitimate authority patterns as a background. In *Economy and Society* ([1968]/1978:212-299), Weber presents a typology of legitimate political domination in society, viz: (i) the rational-legal authority; (ii) charismatic authority; and (iii) patrimonial/traditional authority and discusses their characteristic features and how the logics of charismatic, and patrimonial political domination in particular, can be obstacles to modern capitalist development.

To begin with, Weber maintains that the organization of the modern state is premised on the logic of rational-legal authority. The nature of this authority is that individuals in public office perform their roles and exercise power in accordance with legally defined hierarchical structure of roles and authority directed towards a publicly acknowledged goal. These individuals' public roles are clearly differentiated from their private ones, a distinction normally encapsulated in the idea of "office" (Clapham, 1985). Those occupying public offices behave strictly, at least in principle, according to the official rules and treat these offices as a public domain. Equally, individuals who come into contact with these offices are treated impersonally. Outside the offices, however, the officials are private persons with private roles. Public offices and their organization are highly routinized within a framework of structured formal rules that do not budge to individual idiosyncrasies (Habermas, 1987). Because of this public-private dichotomy, the appeal to universal principles and the efficiency that characterizes its operation, the rational-legal authority that underlies all modern forms of organization is accepted as morally right, and, consequently, its power is legitimated (Gerth and C. Wright Mills, 1977).

Bureaucracy is inevitable in modern societies because it provides the most efficient mechanism for coping with the requirements of large-scale or modern organization. This is because as an institutionalized pattern of procedures, bureaucracy provides rational rules, certainty, calculability, and predictability that are critical for organizational efficiency. An effective state intervention in socioeconomic process requires that the modern state be organized as a bureaucratic *entity* or legal institutional order. As Rueschemeyer and Evans (1985:50-51) note in their contribution to understanding the role of the state in modern development, the "existence of an extensive, internally coherent bureaucratic machinery is the first prerequisite for effective state action... An effective bureaucratic machinery is the key to the state's capacity to intervene".

The second type of political domination is charismatic authority, which is best crystallized in the activities of messianic religious leaders. In charismatic domination, the led attribute special or supernatural qualities to their leader, who legitimizes his rule by virtue of claims to possess such qualities. Unlike rational-legal authority, charismatic authority is not a routinized phenomenon anchored in social and political order. Charismatic domination flourishes best in *anomic* social situations, such as the immediate postcolonial African context, when traditional authorities were rapidly declining in the wake of modernization, or when traditions were "dissolving" and societies were experiencing systemic crises. From this perspective, it is often argued that it is circumstances rather than the individual's innate qualities that create charismatic leadership (Clapham, 1985:47-7). But whatever its sources and pretensions, charismatic domination does not conduce to long-term development because, as a personalized political attribute, it is ephemeral, often collapsing with the death of its holder. Moreover, situations giving rise to charismatic domination may change, or the charisma itself may wane. For charismatic authority to survive the test of time, it must either be routinized into bureaucratic authority or be allowed to "regress" into traditional authority, or become a mix of both.

The third in the typology of authority is patrimonial domination, which is found in feudal Europe during the era of dynastic princes (see below) and is also thought to be found in some precapitalist societies. The word patrimony denotes authority of a father over a son and the inheritances thereof; and a patrimonial leader's ideal, according to Weber (1978:1107), is to be regarded as the "father of the people". The salient feature of patrimonialism, in contrast to rational-legal authority, is that its authority is vested in the person rather than the office, and, in contrast to charismatic authority, it is routinized into social and political order. As Weber notes, patrimonial office lacks above all the bureaucratic separation of the "private" and the "official" spheres. Political administration is treated as a purely personal affair of the ruler and political power is considered part of his personal property. His exercise of power is entirely discretionary. Thus absent in patrimonial political systems are the binding norms and regulations of the bureaucratic administration (cf. Weber, 1978:1028-30).

In a political system held together by patrimonial logic, those lower on the political hierarchy are not subordinates, in the sense of officials with defined powers and functions of their own, but rather vassals and retainers whose position depends on the leader to whom they owe allegiance (Clapham, 1985). In his discussion of the positions of officials in a typical patrimonial state, Weber maintains that "in contrast to bureaucracy ... the position of the patrimonial official derives from his purely personal submission to the ruler ..." (1978:1030). "Patrimonial officialdom may develop bureaucratic features of functional division of labour and rationalization, especially with the expansion of clerical tasks and of authority levels through which official business must pass. But the genuinely patrimonial office differs sociologically from the bureaucratic one ...", because of the fusion of the "private" and "official" spheres (1978:1028). Patrimonial officials treat their office as "a *personal* right and not, as in bureaucratic state, from *impersonal* interests ..." (1978:1029; emphasis original). The ruler is at one and at the same time the source of most laws and above their influence. Weber stresses that, since rulers in patrimonial states tend to create non-formal types of law and are not bound by an "abstract formalism of legal certainty", they may refuse to be bound by formal rules, even those they have made themselves.

Furthermore, Weber argues that, in all political matters, it is the ruler's personal discretion that delimits the jurisdiction of his officials. Thus, when the distinction between public office and private life is fused, and allowed to dominate the state and its institutions, which are constructed on rational-legal principles, not only is efficiency compromised, but the state might fail to rationalize itself and become an obstacle to socioeconomic development.

These three modes of political domination are undoubtedly ideal types, and, in reality, do not exist in such pure forms or in watertight compartments. They must be treated as heuristic devices against which to measure real political systems. For example, the developed Western societies and their bureaucratic institutions that appear in Weber's account as rational are not always entirely dominated by formal rationality. Western societies and political systems are not as rational as the typology of bureaucratic domination suggests, as evident in the episodic incidents of privatization of public office (e.g. corruption of political and bureaucratic officials). Also, elements of charismatic authority often co-exist with rational-legal authority. Furthermore, it is possible to find patrimonial rulers legitimizing their rule by complex and shifting blends of charismatic and legal-rational doctrines (Clapham, 1985). But when all is said and done, a political system dominated largely by patrimonialism is seen by Weber as anathema to modern capitalist development because:

> the patrimonial state lacks the political and procedural *predictability,* indispensable for capitalist development, which is provided by rational rules of modern bureaucratic administration. Instead we find unpredictability and inconsistency on the part of the court and local officials, and variously benevolence and disfavour on the part of the ruler and his servants. It is quite possible that a private individual, by skilfully taking advantage of given circumstances and of personal relations, obtains a privileged position which offers him nearly unlimited acquisitive opportunities. But a capitalist economic system is obviously greatly handicapped by these factors, for the individual variants of capitalism have a differential sensitivity toward such unpredictable

> factors. Wholesale trade can tolerate them easily, relatively speaking, and adapt itself to all changing conditions...
>
> It is different with *industrial* capitalism... [which] must be able to count on the continuity, trustworthiness and objectivity of the legal order, and on the rational, predictable functioning of legal and administrative agencies. Otherwise those guarantees of predictability are absent that are indispensable for the large industrial enterprise. They are especially weak in patrimonial states... (Weber, 1978:1095; italics original).

Weber, however, made it clear that patrimonialism is compatible with many different types of economic structure, and, thus, the relationship between patrimonialism and capitalism is not deterministic. His argument is that patrimonialism is inimical to capitalist development because the patrimonial ruler is usually arbitrary and does not provide the necessary institutional complexes and predictability for sustaining capitalist development. As Weber comments, "...arbitrariness affect very deeply the developmental opportunities of capitalism. Either the ruler himself or his officials seize upon the new chances of acquisition, monopolize them and thus deprive the capital formation of the private economy of its sustenance, or the ubiquitous resistance of traditionalism is reinforced by them so as to hinder economic innovations that might endanger the social equilibrium" (1978:1094).

As an illustration of the inability of patrimonial mercantilism to generate capitalist development, Weber points out that "neither the period of Colbert nor of Frederick and Peter succeeded in turning their countries into industrial states" and he gives the political reason (they were others of course) as arbitrariness of patrimonial rulership arguing that "the economically interested groups behind the royal monopolies and the industries which were imported, founded or protected" were often members of the royal family and courtiers, who consumed the resources generated by such statist efforts (1978:1098-1099). Weber also maintains that "in the interest of his domination, the patrimonial ruler must oppose ... the economic independence of the bourgeoisie" and ultimately any autonomous action on the part of the subjects is suspected as hostility to authority (1978:1107). This political phenomenon also receives critical analysis from Marx. In *The Eighteenth Brumaire of Louis Bonaparte*, Marx condemns patrimonialism associated with the rule of Louis Bonaparte, who enriched himself and his associates at the expense of a disorganized peasantry.

Today, however, such patrimonial politics have somehow disappeared from the Western world. Nevertheless, it is important in this study because as it will be argued and illustrated empirically, in the case of Ghana, postcolonial African politics exhibits many features that resemble those associated with patrimonialism, a phenomenon that impinges negatively on the promotion of development. But, if even the European state was once patrimonial, how and why did it become not only bureaucratic but also a "positive" force in the modern capitalist transformation as Weber recognizes? Answering this question can reveal how the state and politics can stunt or aid socioeconomic development.

The Rise of the Modern State and the Development of Capitalism: A Brief Historical Outline[5]

The rise of modern bureaucratic state in many European societies is often attributed mainly to political imperatives of political domination (war and interstate competition externally and domination and extraction internally) and the state formation processes and struggles which they generated (cf. Callaghy, 1988:73; Poggi 1978). Following the collapse of the Roman Empire in the fifth century A.D., no successful over-arching theocratic federation ever emerged in Europe again. Rather from about this period to somewhere in the fourteenth century, Europe "regressed" politically to be governed by multiple feudal-patrimonial centres. This system was later supplanted by a single system of states dominated by princely or absolutist rule in which change in one cell affected the others. This change is crucial to the emergence of the modern-state and its role in laying the institutional foundations for modern capitalist development.

Certainly, as Jones (1987:105) correctly observes, there is no *a priori* reason why a state system alone should be responsible for modern capitalist development. An empire could equally do the same job—perhaps even better—since it could control more resources and generate greater economies of scale beyond the capabilities of a disarticulated state system. But an empire lacking internal tensions can retard modernity, and "cultural homogeneity" could also make for political autocracy. China, for example has had a relatively homogeneous culture since 200 B.C. and did not undergo the Western-type economic transformation until now. Whatever cultural homogeneity existed in Europe was undermined by the demise of the Roman Empire.

In the absence of an imperial centre, the system of rule relied, both for order-keeping and for the enforcement of rights, on self-activated coercion exercized by a small, privileged class of warriors, the nobility and retainers who dominated the feudal system. The feudal system was one in which lords controlled the land and allowed peasants (serfs) to work it so as to extract rents to support their aristocratic indulgence. The peasants, however, often did not have security of tenure. The lords shared the surplus from their estates with large numbers of retainers, who constituted their private armies. Consequently, the lords were very keen to control rents and defend their respective interests against other lords.

The imperatives of "self-defence" soon led to anarchy whereby feudal lords competed fiercely with one another for rents. According to Hirschman (1977:101), the lords made war according to their own discretion, almost continually upon one another, and very frequently upon the king; and the open country was a scene of violence and rapine disorder. The feudal anarchy came to pose a nuisance to the king because of the clash between kingly interest in taxes and the nobility's preoccupation to control rents. In order to ensure social peace in the kingdom, the king monopolized all the faculties of rule. He obliged the lords to settle their disputes in his court, and also provided protection and justice for the peasants.

The long interlude of rumbling political contest that ensued between the monarch and the nobility led to the development of absolutist-monarchical rule. The gradual centralization of power enabled the king to overawe self-seeking political centres making life "solitary", "poor", "nasty", "brutish" and "short". Writing about this era, which lasted up to the beginning of the Renaissance, Hobbes provided a royal pedagogy that Reason alone was insufficient for regulation, as each man was driven by his selfish interest seeking honour, wealth and glory, a situation that was likely to degenerate into "war of all against all". His solution: the King should have absolute rule, a *Leviathan* to be able to achieve order. The absolutist state was, thus, with few exceptions, to preside over not only the destruction of feudalism but also the establishment of the nation.

From the period of the Renaissance to the time of the French Revolution—almost three centuries—Western Europe was a theatre of rule by princely-dynastic state. In this period, the state was not an expression of the people, the bourgeoisie, or the proletariat, but merely of those who controlled it and its confederates. We are thus confronted with a peculiar political system of a feudal nature based on personal rule that Weber describes as patrimonial in nature, whose functioning he conceives as an obstacle to modern capitalist development. In their recent contribution to understanding the political economy of the princely state, Jackson and Rosberg (1994:292) refer to a simple statement from Ludwig von Mises (supposedly written in 1919), that "the political conception of the princely state is the interest of the ruler", and argue that it summed up the view that dominated West European political economy from somewhere in the fourteenth century to the French Revolution and was also prominent in Central and Eastern Europe until the end of World War I. This view, it must also be added here, lay at the root causes of World War II, and, somehow, albeit mildly, underlined postwar Cold-War politics. In the state-society relationship, land and people were nothing more than objects of princely ownership and demands. They were the patrimonialized property of the ruler. The interests of the state and ruler were assumed to be identical. This is manifested in the famous expression: "L'etat, c'est moi", (the state is me) attributed to Louis VIX of France (1665-1690), (Jackson and Rosberg, 1994:293). Other examples of absolutist rulers include King Frederick William (1917-40), and Frederick the Great (1740-86) of Prussia.

The Bureaucratization of the State and the Promotion of Development

The personalized use to which princely rule was principally put, however, led to protracted struggles towards the making of nation-states. Each prince began to wage war to incorporate new peoples or nationalities, or at least to resist submission to others. In his seminal sociological analysis of the development of the modern state, Poggi (1978:60) observes that "the strengthening of territorial rule and the absorption of smaller and weaker territories into larger and stronger ones ... led to the formation of a relatively small number of mutually independent states, each defining itself as sovereign and engaged with the others in an inherently open-ended, competitive and risk-laden power struggle".

As nation-state making processes began, the European political environment became highly volatile and unsettled. The emerging state boundaries were fluid as groups of people or nationalities were now incorporated into new states and lost to others. When territories were acquired, their populations were automatically acquired with them, with the result that in the European dynastic state, rulers and the ruled often came from entirely different national groups. States were like firms in the so-called free market economy, where larger firms or monopolies tended to eliminate or absorb the smaller and weaker ones. For example, in Central Europe, the Hapsburg had ruled not only Germans but also Bohemians, Moravians, Poles, Hungarians, Slovenians and Croatians (among others) and in the southern Balkans, the Ottomans controlled Slav, Albanian, Greek, Macedonian, Romanian and other peasantry (cf. Jackson and Rosberg, 1994:293). In these societies, the interest of the ruler, the dynasty, and survival of the empire (which amounted to the same thing) were the overriding concerns, and the political skills for administration were those that emphasized these as imperatives. The predominant principle of statecraft was how to get more territories, more revenue and more people; to expand or at least survive. Indeed, in this era, when there was no industrialization, natural resources, land and minerals were considered the prerequisites for development and wealth. This helps to explain European colonization of other parts of the world.

The emerging new states were different from contemporary states, which are governed, at least in principle, by internationally recognized rules. In the early modern states system of Europe, there were simply no international bodies such as the League of Nations, the United Nations or the International Court of Justice that could establish peace among nations or defend the weak states against the strong. The emerging states were self-assertive and did not presuppose that there was a supra-states system with recognized rules to which their behaviour should conform. In his sociological study of the origin of the modern state, Poggi notes that each state was a self-originating, self-perpetuating and self-empowered autonomous unit operating exclusively in pursuit of its interest. Nothing could stop it except another superior state. But each state's definition of its interest and actions was continuously affected and modified in response to those of the others. The state system was fragile, disequiliberating, and perpetually in need of re-adjustment, but which cannot be secured since there were no internationally enforceable norms. The political universe was intrinsically and irremediably "opened at the top": highly contingent, volatile and inherently dangerous. The principle of "might is right" was the order of the day. The system had a homoeostatic or self-regulation property. Every accommodation was conditional, every alliance was temporary, and every claim was enforceable only through coercion, if necessary on the battlefield. As Poggi ironically comments, "within this context, the abstract and essentially fictional notion of equality of all states as sovereign possessed considerable credibility" (Poggi, 1978:88-89).

In this context, there was simply no insurance against unfortunate geopolitical location. Many of the smaller and weaker states that failed to meet the test of competitive state system temporarily and permanently disappeared as a result of conquest or voluntary incorporation into other political units. Poland, for example, was partitioned and divided between Russia, Prussia and Austria in 1795 and ceased to exist as a sovereign entity until 1916 (cf. Jackson and Rosberg, 1994:294). Thus, the failure of one state was simultaneously the success of others, in a zero-sum international political game focused largely on the possession of territory as the principal source of wealth, power and glory. This competitive and combative environment put a premium on the dynastic states to tighten their territorial claims and their respective internal political orderings and to give structure to their rule; thus, as rulers schemed for more glory, the preparations for war indirectly or directly began to improve the economic base and the military and administrative capacities of their emerging nation-states (Jones, 1987:135).

Conversely, by implication, it can be argued that rulers in contemporary Africa, monopolizing the means of coercion and resources but not threatened by hostile neighbours, are not likely to be compelled by competition to adopt new organizational methods or develop scientific techniques to modernize society. The states of Europe were, however, not so fortunate; they were surrounded by actual and potential competitors willing to "swallow" them. If the government of one were lax, it impaired its own prestige, military security, and survival. In this respect, Jones (1987:118-9) argues that the belligerent states system acted as insurance against economic and technological stagnation. Thus not considering ethical implications of war, war itself is a factor of technological development.

The emerging states also needed sturdy economic foundations to survive and fight other competing states. This demanded a regular and reliable source of wealth (taxes) for state reproduction. Thus, a new conception of state and state politics (enlightened despotism) gradually began to evolve; arbitrary rule declined and the political behaviours of leaders became increasingly predictable. Internally, the state moved away from mostly patrimonial domination to make its rule more encompassing, more continuous and more calculable. Gradually, though not always intentionally, the princely-absolutist state began to change into a bureaucratic state. Public rule systems and laws became more universal in application and more transparent; property rights, as well as contracts were enforced by the state. Internal barriers to trade, both physical and institutional, were removed and markets were expanded. Communication was improved and laws were made more formal, more consistent, and explicit; weights and measures were standardized; coinages were unified; banking, accountancy, insurance and financial policies were developed to facilitate capitalist development. The institutional complexes identified by Weber as promoting capitalist development began to emerge. Localism and "anomalies" from the past were ironed out. Laws were codified into the "vulgar" languages (no longer Latin and French of the ecclesiastical establishment and the nobility). The process was, of course, slow and coral-like, because Europe at this time was an ethnic and linguistic mixture, amid considerable topographical diversity. But, gradually, all these worked in diverse ways and in various combinations to strengthen the bureaucratic integrity of the state.

The European state of late eighteenth and nineteenth centuries also pursued distinctive (though by no means, at least initially, always economically progressive) mercantilist policies, with the main objective of building a strong domestic economy to enable it fight its enemies. Furthermore, the state "protected" its national bourgeoisie domestically, and, later, their economic activities abroad under colonialism. In some instances, it deliberately established textile works, iron foundries, and arsenals behind tariff walls and encouraged import substitution industries. Most of these efforts were, however, disasters, but they did constitute rudimentary statist attempts to modernize society and also serve as useful precedents which were later followed although intermittently (Jones, 1987:135). In some cases, the state also supported the bourgeoisie, and they, in turn, sought accommodation and special rights. They recognized national greatness and the desire to build empires. This reinforced the economic base of the absolute ruler. The wishes of the ruler and that of the emerging bourgeoisie were in many ways, confluent. An alliance was thus forged by the bourgeoisie, who possessed critical market resources and needed public guarantees for its resources, and the state, which needed part of these resources for its own reproduction.

But the pact between the absolute prince and a bourgeoisie content with the prince's mercantilist policies did not last. The rise of the bourgeoisie as a product of mercantilist policies had unintended political consequence for the prince. The bourgeoisie, as an *estate*, became a *class*, or corporate actors expressing themselves as intrinsic aspect of capitalist development. As the bourgeoisie acquired economic power, it sought to partake of the absolute ruler's power in order to advance its economic interests. This provided one kiss of death to absolutist rule and paved the way to liberal democracy. As one author comments:

> trade and industry... owed their establishment to the ambition of princes... particularly with a view to enrich themselves, and thereby to become formidable to their neighbours. But they did not discover, until experience taught them, that the wealth they drew from such fountains was but the overflowing of the spring; and that an opulent, bold, and spirited people, having the fund of the prince's wealth in their own hands, have it also in their own power, when it becomes strongly their inclination, to shake off his authority. The consequence of this change has been the introduction of a more mild, and a more regular plan of administration.
>
> When once a state begins to subsist by the consequences of industry, there is less danger to be apprehended from the power of the sovereign. The mechanism of his administration becomes more complex, and... he finds himself so bound up by the laws of his political [economy], that every transgression of them runs

him new difficulties (Steuart, quoted in Hirschman, 1977:83-84).

This quotation underscores the indispensability of the economic activities of the bourgeoisie to the capitalist state, and the political concession the state must make to the bourgeoisie in order to take part of its resources. As Moore argues in his historical-comparative sociological study about the western or liberal route to modernity, representative government is the price the absolutist state had to pay for taking away the economic resources of the bourgeoisie. "No bourgeoisie, no democracy", says Moore (1968:418), as implied in the old adage: "no taxation without representation". The bourgeoisie, then, is or rather can be a threat to dictatorial or personal rule. What this suggests is that the viability of the capitalist state cannot be divorced from the bourgeoisie, who control resources that the state permanently needs (Habermas, 1975). In this respect, some Marxist approaches to the state see it as an instrument of the bourgeoisie (see e.g. Offe and Ronge, 1988).

But the quotation above also suggests that the modern economy itself is the one important effectual bridle ever "invented" against the folly of despotism (Hirschman, 1977:85). A modern economy is highly vulnerable to arbitrary rule because its complexity which makes it fragile requires "ordered rule", thus making despotic rule out of the question. The state creates or shapes institutions to support modern capitalist development or a modern economy, but the state needs to continuously enforce or maintain these institutions if its economy is to survive. This means that the state itself cannot always act arbitrarily or flout its own rules without jeopardizing the economy. This understanding has important bearing on Montesquieu's observation noted above that, in the birth of states, leaders create institutions, but in the long-run, the institutions turn to shape them.

The demise of the absolutist rule was also accelerated by the rise of "reasoning publics" and/or civil society. Increasing industrialization and division of labour accelerated social differentiation in society. Greater freedom of thought and lifestyles emerged. This led to a revival of discussions about the public domain started by Greek philosophers but sequestered by religious dogmatism. A "critical intellectual community" in the renaissance period thus grew in tandem with increasing secularization of society. A distinctive public interested in intellectual and artistic pursuit, as well as the development of a new social identity not totally distinct from the bourgeois class (since in most cases the members of these social groups overlapped) was emerging demanding political reform. Sometimes they were politically radicalizing and energizing and they conducted discourses on society in distinctive settings and media such as scientific societies, literally salons, Masonic lodges, coffeehouses and daily and periodical press.

Such fora for cultural discourse and communicative action were often open-ended, relatively unconstrained, thus arguments could be developed with critical imagination and justification leading to the establishment of "public opinion" about given themes, issues and events (see Poggi, 1978:81-82). Private citizens took to publishing exhortations to purposeful development "growth programmes", which were virtual national plans (Jones, 1987:134). These publics were not always censored. They dealt with political matters: the characteristic civic virtues and vices of the nation; the ways and means to the promotion of welfare; the improvement of legislation; the relationship between the church and state; and the conduct of foreign affairs. These were incorporated into political theory of politics. In this sense, the bourgeoisie and civil society constitute prodemocratic forces in society. In a book dealing with capitalist transformation and the development of democracy, Reuschemeyer *et al.* (1992) also draw attention to labour power's role in consolidating democracy in western societies.

The discussion so far gives the impression of deliberate interfering, intent and easy development of capitalism. It needs to be emphasized, however, that this was not always the case. The positive consequences of state policies in this era were mainly unintended consequences of the search for personal glory. They did not arise from the goodness of the rulers. The national purpose in the *ancien regime* was largely a cloak for ruling interests. In this vein, Jones (1987:237) argues that the effects of forward-looking policies [were] hard to separate from the independent effects of growth itself; that measures to reduce risk were still feeble by modern standards; that any cherishing of human capital often thought of people as no more than tax payers and cannon-fodder; and that the distribution of wealth and social relations, remained preposterously unequal. The brutality of transformation is captured by the Chinese emperor who declared to a British emissary who came to China that China had everything in abundance and did not need anything from the barbarians of the West. Yet, with all these reservations, the "modernizing" effect of state action was clear even though the results were out of all proportion to the motives.

It would also be an exaggeration to claim that feudalism was totally destroyed. In some parts of Central and Eastern Europe, personal rulers were unsuccessful in completely smashing the feudal order. Those which survived had to await destruction by revolutions. Again, as indicated above, there is overwhelming evidence to show that the actions of the absolute leaders, were in many instances, not conducive to capitalist development because they were resentful of their subjects, who could become wealthy and challenge their authority. Arbitrary rule is what the patrimonial leader unleashes to limit the rise of the capitalist class that could threaten his rule.

In his study of European economic history, Jones (1987:93) maintains that arbitrary political behaviour in feudal and medieval Europe included irregular levies, confiscations, forced loans, debt repudiations, debasement, expulsions, and judicial murder, and argues that all of them were productive of uncertainty in the socioeconomic environment (cf. also Collins, 1980). Jones recounts some interesting episodes to illustrate the point. For example, King Edward III of England ruined his Florentine bankers by repudiating his debt, and the Order of the Knights Templar, which became a banking operation, was also destroyed by Philip the Fair of France (1285-1314) because he repudiated his debt to them and subsequently banished the bankers. In France in the fifteenth century, Jacques Coeur, the richest merchant in the country, was arrested on a trumped-up charge. His property was taken and he was forced to flee the country.

Other anecdotal evidence Jones gives are: one Samuel Pepys buried £1,300 in gold in Huntingdonshire orchard when the Dutch came up to Medway in 1667; a hoard was also buried at Bristol when William III was marching up to Torbay in 1688. During the Swedish-Danish war of 1562-9, large amounts of coins were buried in Southern Sweden as a result of troop movements between these countries. Montesquieu, in his *Esprit de lois,* 'How Commerce Emerged in Europe from Barbarism'argues that the Jews who suffered constant extortions at the hands of kings and nobles, invented the bill of exchange to circumvent the arbitrary confiscations induced by personal rule (cf. Jones, 1987:92-96). Indeed, even though some of the overseas wars conducted by the princely states might have brought home profits, the merchant class sought an end to unpredictable and arbitrary politics because they were inducing insecurity and institutional rigidities, and, consequently, holding back productive investment.

Emphasizing the importance of freedom to capitalist development, Jones (1987) reiterates John Locke's view that freedom is "not to be subject to the inconstant, uncertain, unknown, Arbitrary Will of another man". The prevalence of this view (see Hirschman, 1977) was crucial to the emergence of risk-free environment for capitalist development, since economic actors, be they individuals, investors or companies, would not invest or risk their resources if there is uncertainty about the future or there are prospects that they would be robbed by governments when they become successful.

The removal of these political obstacles was in itself a protracted process spanning several centuries. But greater security of property was won, if intermittently, because the kings gradually realized that they could not fight their enemies when their subjects were poor or could not engage in long-term entrepreneurial activities that could generate reliable and sufficient taxes for their kingdoms. Arbitrary behaviour such as confiscation could only harm revenues. Because of royal quest for definite sources of taxes, the rulers gradually recognized the absurdity of random disruptions and confiscations. Slowly, they converted to the advantages of commerce because of the prospect of taxing it (Jones, 1987:88). They also slowly "realized" that their own kingdoms run the risk of poverty if they kill the goose that lays the golden egg. In other words, if they created poverty at home they would not be strong enough to fight their enemies.

The point being made here is that the absolutist state gradually came to reduce cheating, price gouging, episodic and distrustful, as well as predatory economic behaviours that had been the salient features of earlier political societies of Europe (Collins, 1980:931-2). The absolute ruler granted trade agreements under public seals, and gave imperial and royal or municipal authority to the bourgeoisie, though initially mainly to his favourites, to enhance this authority and control over important sectors of the society and economy.

The absolute ruler's scheme for glory and the historical circumstances within which he operated forced or drove him to "think" in terms of how to make his kingdom more wealthy. This entailed some reduction of his "thoughtless" and selfish behaviour, as well as of those who administered his state. Laws and policies gradually became somehow constructive and predictable; this indirectly became a crucial factor in the development of "risk-free" internal environment, which conduced to capitalist development (Callaghy, 1988:75).

But the alliance between the absolute state and the national bourgeoisie within a particular national context is just the tip of the iceberg. By acquiring other lands outside Europe, imperial powers were able to export their bureaucratic states abroad to provide protective cover in the colonies for their national bourgeoisie to accumulate resources such as raw materials, minerals and finance capital from trade. Much of this became vital for the industrial expansion of the imperial powers of Europe. Throughout the colonial period, the metropolitan state did not hesitate to use all means (diplomatic or war) available to enhance the reproduction of its bourgeoisie abroad. The product of this historical alliance is the continuous expansion of western businesses and investment on global scale, or what we now term transnational corporations, many of whose annual turn-overs dwarf the Gross Nation Product (GNP) of many African countries.

It is also important not to exaggerate the speed and intent of these "progressive" developments. Things were still done intermittently, sporadically, and not out of concern for the welfare of citizens. Extreme poverty and social inequality remained characteristic of the population, wars were rampant and arbitrary rule took a long time to disappear. In this regard, according to Jones, probably the greatest *eventual* achievement of the European state was the progress it made in bringing political arbitrariness to heel in decreasing uncertainty and increasing calculability (Jones, 1987:93-94; 103).

That Europe became an economic "miracle" vis-à-vis other societies or nations is attributed to many factors but one of these is this "political development" outlined above. Given the significance of the political factor, Jones submits that the explanation of the origin of modern capitalism in terms of purely market forces, is insufficient if political factors are excluded. As he argues:

> what the rise of the nation-state does seem necessary to account for is the establishment of the stable conditions necessary for expanding development and growth, for the diffusion of best practices in technology and commerce, and in several countries for the actual founding of manufactories where there had only been handicrafts... The self-propulsion of market forces explains much at least in less authoritarian parts of northwest Europe. A full explanation of the generalisation of novelty must also take the nation-state into account (Jones, 1987:149).

In what can be interpreted as a reaction to Smithian analysis, which stresses the "invisible hand" of the market as the main determinant of modern capitalist development, Jones contends that "an explanation of the genesis of the modern economy will surely have to incorporate political power into price theory and thus account for that great bulk of human behaviour and choice which otherwise must remain *ceteris paribus*" (ibid., 226). In a sense, Jones reiterates Weber's "historical institutional theory" of capitalist development, from which Collins concludes that:

> the bureaucratic state is a crucial background determinant for all legal and institutional underpinnings of capitalism... the key to all of the *institutional* structures of rational capitalism... It is [the] bureaucratic-legal state that broke down feudalism and patrimonialism, freeing land and labour for the capitalist market. It is this state that pacified large territories, eliminated internal barriers, standardized taxation and currencies. *It is this state that provided the basis for a reliable system of banking, investment, property, and contract, through a rationally calculable and universally applied system of law courts* (Collins,

1980:931-2; first italics original; second italics added).

The State and Institutions in Long-term and Late Development

In recent years, economists who address the issue of long-term economic change have stressed, in varying degrees, the crucial role of institutions in the political economy of growth, giving further support to Weber's institutional theory (cf. North 1981, 1990; Goodell, 1987; Eggertsson, 1990; Ingham, 1993; Killick, 1995). For North, both formal (e.g. laws and constitutions) and informal institutions (such as conventions and codes of behaviour) are the main underlying determinants of long run economic performance of societies. Through institutions, economic actors and investors structure their actions. According to institutional economists, institutions arise in societies due to the pervasiveness of uncertainty, the imperfect delineation of property rights, and the bounded rationality of agents, (Neuber, 1995) all of which raise the cost of production and exchange (or transaction cost) in the absence of constraining institutions. Institutions, as a framework, reduce uncertainty, structure incentives, constrain human interactions, affect decisions of individuals, groups and organizations, and consequently, bear on transaction cost. Institutions, together with the technology and other factors of production North (1990:16, 118) maintains, determine the "transaction cost" of societies and the long-run economic performance of societies. The major developmental role of institutions, according to institutional economists, is to reduce "transaction costs" in society by providing stable (though not static) structure for human interaction.

Institutional economists thus argue that the existence of well-specified and enforced property rights systems and institutions, which enable markets to function with relatively low transaction costs, are the key to economic growth. In this respect, Goodell (1987) maintains that the most destructive force in any form of economic development is arbitrariness. In one earlier work of his, North (1981) argues that it is the high costs of transaction induced by political arbitrariness, wealth confiscation and institutional rigidity that prevented certain economies and societies, such as sixteenth and early seventeenth century Spain (unlike England), from realizing their potential. North's contribution is an argument that a capitalist economy has to be supported by laws of contract enforceable in courts by impartial official judges (i.e. the rule of law) by a common understanding of property rights, agreed weights and measurements and other rational practices. For:

> economic development requires increasing rationality or predictability in the

> allocation of productive resources. If factors of production are to seek out their most exact productive use on their own (capitalism) they need conditions of ample mobility... But except in desperation—and then not rationally—people are unlikely to venture their resources beyond their existing situation (whose risks are at least known) and certainly are unlikely to commit themselves to *new combinations,* unless two conditions are met: predictability and the security of bonding. Inasmuch as economic development entails an increasing division of labor, it *depends on* an ever wider and firmer guarantee of bonding, which assures resources a safe linking with their economic complements and enables accumulation through time. Of these three requisites for capitalist development—rationality, predictability, and bonding—predictability is the foundation, because just as there can be no rational allocation without it, neither can there be bonding (Goodell, 1987:102; italics original).

Like these institutional economists but writing from the Marxist perspective, Warren (1980) also points out that European capitalist development was facilitated by:

> the permeation of Western European society by the sanctity of individual property irrespective of rank, and thus by a cultural climate conducive to maximum scope for private enterprise. Supplementing this was the rule of non-arbitrary law, which permitted and encouraged the development of contracts, implying agreement between persons of equal rights and not depending upon relations of force between economic transactors. The more force decline in economic life, the greater was the weight of rationality in resource *allocation* (Warren, 1980:29; italics original).

Goulbourne sums up the intimate pact between the state and capitalist development and its later modification this way:

> the widespread practice of the capitalist state has been first to establish the basic ground rules for peaceful market relations (free labour power, the growth of commoditisation, etc.) as well as to assist the construction of material infrastructure necessary for capital accumulation. Later on the capitalist state became more directly involved with both social arrangements between individuals as well as the productive capacity of its territorial base. Thus, in the early phase of capitalist development the state was the principal protector of the emergent bourgeoisie, and as capitalism evolved a highly complex social division of labour along with an integrated social and economic system, the political became more and more intertwined with these (Goulbourne, 1987:28; emphasis added).

Given that economic behaviour is shaped by institutional structures and incentives, without rationality and predictability in the rule systems, investments are likely to suffer because entrepreneurs would choose not to commit resources into highly uncertain and volatile socioeconomic and political environment, especially if those resources would be difficult to redeem should this environment turn unfavourable. Indeed, it would be entirely irrational to engage in long-term projects in an environment that is uncertain and riddled with political arbitrariness. Governments that formulate, modify or change rules play a leading role in setting up the broader institutional framework for economic behaviour: the incentive structure to which economic actors such as individuals, corporations and organizations respond and within which they engage to undertake productive activity. For the purpose of establishing certainty and predictability, when governments make rules, they need to make sure that the rules they establish are enforced consistently. Governments or states that change rules frequently and unpredictably signal that the environment is risky and chaotic. Perhaps the greatest injury a state can impose on its economy is to create uncertainty through its actions and policies. It is thus not sufficient for the state to establish the rules for economic actors to abide by; the state as an organization must also adhere to those rules. In other words, socioeconomic development cannot be sustained in an institutional vacuum or in an environment that institutionalizes anarchy.

Weber's institutional theory and the historical material discussed above provide an important insight into how the modern state "promoted" capitalist development and the growth of markets in western Europe. This Weberian analysis challenges the general belief that nineteenth-century Western industrialization was led mainly by powerful, rising commercial bourgeoisie who having acquired state power, were then able to remove the remnants of feudal obstacles to pave the way for modern capitalist development. Rather we find some "collaboration" between the modern state and the bourgeoisie in early capitalist development. It thus a misconception of the development process in dependency theory to assume *a priori* that local capitalists in twentieth century Africa are incapable and ineligible to make some kind of positive contribution to the economic development of their countries and for that matter the state need not aid their ascendancy (see also Kennedy, 1990:3).

But the relation between the state and development becomes even clearer in the case of "late developers" such as Germany, France, the former Soviet Union, Japan, South Korea, Singapore and Brazil in their transition to modern development, whether capitalist or socialist. In this late context, the state assumed an even more active role in development than in the early

European context outlined above, through the "revolution from above". The explanation offered by List for this phenomenon is that "late developers" require "artificial means" or what he calls "developmental states" to catch up with the advanced nations (List, [1841]/1966:175).

Recently, Hettne (1990) makes the point that, after England, all other countries had to make their decision in a different, more competitive and economically complex international climate. In Western Europe, the post-war Keynesian strategy of reconstruction made the state the controlling agency in the process of economic growth and development rather than simply the market. Gerschenkron (1962), whose study deals with the experiences of nineteenth-century late developers such as France, Germany and Russia, brings out the centrality of the modern state in the development process by stressing effective state intervention in these European countries. The Gerschenkronian thesis posits that "economic backwardness" or the context of "late development" necessitates a strong economic role for the state. He notes that, particularly in the case of Russia's late development, (at least up to the 1980s) it was the state that assumed the role of the primary agent propelling the economic progress in the country (Gerschenkron, 1962:17; but also Moore, 1968; Gregor, 1979). In this sense, Hettne suggests that we can regard the development projects of the Soviet model as part of substituting process, whereby alternate means were made use of in order to reach similar ends as in Western Europe (Hettne, 1990:42-6). The state's role in development then is not so much the pronouncement of ideological preference, but rather, the development of state capacity to undertake development which is much more of a technical matter than purely ideological preference (Leftwich, 1994). Conversely, however, in the former Soviet Union and Central and Eastern Europe, it was the state's long-term failure to fulfil its promises that led finally to its overthrow.

Within the context of Asian "economic miracles", a number of researchers have also shown that the character and especially, the capacity of the state has been a critical variable in the developmental success of many countries through the so-called "revolution from above". Even long before contemporary times, Leftwich (1983) argues that "state politics" had been important in Aztec empire, Song China and post-Meiji Japan. The idea that the state has played a vital and catalytic role in the development and growth of markets of the Asian "miracles" and other relatively successful countries in Latin America has been repeatedly stressed in the literature (see Fukui, 1992; White and Wade, 1985; Wade 1990; Evans, 1979; Amsden, 1985, 1988, 1991, Girling, 1981). For White and Wade (1985), as well as, Wade (1990), the dominant political explanation for successful interventionist strategy lies in the authoritarian, corporatist state that emerged since 1949 in Taiwan and from 1961 in South Korea. These analysts draw attention to both countries' clear and urgent attention to industrialization, which was pursued in a comprehensive rather than piecemeal perspective and framework. The industrialization programmes of these countries were also said to be sustained by the "threat" from powerful external enemies (mainland China and North Korea), which is reminiscent of the external threats the states of pre-modern Europe, and the Meiji state of Japan experienced in the mid-nineteenth century.

Also mentioned is what is termed a "corporatist" alliance between the state leadership, the bureaucracy and business interest, although not labour unions. Some authors, however, emphasized the institutional reforms launched in the Korean case by military leadership in the 1950s, which purged and reformed the civil service, enhancing the status and power of the technocrats in coherent policy formulation and the quality of the civil servants at lower levels who implemented policy. They stress that the shift from import-substitution policies to more export-oriented strategies in the 1960s was made possible by this key administrative change which derived from military autocracy (cf. Leftwich, 1994). Evans (1979) has also explained recent development of Brazil in terms of the triadic alliance between the state, multinational corporations and the local bourgeoisie. China's phenomenal economic performance in recent years, is also occurring under a strong "undemocratic" state.

In all these cases of "late development", the capacity of state to directly facilitate production through the provisions of roads, railways, ports electricity, water, telecommunications, etc. and indirectly, to assist production by first providing a predictable and calculable environment for investment and schools, technical education, public housing, sanitation and health facilities through which skilled, healthy, motivated labour force is developed, has been crucial. Governing elites are using or have used public institutions at the early or intermediate stages of their countries' development to foster healthy relations between them and the local bourgeoisie in order to expand manufacturing capacities of their countries.

In summary, the state (though not always democratic in the western sense) has everywhere facilitated modern development; be it capitalist or socialist. What then, is the state-development nexus in the Ghanaian context? This is a question to be answered empirically in the next chapters. But before proceeding to the empirical section, it will be useful to provide some background to the origin, nature and functioning of the African state, because it helps to understand why the socioeconomic and political environment has so far not been very conducive to development.

The Making of the African Bureaucratic State: Contradictions and Problems

The states of Africa are largely the products of European colonialism, and many of their features and, consequently, some of their development problems, can be said to have colonial roots. By the latter half of nineteenth century, when European powers formally colonized much of Africa, their own societies had undergone considerable economic and political transformation: the "bourgeois-democratic revolution", so to speak. Institutions such as democratic governance, political representation, civil society, political citizenship and concepts of nation-states, had taken or were at least taking hold. Capitalist development, mainly through the nineteenth century Industrial Revolution, grew in tandem with the intense bureaucratization of the modern state and territorial consolidation. But although states with democratic culture had emerged at this time, the mentality of acquiring land and people did not die down. As history was to show, it took two World Wars, the establishment of the United Nations, and the imposition of democracy on nations such as Germany, Italy and Japan to put an end to such mentality. In a sense, European conquest of other parts of the world was European barbarism exported abroad, as the main objective was to acquire foreign resources to build empires or states in Europe. Poggi (1978) writes that, during this period, Europe was to experience one hundred years of peace.

The acquisition of lands and colonies, however, went against the democratic principles of social mobilization and political citizenship that had transformed European countries into national states. Colonialism, the direct subjugation and control of other people by outsiders, was infused with a dynastic mentality. As regards Africa, the continent was parcelled out at the Berlin Conference of 1884-5 by the imperial powers without regard to the natural boundaries of ethnic, linguistic and religious groups. Most colonial boundaries were arbitrary and consequently divided similar, or brought together different ethnic groups. Colonial boundaries were often redefined to suit the interests of the imperial powers (Jackson and Rosberg, 1994). By establishing jurisdictions over African colonies and by recognizing such colonial boundaries, the imperial powers abruptly aborted endogenous state-making processes that were incipient or potential. This political "aberration" was later to be a source of ethnic conflicts and political strains in the postcolonial nation-state projects, which was a factor that compelled new leaders to reinforce the inherited coercive colonial political structures in order to survive, rule and build nation-states (Ake, 1996).

The imperial powers, like the dynastic states of early modern Europe, acquired Africans not as citizens, but as subjects. Possessions of land by European powers were mainly through unequal treaties signed with traditional rulers who, in many cases, did not understand their implications; where they resisted, they were conquered. Africa's state-making process and "bourgeois-democratic revolution" had barely begun and no militarily powerful nation-states had emerged on sufficiently large scale to ward off foreign intruders.

The colonial conquest led to the imposition of the colonial state on African societies, consequently weakening pre-existing African self-governing institutions and "constitutional" practices defined by reciprocity, public accountability and political legitimacy challengeable through cultural and political norms (Healey and Robinson, 1992). Whether it was the colonial administrative policy of *Assimilation* (to culturally make the colonized people French) or *Indirect Rule* (British policy to rule colonies through their own political institutions), the political consequence was the same, in that, the powers and functions of Africa's traditional political institutions declined as they came under the control of the new colonial centre through regional and/or district commissioners (Hodder-Williams, 1984). The colonies resembled postfeudal and preindustrial states of Europe and their administration was reminiscent of the dynastic states of Europe. For example, the Congo or Zaire, now the Democratic Republic of Congo, a country roughly the size of Western Europe became the personal property of King Leopold of Belgium and its administration resembled that of Colbert's France (cf. Jackson and Rosberg, 1994).

The colonial state was an appendage of the metropolitan bureaucratic-democratic state and was manned by a small core of expatriate bureaucratic officials answerable to the metropolitan political centre. But crucial aspects of the recently acquired democratic practices and institutions in Europe that could integrate the colonial state and African societies (for example, political mobilization, representation, and public accountability) were not disseminated, at least not initially. The colonial state provided a Hobbesian framework of political order to enable the metropolitan bourgeoisie to extract resources from the colonies and, at the same time, use the colonies as markets for the manufactured products that originate from the imperial centres.

The state developed those instruments that would enable it and its national bourgeoisie to extract resources from the colony; it was a predatory state. Thus it subjugated the colony from outside and represented primarily external interests (Healey and Robinson, 1992:22). Only its bureaucratic and authoritarian structures and coercive institutions (the military and police) were exported to Africa. The colonial state had to be coercive, because there was no other way through which a small number of aliens could rule other nations. In this regard, Jackson and Rosberg 1994:297) observe that the acquisition and holding of colonial territory was a throwback to the era of dynastic states. The nature of the state has also led to the development of the notion of the "over-developed state" (cf. Leys, 1976), by which is meant that the state apparatus inherited at the time of independence was larger and had greater administrative and coercive powers than would have been the case if it had not been necessary to subjugate locally dominant classes.

Thus, from its very conception, the colonial state was an alien, undemocratic and coercive institution to the colonized people. To use Hyden's (1983:19) metaphor, the colonial state in Africa was like a balloon suspended in mid-air; an artificial institution having little or no organic relation to the society. As the state was detached from society, it could not win popular support and its effectiveness rested on its ability to back its orders with force. Colonial state-making process was contradictory, in the sense that those being governed did not ask for the state and were not allowed to influence it. As Hodder-Williams (1984:32) notes, "imperial rule was bureaucratic, rather than political; it was authoritarian, rather than participant; it was centralist, rather than devolved". In other words, "good governance", originating from the society below mediated by civil associations, political contestation, and other democratic practices were not encouraged.

Colonialism was also conceived as a civilizing mission, largely paternalistic in the form of the "white man's burden". This conception was also necessary because there was really no way that the colonizers could have come to terms with their dehumanizing treatment of their victims except by regarding them as somewhat less human. The only problem was that, in reality, the victims were no less human. Therefore, they had to be rendered less so, and this was what colonialism proceeded to do in its cultural and political policies. This led to the systematic "abolition" of African history, the imposition of foreign religions, the "debasement" of African cultural practices and values, and the creation of a centralized colonial state that concentrated political power in the hands of European administrators at the expense of existing political centres. By "brushing aside" Africa's time-tested indigenous political institutions, colonialism usurped the spirit of political self-determination. Since the colonial state was not constructed to advance upon existing African politics its one major problem is that few Africans, if any, had a strong *gemeinschaft* attachment to it.

Colonialism was oppressive, but freedom from it was seen as an opportunity to take control of the colonial state rather than reforming it and salvaging what good traditions remained of Africa's indigenous politics. Instead of searching for, or building upon political structures that truly reflect Africa's own peculiar circumstances and traditions, the colonial state's administrative framework was accepted in toto for postcolonial state business. Where colonial ideology, usually capitalism was rejected, it was not because elites sought to replace it with something peculiarly African but rather with other foreign ideologies. By so doing, they indirectly accepted the colonial self-serving view that Africa should look to the outside world for models of development. Thus Africans were not to remain true to themselves but were to be changed. In so doing, leaders alienated their people from their social and cultural environment and all that they believed in. Consequently, state policies were divorced from the aspirations of the people and local experiences. It was as if Africa had nothing of its own in terms of institutions and cultural traditions that could provide frameworks for its development.

This approach to development which "denies" Africa realities, can be read to mean that the Japanese or Chinese for example would have developed better had they totally abandoned their ways of life and embraced Western capitalism or Soviet socialism. Indeed a great deal of attention has been devoted to the detrimental economic or material impact of colonialism (such as the orientation of African economies to produce what they do not consume directly and their dependence on manufactured goods from the North), but as Ake (1985) correctly contends, this might not be necessarily the only or the perhaps the most damaging effect. In all probability it contributed rather less to the African predicament than the political and cultural policies propagated in the developed world which leads to the "monopolization" of African development thinking by Western development establishment, which is manifested in all kinds of outside ideologies "imposed" on Africa or accepted by African leaders with the implicit assumption that Africa can only develop by alienating itself from its own institutions and traditions.

This orientation in development discourse and practice has ruled out any African initiatives that could secure development through African people (the means) and for African people (the ends). Europe's confrontation with Africa was not based on Africa's terms, but, rather, led to Africa's subjugation in all aspect of social life. By maintaining right from the colonial period that Africa had to be civilized in the image of Europe, the colonial masters implicitly laid a foundation for future development discourse that assumes that Africa's development should be imitative of the other societies elsewhere and should not grow practically out of African social experiences.

Colonialism was a factor of change in African modern history. Directly or indirectly the colonial state induced socioeconomic change in society. The state mobilized the people and constructed roads and railways to facilitate the exploitation of colonial resources, such as gold, diamonds and copper. It also built schools to provide literate work force for its public sector. It introduced modern money and banking institutions; it set up public sector employment and administration and thus provoked migration and urbanization. Villages were turned into towns, and towns into cities. Colonial economic policies compelled Africans to specialize in the production of new crops like cocoa, coffee, sisal, cotton, groundnut and palm oil to meet the demands of the burgeoning industries and markets of the Western societies. Because each territory was seen as a cog within an empire, and not as a separate economic unit that should aim at self-sufficiency, the tendency has been to make the colony specialize in one crop or resource that fitted its geographical circumstances (Hodder-Williams, 1984). Often this resulted in the development of monocultural economies. The externalization of productive activities and lack of economic diversification were, however, to expose African countries to the vagaries of the international market which, one way or the other, continue to undermine postcolonial domestic planning and economic stability.[6]

Colonialism was a carefully guarded project. Control over Africa was to be exercised as long as possible. But events have a notoriety for upsetting our best laid plans. Colonialism brought about socioeconomic transformations, the consequences of which its architects probably never anticipated. Newly educated elites, merchants and professionals with skills for participating in the formal economy, emerged on the historical stage. These groups became sharply aware of colonial injustice, racism and exploitation, as well as internal political contradictions. They sought, through constitutional means and civil disobedience, the democratization of the colonial state and even an end to colonialism. After World War II, the British and later, more reluctantly, the French and the Belgians, embarked on constitutional changes to accommodate these internal tensions. In Guinea-Bissau, Mozambique and Angola, where Portugal refused to let go easily, arms struggle ensued. Finally, decolonization movements (peaceful or violent) could not be contained. The colonialists conceded defeat and hurriedly assembled for their colonies liberal democratic political systems reflecting the experiences of their own societies, granted independence and left.

The anti-colonial struggle drew diverse social groups together to present a unified front against the colonialists. Although the members of the nationalist coalitions closed ranks to overthrow the colonial power, there was much political anxiety about the enormous power, privilege and coercive instruments of the state nationalist leaders were to inherit. The normative, institutional and ideological constraints, accountability, negotiation and compromise that are the hallmark of liberal democracy, had not yet crystallized. With independence and the colonialists gone, the front began to decompose as various social forces and elites competed with each other to capture the state and exclude others, usually their rivals.

One major political problem that confronted governing elites, was that the states they inherited were not sturdily rooted in society. The economic transformation induced by colonialism did not produce capitalist classes or strong civil societies that could anchor the representative governments that the colonialists bequeathed to the new leaders. The colonialists failed to effectively institutionalize their political structures and norms when they swept aside those of Africa. It was not long before the moral basis of the new leaders' authority was questioned through ethnic resistances to the national centre. Indeed, few of the new states evolved from a single dominant ethnic group that could constitute a nation. Many African countries are artificial entities and not nation-states in the true sense. They were agglomerations of diverse ethnic groups with their own political institutions. The modern state (a political organization claiming jurisdiction over a particular territory) has preceded the nation (a social group that develops solidarity on the basis of shared customs and institutions based on some common belief, real or imagined, in a common descent).

As the new states and nations did not coincide, the states were assigned the task of making the nations. But the citizens of the new states still preferred to divide their loyalty between the local and national political centres. The need to garner support to build the nation-states the colonialists had began and pursue socioeconomic development remained crucial. Thus, leaders sought to control their people and resources to channel them into development and nation-building. Given this political configuration, there has been a certain tendency on the part of African leaders to intensify and expand the coercive features of the colonial-inspired political structures and to steer politics in authoritarian directions to enable them consolidate the state and territory in order to build nations. This was a practical or easy option because they have already inherited autocratic states with highly developed coercive instruments that could enable them control their societies.

Nation-Building, Development and Political Centralization

For African countries, political independence was a great moment to take care of their own affairs on their own terms. Expectations for a better future were high. Leaders made great, if not sweeping promises, to those who supported them to articulate the anti-colonial struggle. They became instant heroes and exuded great charisma. Their promises, however, defined a pact between them and their followers and were to shape in no small way the character and direction of policies and politics. The colonial state itself was accepted as the framework in which to do political business, but was somehow to be re-cast to serve as an instrument for economic development, national integration and social welfare (Clapham, 1991:92).

As the new African governments yearned for rapid socioeconomic transformation, the establishment of "developmental states" soon became the hegemonic issue, and the postcolonial development strategy came to be mainly statist. The preference for developmental statism reflected the postwar international development consensus or ideological climate. In the Western world, the post-war reconstruction ideology—Keynesianism—accorded the state an active role in socioeconomic transformation. In the former Soviet Union and Eastern Europe, state socialism also accorded the state a prominent role in development, although for different reasons (Hettne, 1990). The view that the state is not only the guardian but also the motor of development was well established in the development literature by the time African countries got independence (Young, 1988). Logically, developmental statism was extended to Africa.

Yet, besides this external influence, Africa's own historical and peculiar circumstances were said to have conspired to make a strong case for the state-led development. Analysts have advanced a number of reasons for this (see e.g. Sandbrook, 1993:22; Goulbourne, 1987; Healey and Robinson, 1992). Principal among these are: (i) the need to provide public goods, such as roads, power, transport and communication, defined as those from the use of which no one could be excluded and which are strategic to modern development, but whose provision could not and would not be undertaken by private individuals; (ii) the need to industrialize and modernize African countries in the absence of a vibrant, indigenous, productive bourgeoisie which could fulfil their historic mission of advancing the means of production; (iii) the need to create social services such as health, education and sanitation to meet much broader developmental objectives; and (iv) the need to form specialized bodies entrusted with the task of promoting rural development such as agricultural extension services, rural credit agencies, and departments responsible for managing natural resources.

Given the European approach and the above African circumstances, the idea that the African state should become the vanguard of development was hardly contested. Within the modernization mode of interpretation dominant at the time, development seemed more easily surmountable—largely technical—challenge. Social scientists, mainly economists and technical experts would formulate good policies that good governments would then implement for the good of society. Implicitly, the African state was "accorded" Weberian features of rationality and bureaucratic competence for not only providing the basic ground rules for economic actors but also catalyzing the development process.

The state was supposed to provide rules, institutions, public goods and services. It was also supposed to mobilize foreign finance capital and technology to supplement domestic savings (mainly through surplus extraction from agricultural producers, either as direct taxes on crop sales and excise duties) to penetrate traditional societies, aid the emergence of capitalism and to pursue broader socioeconomic goals so as to break the "vicious circle of poverty" (see Rostow, 1960).

Analysts from all sides of the ideological spectrum reasoned that, since underdevelopment was so grave, the promotion of economic growth and development would be the overriding objective of political leaders. African leaders, it was thought, would be "philosopher kings", leading their nations to greater heights of socioeconomic transformation and guiding them towards political pluralism. In this regard, Bottomore (1985:99) maintains that "in many respects, government officials are to the economic development of the new nations in the twentieth century what the capitalist entrepreneurs were to the economic development of Western societies in the eighteenth and nineteenth centuries". Indeed, some independent leaders sought the attainment of such "noble" goals. They responded to the early modern nature of their societies and economies by relying on centralist, corporatist, or neo-mercantilist strategies to promote development.

Beside the strong optimism shared by many scholars and development practitioners about the capacity of the state to effectively undertake social engineering, there was also a belief that members of society would shed ethnic identities and relationships and become national citizens under a secular postcolonial state. According to Doornbos (1990:182), however, this optimism and belief with the benefit of hindsight, markedly ignored two crucial things: (i) how uncertain and precarious the efforts of the African state would be to define and develop its role and position vis-à-vis the larger society; and (ii) the occurrence of much wider socioeconomic differentiation in society that was to throw into prominence social movements and state reactions during the immediate post-independence period, which would curtail the social basis for long-term national (re)construction. To these must be added, the personal ambitions of leaders lacking independent economic base and wishing to stay in power forever and thus politicizing and undermining the development process. Also, analysts readily assumed that the state was an effective institution, forgetting that it would take time for young countries and African leaders to build effective states in the Weberian sense, something that has taken the West several centuries of wars and protracted state-making process to develop.

The African state's role in development was taken for granted. It was to pull society along in an all-out assault on underdevelopment. Whether in agriculture, industrialization, education, or other sectors, the burden of formulating and implementing policies that would transform society was laid squarely upon governments' shoulders. It is important to note here that all the hegemonic paradigms—(neo)liberal and radical and the discipline of development economics—shared this view of development. Indeed as one analyst observes, "in the 1960s the development strategy shared by African governments, most Western advisors and international development agencies pointed to the importance of big investments and the need to achieve large-scale shifts in economic practice and thinking. Individual, micro-level, private initiatives were identified as essentially peripheral to the main thrust of development strategy. They might be tainted by traditional attitudes which involved elements of resistance to real change. It seemed almost axiomatic, therefore, that state regulation and extensive public enterprise, alongside a greater or lesser contribution from foreign capital, would shoulder the main burdens of economic management" (see Kennedy, 1990:2). Interestingly, as a result of this "grand approach", some scholars became ambivalent, if not actively hostile to the role that local capitalist classes can play in their countries' economic transformation.

The state thus came to be conceived as the motor of development, whether under capitalism—as in Cote d'Ivoire, Kenya, and Malawi—or socialism—as in Ghana and Tanzania, for example. But the countries that opted for state capitalism pragmatically asserted a continuation of historical and socioeconomic antecedents in Africa, or rather, a continuation of colonial capitalist legacies. But in the "socialist" countries, new leaders often conceived of independence as an historic watershed in the evolution of the African state and society. They sought new strategies of administration and development often based on socialist ideology. For these leaders independence was an anti-colonial revolution and the late colonial legacy of liberal democracy was rejected substantively (if not in rhetoric). They established legal or *de facto* single party regimes, often spiced with eclectic blends of nationalism, populism, socialism, Leninism, and anti-imperialism, with strong traditional, quasi-traditional, personalistic qualities (Callaghy, 1987:95). These actions and ideologies were crucial in creating new states out of diverse ethnic groups. But they were also occasions for ambivalence, if not an outright refusal, to continue previous capitalist economic processes already unfolding in their countries because colonial capitalism was seen as exploitative of the colony. In this respect, McCarthy (1994:98) observes that few of these leaders could say "things must be done this way because that is what the colonial administration did". A new often radical imprint was sought for society.

While this no doubt was important to restore Africa's balance after colonial domination, the imprint was based on foreign ideology: usually socialism and not on Africa's unique social experiences and indigenous institutions. The "spirit of capitalism" which was embryonic in precolonial Africa, was associated with imperial colonialism and defined in some countries as either alien to society or injurious to the new egalitarian society to be constructed.

Private enterprise and the search for profit were associated with Western domination and exploitation. And just as in Western European development history there was doubt about the ability of capitalism to achieve a humane socioeconomic development and social justice, so was it in many independent African countries. Many newly independent governments of Africa viewed the prospect of capitalist development achieving a self-reliant development and social justice with suspicion (Kennedy, 1990:1). It was thus officially "negated" to pave the way for more humane socialist transformation which, most leaders perceived as successful in the former Soviet Union and Eastern Europe. But generally too African intellectuals and informed public at this time also regarded, to a greater or lesser, some kind of socialist development as infinitely preferable to capitalist development.

With this official animosity to capitalist development and as Kennedy (1990) notes, when Neo-Marxist theories of underdevelopment became influential in the late 1960s and 1970s, the theoretical basis for believing that the local bourgeoisie have some useful role to play in [capitalist] development that could complement the state's own efforts lost all credibility in many parts of Africa. Consequently, in a sense, Africa's postcolonial development was put on a *false start* (see Dumont, [1966]/1988) in that, most development policies were divorced from social realities, and the indigenous economic values and aspirations held by society at large. This was disturbing of the positive balance that must exist between the material development of societies on the one hand and the prevailing ideas and values that anchor such a development, on the other.

At independence, African leaders also felt the need to develop sovereignty and true symbols of independence to reflect their newly acquired status among the comity of nations. But they also realized that their states operated in fragile social environments. As such, they took pragmatic actions and also constructed autochthonous institutions and/or adopt a variety of the centralist corporatist traditions set by the colonialists to reflect what they judge to be the political realities and aspirations of their society and people in order to hold their societies together (Hodder-Williams, 1984). For example, they proclaimed single party states, proscribed opposition parties, co-opted opponents or sent them into prison and sometimes exile, cancelled constitutional "checks and balances" on executive power, amended constitutions and placed restrictions on the autonomy of civil societies and political associations.

In the late 1960s, students of Africa's political development came to terms with the dilemmas that confronted these leaders seeking to create or consolidate new nation-states and suggested that "mobilizing politics" and/or "political authoritarianism" rather than western "reconciliation political institutions" might be more appropriate to the circumstances of the African state (cf. Apter 1965, Huntington, 1968). Recently, Amin re-echoes the sociological necessity of authoritarian states by noting that:

> the role of the state as the focal point for social cohesion and social engineering is particularly important in societies that have not had time to melt and merge into solid nations where people have some historical and collective memory which may help explain away certain conflicts and rationalise or justify political and social domination. Weak and young states, lacking time-tested legal frameworks and political cultures for conflict regulation, are likely to be very brittle in attempts at conflict resolution (Amin, 1987:17).

The inherited liberal democratic politics was defined by most new politicians as a luxury that poor countries could forego, because the goals of development were clearly defined and governments needed to act quickly. There was no time or need for "unnecessary" competition, politicking and division that democracy "breeds" so the argument went. African leaders, in the defence of their authority stressed the "consensus building" and lack of expressed opposition that they claimed characterized many indigenous political institutions and argued that political opposition was an aberration in the African context (cf. Markotvitz, 1977; Austin, 1964; Hodder-Williams,

1984). Where opposition was recognized, it was supposed to be conducted within the single-party apparatus, as opposed to within separate political parties.

Thus while the ideology of development remained a rallying point for politics, the political arena began to shrink. Consequently, leaders re-asserted centralizing and exclusionary colonial political structures and de-participated a large segment of their populations. Highly personalistic forms of rulership, with shifting eclectic blends of legitimating symbols and doctrines, came to be some of the defining features of postcolonial politics (Callaghy, 1987:89). In one country after the other, the executive branch of government asserted superiority over both the legislature and the judiciary, instead of the reverse as it supposed to occur in liberal democracies (Hodder-Williams, 1984:140). The economic change induced by colonial capitalism, however, did not produce a strong modern capitalist class. At best, it was incipient and too weak to anchor representative government, demand political accountability, and consequently, to check the patrimonialization of the postcolonial state by the politically dominant class. Constitutions described the formal arrangements of governance, but, in reality, power usually resided in the leaders. Whether there would be stability, peace and socioeconomic development depends to a great extent on the will, skill and wishes or ideologies of the leaders (cf. Jackson and Rosberg, 1994). Most leaders lack an autonomous material base, and the paucity of economic opportunities outside the state apparatus makes the loss of office a serious business; hence the desire on the part of those fortunate to be in power to rule for life. Politics became a matter of life and death (Healey and Robinson, 1992:18-19). This often resulted in inflammatory inter-elite conflicts dividing elites over the long-term national purpose.

As elites became increasingly resentful of each other, they pulled apart and placed a high premium on capturing power for themselves. They also became antagonistic to other societal groups that could help ensure the orderly "circulation of elites" in the political system. This led to a "winner takes it all" political game, such that development remained a rhetoric for garnering political support and for monopolizing the state, but in reality it received only marginal attention. The monopolization of state and state control of resources enabled leaders to extract resources for developmental purposes but interestingly, it also made the ruled perceived their interests as inexorably dependent on rulers or the dominant political party in control of society. This created false consciousness among the people and preempted any resentment or actions from civil societies and organizations that possibly could engender fundamental reform of the state in a way that would conduce to socioeconomic development. Gradually, what began as attempts to construct "developmental states" in Africa in the late 1950s and early 1960s can be said to have now turned into something resembling the European patrimonial dynastic states with all its attendant consequences for promoting development.

Recent discourse on Africa's economic crisis is replete with consequences that point to the view that the performance of the state has declined markedly over the postcolonial period (see World Bank, 1997). Various terms have been used to designate the state's ability to control society and extract resources, and, paradoxically, its weakness to provide what can be described as the Weberian conditions that drive development forward, let alone spur socioeconomic development. These terms include "authoritarian", "overextended", "overdeveloped", "underdeveloped", "weak", "soft" (cf. Callaghy, 1987, 1988), "anti-developmental" (Sandbrook, 1986), "predatory" and "vampire state" (Frimpong-Ansah, 1991), "the lame Leviathan" (Rothchild, 1987), and so on. The state's weakness is also evidenced in diminishing control, political repression, inefficient administration, endemic political instability, societal resistance and withdrawal, enormous waste, policy errors and corruption. All these features, mirror the decays associated with personal rule, which has emerged in much of postcolonial Africa. They constitute new forms of patrimonialism, which Weber considers inimical to modern capitalist development.

Certainly, few if any African societies are feudal, in which patrimonial politics could thrive. Being mainly an outgrowth of the colonial administrative state, the African state is, strictly speaking, not a patrimonial one because it is supposed to function as a rational-legal institution, a situation that leads some analysts to wrongly equate its form to function. The African state is, however, less bureaucratic than often thought; it is patrimonial. Weber's classic term of patrimonialism is certainly not meant to be applied in this context, but it is legitimate to apply Weber's term to the modal African state if it manifests patrimonial tendencies. Clapham, one of the few social scientists who have recognized the retrogression of the postcolonial African state in this term describes it as

> a form of organisation in which relationships of a broadly patrimonial type pervade [the] political and administrative system which is formally constructed on rational-legal lines. Officials hold positions in bureaucratic organisations with power which are formally defined, but exercise those power, so far as they can, as a form not of public service but of private property. Relationships with others likewise fall into the patrimonial pattern of vassal and lord, rather than the rational-legal one of subordinate and superior,

and behaviour is correspondingly devised to display a personal status, rather than to perform an official function (Clapham, 1985:48).

The Anatomy of Neopatrimonial Rule in Postcolonial Africa[7]

The main imperative of patrimonialism is the survival of the ruler, his regime and its hangers-on. That has also become the concern of many African leaders in recent times, and has implications for the conduct of politics, and, more significantly, the promotion of socioeconomic development in Africa. To meet the imperative of neopatrimonialism, the state is used in a variety of ways by its major occupant, the main ruler and his associates in power. To begin with, the leader seeks to keep power as long as possible (to rule for life) by refusing to seek re-election or to relinquish power to others peacefully. If he subjects his rule to democratic process at all, he carefully sets the framework for election in a way that ensures his own victory. In pursuit of this objective, the state operates like a secret or maffia-like organization with its own security networks and own agenda, which, in many cases have little to do with long-term societal development, although this is often the main official reason for rule.

All rulers desire legitimacy. Since legitimacy for governments is rare or rather weak in many African countries because, loyalty is divided, modern politics is not yet institutionalized, regimes are often toppled by military coups, etc. political compliance is secured in pragmatic ways other than through political culture, ideology and norms. While some patronage is evident and perhaps essential in every polity, in the African context, however, it is not unusual for governments to maintain a patronage system even to a point where it undermines productive activities of groups, thus bankrupting national economies.

Often, the political leader confirms himself or behaves as the "father of the nation" and treats state officials like the officials of Weber's patrimonial state. Usually, they are from the leader's ethnic group, or are relatives, friends, or "old boys" from school and they hold public office with the ruler's approval. The ruler's personal conviction and discretion delimits the jurisdiction of state officials (Sandbrook, 1985). A highly developed but shifting patron-client relationship, held together mainly by mercenary considerations, are the major means through which political leaders deal with their opponents in order to survive. The state acts as a patron for persons and civil associations by supporting them financially and expects loyalty from them in return. Legal protection and material privileges are also distributed in exchange for political loyalty.

At this early stage of their evolution, many African countries have not evolved basic institutions that constrain political behaviour. Politics is a game played mainly for self-interest, expediency and necessity, without legitimate and restraining rules. Actors improvise roles, change offices and turn politics into "makeshift" systems marked by feuding, nepotism, conspiracies, purging manoeuvres and factionalism (Marenin, 1987). Given the imperative to survive in such an intrigue-ridden political environment, leaders can no longer come to act in ways that raises the fortune of their societies. The rulers who dominate states may "allow" their officials illegitimate or preferential access to state resources. For instance, agricultural marketing corporations will draw resources from taxes imposed on export commodities produced by peasants, yet allocate their loans mainly to the government officials and rich men, the executive of the state can block the auditing of public institutions, can instruct the central bank to make available foreign exchange to certain individuals, and corruption in high places can go unpunished while petty theft by ordinary people results in heavy prison sentences. Leaders can impose rules that restrict the importation of "luxury" cars as a way to limit import bills on oil, while they and their ministers use such cars with air-conditioners and free fuel, even during their private moments. There is thus little in the way of leadership by example. As a result, most economically weak African countries are saddled with enormous cost of state administration.

For these officials and other followers of the leader who enjoy these privileges, the expectation in sharing the spoils of office reinforces personal links with the ruler. The personal ruler ignores or frustrates institutions that could ensure public accountability. Institutions, legal norms and administrative regulations that define the uses and abuses of political power are at best embryonic, ineffective and without legitimacy. Officials hold power and office formally, but the distinction between public trust and private gain is often neither accepted nor acted on. The public and private spheres are often blurred. In the Weberian sense, office is privatized. Corruption, in particular, in most countries has now become so widespread—almost institutionalized—that nothing gets done without payment of a hierarchically determined bribe or kickback as a matter of routine (Ergas, 1986:305). Leaders often acknowledge the extent of corruption and other evils and their corroding effects on the economy and civil sensibilities and engage in episodic actions to curb them but soon resign themselves to inertia.

In the African neopatrimonial political systems, important public posts, which should formally go to bureaucrats, are assigned on grounds of "connection" to the chief executive. Jobs go to the "right" guys or party stalwarts, or "the boys" for political support. Contracts are given to political favourites, and these favourites are expected to return kickbacks to the approving minister. These features are certainly not unique to African political systems. Indeed, the politics of the US, Italy and Japan, for instance, can sometimes be seen in similar ways, as businesspeople often "invest" in politicians and expect economic and political rewards. But in Africa, politics has become more instrumental than programmatically geared towards the promotion of national wealth. Economic and social life is much more politicized. Consequently, the social division of labour that should accompany societal advancement is inhibited, since, for example, the businessperson has to become politically involved to get contracts. Often, particularistic rather than universalistic criteria are used for recruitment into public office. Pork-barrel politics (the provision of services or development projects for political and not economic considerations) ensues, making the development process uncoordinated, disjointed and ad hoc.

As the positions of officials are largely bestowed by the personal ruler, state officials do not "rock the boat" by initiating alternate or contrary policies, because to do so is likely to be seen as "sailing against the wind" or challenging the leader which could result in a loss of position. Almost every important decision is left to the leader, and, since the leader is practically not available and humanly incapable of being competent on all matters, policies are often not made, instead government officials take the less politically-risky avenue by appealing to citizens to act morally and properly. If policies are made at all, they are either wrong or not clear in terms of their development objectives because very often, they are not subject to rigorous debates that could lead to consideration of alternative policies and views.

Furthermore, internal criticism is suppressed because the leader is "supposed" to know best what the nation needs at all levels. Cabinet meetings are often ceremonial to approve executive decisions. Adulation is heaped on the leader, and a hero-worshipping cult ensues as a way of assuring the leader of one's loyalty. The personal ruler dominates the party, the regime, state and society. He defines his role as the key to the nation's destiny and his views often come to dominate what policies should be pursued. He surrounds himself in varying and often uncertain ways by a cadre of state administrators and intellectuals who share his views or are willing to substantiate his ideas however bizarre they are. To challenge the leader, even if his ideas are detrimental to economic viability of the nation, is to risk ostracism and one's political position, which normally comes with privileges. Lacking independent economic base, politicians are unable to sacrifice their well-being for principles that advance national interest.

Consequently, competence, probity and credibility in matters of national administration are sacrificed, as many times "square pegs are put in to round holes". In many African countries, it is not unusual to see mismatching of competence and positions. For example, a person trained in biology can be directly appointed to become administrative head of a region or a university lecturer, a head of state organization, or to find a Minister of Agriculture who has no agricultural academic or rural background. Positions have more often than not, little or no connection with professional qualifications. This often results in administrative incompetence and policy blunders as well as waste of public resources. By confining top professional jobs to mediocre or incompetent followers, leaders under-utilize existing competence or human skills and demoralize professionals and even indirectly drive them out of their own countries. The IMF and World Bank have time and again urged governments to adopt policies to rectify this anomaly since the early 1980s when they assumed greater management control over many Africa economies. Today, much of Africa's trained manpower is outside the continent, though the trend might be on the decline as a result of programmes being initiated by the European Union and other North American governments to re-settle skilled Africans in Africa or to curb their exodus.

Since the political environment is so uncertain, the crucial political concern for leaders is not to lose guard so as to be toppled. Non-governmental actors or the "critical public" with a direct interest in promoting debates on economic and political matters notably the professional associations, trade unions and business associations are often muzzled, if not eliminated. Restricting debates on vital national issues helps to limit criticisms and opposition against regimes, but this undermines public accountability and public-spiritness of the state. It also discourages the generation of innovative ideas that are crucial to move societies forward. Leaders have a propensity for expanding the state sector activities to satisfy public demands for employment, but this is often done at the expense of the private sector. This is because through the state sector, leaders can satisfy the demands of the constituencies that support them and further control society. Yet still, the expansion of the state sector is not always guided by considerations of efficiency, and they soon impose huge recurrent expenditures on the state and blow budgets out of proportion that are often resolved through deficit financing. On this issue, the World Bank (1989a) notes that Africa has levels of "public employment" substantially higher than that of Asia and Latin America.

Also, following a tradition set by the colonialists, the African state monopolizes the marketing of primary products, and thus extracts substantial resources from society in the form of revenue from oil, export commodities, and foreign aid. Its activities are wide, as evidenced in the large bureaucracy it operates and the controls and frustrations it puts on private sector development. This means that the state also monopolizes much opportunity and distributes privileges; thus, participation in it is very attractive. Given this, the state offers one—or perhaps the only—quick avenue for upward mobility, wealth and power. Office carries incredible power over others, prestige is high; and the material privileges are inestimable. Such a situation can hardly compel state officials to reform the state in a way that limits the material and intangible privileges they enjoy. This helps to explain the phenomenon in many African countries whereby an elite group claims to be radical "reformist", denounces the privileges of incumbent elites and captures power only to refuse to reform the state.

Perhaps the rich prose of literary scholars is better suited to describe this political situation than that of the social scientist. For example, in *A Man of the People,* one celebrated novel satirizing postcolonial Nigerian politics, Achebe (1967) deals with three main social groups in society: (i) the corrupt politicians (who took over power from the colonialists), represented by Chief Nanga (who know how to appeal to the electorate with promises so as to be returned to power); (ii) the intellectuals, who want positive change and end up unable to distinguish between their private ambitions and public goals; and (iii) the cynical masses, who do not care to change their leaders so long as they are not dying. Through the teacher, Odili, Achebe provides an acid indictment of Africa's new rulers. The plot of the book resolves around Odili's excursion into politics and confrontation with Chief Nanga about what, in this analysis, is neopatrimonialism. Odili explains why politics is so in Nigeria:

> We ignore man's basic nature if we say, as some critics do, that because a man like Nanga has risen overnight from poverty and insignificance to his present opulence he could be persuaded without much trouble to give up again and return to his original state.
>
> A man who has just come in from the rain and dried his body and put on dry clothes is more reluctant to go out again than another who has been indoors all the time. The trouble with our nation ... was that none of us had been indoors long enough to be able to say "To hell with it". We had all been in the rain together until yesterday. Then a handful of us—the smart and the lucky and hardly ever the best—had scrambled for the one shelter our former rulers left, and had taken it over and barricaded themselves in. And from within they sought to persuade the rest through numerous loudspeakers, that the first phase of the

> struggle had been won and that the next phase—the extension of our house—was even more important and called for new and original tactics; it required that all argument should cease and the whole people speak with one voice and that any more dissent and argument outside the door of the shelter would subvert and bring down the whole house (Achebe, 1967:42).

Achebe's satire comes out when Odili accuses the masses of resignation and cynicism, and, at the same time, turns to exhibit many of the faults of the people and politicians around him. Disillusioned by corrupt and mediocre politicians who seem to have betrayed the promise of independence, a tiny group of intellectuals decided to form a new political party to dethrone the corrupt regime of Chief Nanga and the like. But the masses seem to have resigned themselves to their lot, and do not care as much as the intellectuals about the corrupt and abusive behaviour of the politicians. Towards the end of the book, when the "mess" of the politicians was being analyzed, Odili had no second thoughts about the masses. He remarks thus:

> The people themselves ... had become even more cynical than their leaders and were apathetic into the bargain. 'Let them eat', was the people's opinion; 'after all, when the white men used to do all the eating did we commit suicide?' Of course not. And where is the all-powerful white man today? He came, he ate and he went. But we are still around. The important thing then is to stay alive; if you do you will outlive your present annoyance. The great thing, as the old people have told us, is reminiscence; and only those who survived can have it. Besides, if you survive, who knows? it may be your turn to eat tomorrow. Your son may bring home your share (Achebe, 1967:161-2).

Odili presents the shortcomings of the politicians, but only to exhibit many of the faults of the people around him. Achebe is very well-grounded, painting a picture of how one elite group tends to perpetrate predatory colonial rule and the people themselves have resigned to their fate. There seems to be no way out of this political predicament many African countries face.[8] Some analysts now perceive the postcolonial state as a "curse" and "burden" on ordinary Africans (see e.g. Davidson, 1992).

Achebe wrote about this political condition some three decades ago, but it is only now that this scenario appears to be changing mainly as a result of recent external pressures for democratization. Most African political leaders, like the absolute rulers of premodern Europe, consider themselves above the laws and constitutions of their countries, even those they themselves make, and generally demand to be treated like divine princes. The police, the judiciary and parliament, as well as the state-controlled mass-media treat them that way. The media for instance reports every action of leaders, however, insignificant. The rule of law is circumvented by decrees or refusal to follow established institutional procedures. Personifying the state, some of these leaders build themselves palaces, bring traffic to a standstill when they drive, hold conspicuous parades, impose numerous curfews and checks on their societies and keep private armies to harass political opponents, industrialists and businesspeople, irrespective of the cost such actions impose on production, business and the general functioning of their economies and societies.

Also, some of these leaders and state officials can appropriate public funds at will. In many African countries, the incomes of state officials are a mystery, and can only be estimated from their living standards or life-styles and the resources they "squander" in private. In some cases, a substantial share of revenue generated from the sale of export commodities is deposited in private foreign bank accounts. The amount of public funds that has been deposited abroad is of course difficult to determine, but the World Bank says hundreds of millions of dollars have been siphoned off to private bank accounts outside Africa. And one of the arguments advanced by Western bankers for why they would not offer Africa debt forgiveness is that their previous loans to Africa ended in the West in the private accounts of political elites (cf. Jackson and Rosberg, 1994). While this fact is accepted, Western bankers, however, are unable to give such monies back to Africa when most of these leaders leave or are removed from office.

Furthermore, the policy-making circles are often restricted to informal caucuses surrounding the political leaders and the policy-making process is opaque. Little or no public debate informs crucial national issues. Policies are often chaotic, uncoordinated and do not always follow what is formally laid down. Basic economic data, such as national output trends, unemployment rates and rates of inflation that should guide national policy formulations are ignored until the economy collapses. Criticisms and alternate viewpoints are construed as attacks on the state coming from "enemies" of the regime or an attempt to topple the regime or as detrimental to "national interest", whatever that means. This helps to explain why the African state suppresses press freedom, abuses human rights and arbitrarily incarcerates political opponents, real or imagined.

In addition, the ruling class imposes discourse closure, defines itself as the only group capable of representing the national interest and promoting development. In the Habermasian (1987) communicative terms, truth and rational discourse are impaired by power relations. Consequently, policy errors are often committed, but little punishment and cost exist in the system to check bureaucratic incompetence and deter potentially deviant behaviour.

Likewise, there is little or no consensus formation around policies, nor "going by the books" that could provide legitimacy for government itself and policies. Also, no credible system exists for evaluating how specific national resources and foreign loans are used. Money can be printed and demonetization policies initiated at will, often against expert advice. Governments can instruct banks to refuse or grant loans to individuals, and confiscate bank accounts and means of production regardless of the damage these actions do to the confidence in the national economy, investment and wealth creation.

Furthermore, power is applied to a large extent, in an arbitrary manner. For example, courts can be ordered to convict innocent people, public treasuries can be expropriated almost at will for political reasons and laws and public institutions can be changed by simple executive decree. Basic constitutional principles, such as the separation of powers, "checks and balances", fair and representative elections and alterance in government do not, in general, carry much weight (Ergas, 1986:308). This context parallels the arbitrary premodern European context described by Jones (see above) as inimical to capitalist development. Analogously, it can be hypothesized that this context encourages disinvestment and capital flight, and, consequently, undermines long-term investment and productive activities in Africa.

On development planning and defending national interest, the state's rhetoric is almost infinite. Paradoxically, however, organizational and implementation capacity is often weak (Hart, 1982). Plans and policies are at best only intentions, published widely and elaborately launched; continuous implementation or evaluation is, however, another factor that does not feature adequately in the equation (Ake, 1996:9). For example, year after year, governments offer solemn promises to increase agricultural productivity, to improve health and educational opportunities, but food shortages and social problems are not diminishing. Often, this is not due to lack of resources, but effective organization and implementation capacity as well as optimization of the connections between ends and means. As developmental and organizational capacity is weak, credibility of governments sinks further. A general inertia permeates the political system. In the end, government itself is overwhelmed by national failures and perplexed as to what actually should be its tasks, and waits for the so-called foreign experts, who are eager to define the problem and how it should be tackled.

In this situation, every fashionable ideology or fad in the industrial world and misconceptions has been imposed on and applied to Africa, by local elites sometimes with or without the collaboration of foreign powers and development agencies, no matter how irrelevant and inappropriate they are in solving specific local or national problems. Africa now exists as a vast social laboratory where various ideas emanating from the advanced countries can be practised.

This increasing appropriation of Africa's development process by foreign development agencies has undermined the confidence of political leaders to lead their peoples into self-sustaining development processes based on African social experiences and institutions. A psychology of helplessness and looking up to the advanced North for salvation and development now permeates all levels of society. A large reservoir of active labour force and local experts live in despair and wasting their potential and human capital, as a result of leaders' submissive preoccupation with foreigners to solve national problems.

The proliferation of independent development actors, such as NGOs (Non-Governmental Organizations) and other charity organizations in recent years, by-passing the state, in the management of Africa's development process is a clear testimony of the frustration not only with the failure of the African state to play a positive role in socioeconomic processes, but also to have effective control over its own society and economy. But it is also that various self-serving movements, organizations, consultancies, professing to be doing development now besiege the continent, setting up offices here and there and doing whatever pleases them. In the end, Africa's own development initiative is sacrificed and development has turned into charity.

But, above all these internal contradictions or pathologies, perhaps the most devastating aspect of neopatrimonialism is uncertainty, evident in the state's relationships with economic actors and the larger society. Laws, public rules and policies that should provide predictability and act as enabling structures for social actors (peasants, businessmen, investors and corporate actors) are changed at will without adequate provision of time to allow citizens to adjust to them. Constitutions, which should provide a universal framework for political actions are often couched in particularistic terms to protect leaders and their vested interests. Policies are determined and driven mainly by the logic of patrimonial political domination and are implemented instantaneously without adequate public debates and considerations that enable citizens to understand and respond to them. Governance is many times a surprise and an unpredictable phenomenon often inciting panic in society and driving citizens helter-skelter. For example, any hint of governments' intention to read a budget throws the societies into confusion. Goods disappear until the budget is read. While state politics can generate arbitrariness and unpredictability in the socioeconomic life of society, the players of state politics and those people responsible for such capricious policies often shield themselves against their negative impact, because they consider themselves as above the laws of the land and consequently they flout them. Given this state of affairs, McCarthy (1994:106) notes that "behind every decision lurks a hidden agenda", and one can add intentions that do not advance the welfare of general society.

Consequently, over the years, institutions and rule systems that should guarantee individual rights and property, investment and growth have all diminished. Predatory politics and arbitrariness are what characterizes government-business and government-citizens relationships. For example, the rise of local businesspersons or an economically dominant and politically independent private class is perceived as a threat to those who dominate society by virtue of political power and thus opposed.

The response of local citizens to this volatile environment is to avoid the state and devise institutions to mitigate the risk induced by politics. Citizens have waged a silent war on the state by retreating into nontaxable economic activities, subsistence, and informal economy, or what Hyden has referred to as "economy of affection". Africa's "reasoning publics" are also repressed and intellectual discourses that should inform key national issues and lead to innovative ideas that should drive development forward are not allowed. Some educated elements with internationally tradeable human capital have adopted the exit option and left their countries for greener pastures. While this exit option might work to the advantage of the rulers, it, however, deprives African countries of much of the human capital they have developed over the postcolonial era.

The African state has increasingly become autonomous of hegemonic class forces, dysfunctional and anti-developmental. Yet, however weak the state may be in terms of achieving its developmental aims or laying a calculable socioeconomic framework for society, those people controlling it are sufficiently powerful to smash their opponents. The coercive instruments of state inherited from the colonial period have not changed but have been intensified to deal with political opponents. The continued success of many rulers to rule for a long period of time produces the appearance of stability, and, perhaps, the formation of a dominant class. If such rulers are fully able to crush their opponents, perhaps they might enhance the consolidation of dominant class, but there are always new rival groups ready to replace the threat just broken.

In some cases rulers are unable to stave off threats to their political domination. In this context, Kasfir (1987:56-7) notes that, class deformation, rather than class formation, is likely to result. This further leads to the development of intensive security networks, patron-client relationships, ever shifting alliances (Sandbrook, 1986:88) and the development of a panoply of coercive instruments to destroy competing incipient class fractions or elite groups. Governments often claim to represent the national community and unity is made paramount. All politically and economically significant cleavages—particularistic, pluralistic and class related—and all conflicts are defined as illegitimate and society is given a sociological uniformity (Callaghy, 1987:95). In such an environment, one is tempted to see the state as unrepresentative of any hegemonic group except those who currently dominate it or have privileged access to it.

Thus, if the political skills of dynastic rulers are to build strong national economies so as to defend themselves against other powerful external rulers, the political skills of most personal rulers of twentieth century Africa are simply to develop ruling class closures to prevent others from unseating them. In the end, a monolithic patrimonial behemoth nakedly confronts society and whose functioning suffocates the development process. As far back as the early 1960s, the French agronomist Rene Dumont in his insightful book significantly entitled *False Start in Africa* ([1962/1966]/1988) was warning about the impending erosion of developmental capability of the African state when he accused new governing elites of inheriting predatory and coercive colonial state structure only to create self-serving modern versions of Louis XV's court. Given this, a major problem confronting many African states today is how to establish "development administration" as bureaucratic norms and principles incipient at independence seem to have been eroded over the postcolonial period mainly as a result of endemic political instability, neopatrimonialization of the postcolonial state and military praetorianism.

Military Praetorianism and Re-patrimonialization of the African State

In addition to neopatrimonialism, many African countries also experience frequent political instability. Given the advantages to be derived from associating with the state, its capture is attractive to various groups, particularly the military, who control the instruments of coercion. Political instability, induced by the constant incursion of the army into politics, has become so rampant that many African states can be described as *praetorian*. A *praetorian* state, according to Perlmutter (1969:383) "is one in which the military tends to intervene and *potentially* could dominate the political system. The political processes of this state *favor* the development of the military as the core group and the growth of its expectations as a ruling class; its political leadership (as distinguished from bureaucratic, administrative and managerial leadership) is chiefly recruited from the army, or from groups sympathetic, or at least not antagonistic, to the military. Constitutional changes are effected and sustained by the military ...". In a *praetorianist* society, the military class exercises independent political power within the society by virtue of an actual or threatened use of military force. Today, the African state is often a battleground for soldiers, intellectuals and civilian politicians who can form a powerful coalition to capture it.

The persistence of military coups, however, has called forth numerous attempts to explain why the military intervenes in politics and the role they play in the socioeconomic transformation of their societies. Some have sought statistical correlation between those states that have experienced coups and military rule and those that have not; some, particularly neo-Marxists have sought explanations at the global level: economic dependence (e.g. the drop of value of exported primary products and coups), and some have argued that they are due to the lack of political culture (see Finer, 1962). This political culture according to Clapham (1985), can be liberal-democratic or Marxist-Leninist or traditional religious, that enables a broadly-held belief in a certain political structure in which leaders are selected according to a given procedure or exercise power in accordance with given standards as to what is legitimate or right. Yet still others have seen praetorianism as lack of political institutionalization (see Huntington, 1968) which mirrors socioeconomic decay rather than political culture, per se.

From this latter perspective comes what Decalo (1990) refers to as "systemic-failure hypothesis": that socioeconomic decay is associated with military coups. Adherents of this perspective have looked up to the military to play a positive role in socioeconomic development. Indeed, one often repeated argument in the sociology of military rule is that the military is a modernizing agent. This view is informed mainly by organizational theory, which stresses the organizational structure, discipline and efficiency of the military institution as perhaps the closest approximation to the Weberian bureaucracy. The view also sees the military as a disinterested nationalistic organization. Therefore, the argument goes, when the military enters politics, it brings its bureaucratic characteristics and orientations to bear on the ruling process, and this, in turn, facilitates honest administration and economic development.

One major problem with this "systemic-failure hypothesis" is that it easily leads analysts to neglect agency and actor perspectives in the genesis of military coups. As Decalo again argues, such an intellectual predisposition and analytic approach can result in the gross reification of armies and also in the uncritical acceptance of the view that armies are modernizing or developmental agents. Decalo draws attention to two major pitfalls in this approach to military rule that are relevant to this study: (i) the general tendency to accept officially enunciated reasons for military take-overs as valid, particularly when the toppled civilian regimes have only too obviously been manifesting the weaknesses (e.g. corruption) for which they have been attacked; and (ii) the propensity to ignore the "personal element". As he further elaborates, the main weakness of attempts to explain military interventions by pinpointing areas of systemic stress is in not placing sufficient weight on the personal and idiosyncratic element in military hierarchies, which have much greater freedom and scope of action within the context of fragmented and unstructured political systems (Decalo, 1973:113).

Perhaps the military can be a positive force in socioeconomic development. But this does not seem to be the case in many African countries where frequent military coups and changes in military rule have been accompanied by abandonment, changes, or reversals of development plans, projects and imposition of new macroeconomic management systems or rules of the game in society. This has resulted in a high degree of oscillation of development plans, ideologies and a lack of continuity in development of any kind. This is so because, by their very nature, military coups are illegitimate. As such, they must be justified and consolidated. Usually, this begins with the ideological manipulation of civilian society into believing that the previous leaders are the main culprits of the nation's problems. Often, this necessitates the abandonment or reversal of ongoing plans and policies and intense media campaigns to discredit previous leaders.

Praetorianism constitutes a setback to institutionalization in the political, economic and social spheres of many African countries (Opoku-Agyeman, 1988). When coups occur, constitutions that provide formal institutional frameworks for political leaders are almost always thrown away and society is administered by decrees. Military leaders are always keenly aware that their colleagues in the army would emulate their success; thus, national security is given priority. The army is also taken "good" care of, with the hope of minimizing dissent and preventing possible future coups. Soldiers are often promoted, their remuneration revised upwards and their amenities and privileges expanded. They are often given jobs in the public sector and new uniforms and equipment are acquired which helps to explain the sight of MIG-fighters and other sophisticated planes flying over fields ploughed by rudimentary agricultural technologies. The external enemy is, however, virtually absent, and the military then deploys its coercive instruments upon the state and society.

Each new ruling group that successfully captures the state denounces earlier rulers, policies and development projects to justify its own. Ideology and development strategy become contentious issues, leading to discontinuities in the development process. Thus like the political system itself, the development process is also characterized by perpetual renewal and decay. The formulation and implementation of development policies focuses on short-term gains that directly benefit governments and leaders. A coup begets coups, because the military is politicized and civilians want to re-assert control over the governing process. Politics becomes a game of musical arm-chairs or a revolving door through which soldiers and civilian politicians enter and exit. Increasingly, a militarization of postcolonial politics has emerged to a point where in some countries civilian politicians cannot re-assert control over the military institutions. If they have to rule at all, they have to do it under the threatening shadow of the military.

As hyper-instability paralyses the governing process, few development plans and ideologies have worked to their logical conclusion. Ghana, which provides the context case study, is one country that has suffered from such military praetorianism, and consequently lack of continuity of plans, policies and development models. The figure below, presents a synopsis of the type of regimes that have so far ruled Ghana, their leaders' names and dates of rule, as well as the main elements of their respective development strategies.

Figure 1: Regimes and Competing Models of Development in Ghana (1957-1999)

Name and Type of Regime	Date	Leader	Main Elements of Development Strategy/Policies
Convention People's Party (CPP), liberal democratic turned authoritarian, pro-East and anti-West	1957-1966 9 years	Kwame Nkrumah	Structural transformation and economic nationalism: state-led industrialization and large-scale capital intensive agricultural schemes, distrust for peasant agriculture, state regulation and control of the economy, social redistribution and egalitarian system, anti-local capitalism
National Liberation Council (NLC), military dictatorship, pro-West and anti-East	1966-1969 3 years	General A. Ankrah	Gradualism and free-market economy: state sector activities restricted mainly to infrastructure and public utilities, a balance between agriculture and industry, stress on growth and efficiency, promotion of private foreign and domestic investment
Progress Party (PP), liberal democratic regime, pro-West	1969-1972 2 5 years	K A. Busia	Gradualism and free-market economy: economic growth and efficiency, the promotion of private foreign and domestic investment but with emphasis on the development of indigenous entrepreneurial class
National Redemption Council and later Supreme Military Council, (NRC/SMC), military dictatorship, pro-East and anti-West	1972-1978 6 years	Col Acheampong	Re-etatization of society: economic nationalization and "non-policy" regime, social re-distributive policies
Supreme Military Council II (SMC II), military dictatorship, pro-West	1978-1979 1 year	General Akuffo	Gradualism: free-market economy
Armed Forces Revolutionary Council (AFRC), military dictatorship, pro-East	1979 (112 days)	Fl. Lt Rawlings	No development strategy, moral revolution: anti-corruption crusade, "anti-commerce" and re-distributive policies
People's National Party (PNP), liberal democratic, regime pro-West and East	1979-1981 2 years	Hilla Limann	Gradualism and free market economy: promotion of private foreign and domestic investment
Provisional National Defence Council (PNDC), military dictatorship, pro-East and anti-West	1981-1983 2 years	Fl Lt Rawlings	"Dependency"-inspired actions: class struggle, anti-business policies, confiscation of means of production and property as a prelude to socialism
PNDC, military dictatorship and civilianized as National Democratic Congress (NDC)	1983- so far 16 years	Fl Lt Rawlings now Mr Rawlings	Economic liberalism: free-market economy via a structural adjustment programme

The elements of these different development strategies can be summarized as follows: from 1957 to 1966, Ghana practised socialism; from 1966 to 1969, capitalism; from 1969 to 1972, capitalism; from 1972 to 1979, quasi-socialism; then, a populist revolution in 1979; from 1979 to 1981, capitalism; from 1982-1983, socialism; and from 1983 to the 1990s, capitalism.

There has been a considerable debate about whether the development policies pursued in Ghana can be neatly classified as socialism, capitalism, etc. The categorization of the regimes' orientations in this study is based on the popular understanding of what their leaders proclaimed and perhaps sought to do, as to whether some regimes were truly socialist or capitalist or even revolutionary or not, is a matter that cannot be settled in this book. What is important for our purpose is that these development ideologies or strategies were not only different and conflictive, but were also abruptly discontinued as a result of inter-elite rivalry and incessant military coups.

This Ghanaian case is, however, not unique; it is repeated more or less in many African countries. Political disorder in many African countries often induces policy discontinuity, unpredictability, and uncertainty into fields of social interaction and economic activities. As praetorianism becomes endemic, it reinforces neopatrimonialism which feeds administrative inefficiencies and corruption. Given this, the gradual depersonalization of political power and system and the progressive development of a bureaucratic state apparatus with a development-oriented administration, which could pave the way for modern development, is yet to seriously emerged in the African context.

Chronic political instability and uncertainty about the future of political leaders and civil strife (as evident in frequent military coups and counter-coups as well as internal wars) also compel rulers to divert scarce national resources to the national armies for security reasons. In this present era of structural adjustment programmes when much of the public sector employment is being down-sized, in many countries the army has remained unaffected by economic retrenchment policies and army recruitment continues unabated. The enormous cost of maintaining Africa's "idle" armies are not lessening. Political instability also reduces political time horizons so that rulers focus their attention on consuming state resources quickly and also personally appropriating the state.

In the immediate postcolonial period, it was understandable to see Africa's underdevelopment problems as originating mainly from the lack of modernization or as a result of (neo)colonialism. But almost half a century-old independence has increasingly revealed the nature and internal dynamics of African political systems more than before. What has become clearer now is that the African state might not be as bureaucratic as originally thought and the state's ideological proclamations do not mean much when one tries to understand African politics and how national development projects are pursued. Indeed, one can argue that socialism, capitalism and other ideologies do not say much about how the empirical African state functions. Whatever the state's ideological complexion, whether it is dominated by a party, president, military leader or other ruler, the main underlying form of most postcolonial African politics is the concentration and personalization of power (Hodder-Williams, 1984), and Weber's views about patrimonialism promises a perspective that can better unlock the nature of the Africa state and its relation to development.

These internal contradictions, seem just as important as the external factors that militate against Africa's development. These contradictions, it can be argued, have retarded the rationalization of many spheres of social life in much of twentieth century Africa and concomitantly have also undermined the development of the progressive aspects of rational capitalism, albeit capitalism remains prevalent at the global level and is the empirical economic system operating in many African countries. Thus Africa's transition to fully developed modern capitalism as at now remains partial and distorted, if not perverted.

If Africa's postcolonial state politics has not effectively served the interest of capitalism, the development experience has also shown that it has not served socialism either. The attempts to construct genuine socialism in several countries were feeble and contradictory and have resulted in abysmal failures. Given this and Africa's precolonial capitalist traditions, as well as the impact of capitalist colonial rule, it is capitalism that is left struggling for ascendancy, but as noted above, postcolonial political conditions put an enormous break on its development. Consequently, majority of Africans have suffered and are still suffering from a peculiarly burdensome type of capitalism in which they endure most socioeconomic deprivations associated with capitalist development without benefiting from its progressive impact.

Contemporary African economies thus possess characteristic features incompatible with or remote from those that Weber identifies as typifying or facilitating modern capitalist economies. To be sure, capitalist colonialism was unable to tear asunder Africa's pre-capitalist relations. These relations have remained resilient, although they appear to be on the decline with the passage of time, as a result of forces of globalization and modernization. Consequently, African economies are syncretic with mixed modes of production in which capitalist and non-capitalist social logics coexist, neopatrimonial states pursuing weak and incoherent mercantilist economic policies thus producing not genuine modern capitalism, but rather patrimonial capitalism.

In one insightful analysis of the capitalist development so far attained in Africa, Callaghy (1988:79) argues that most African economies are characterized by:

(i) lack of integration and interdependence in the economy from geographical, sectoral, and input-output viewpoints;
(ii) sectoral inequality in terms of productivity and specialization of material production;
(iii) partial commodification of (socio-)economic production and stagnant reproduction of economic activities;
(iv) exiguity of the internal market; and
(v) external economic domination and strong dependence of most of the economic activities on the world market.

Furthermore, Callaghy maintains that in these economies there exist:

(i) oscillation of social classes and groups between capitalist relations of production and precapitalist relations;
(ii) the detachment of most social behaviours and activities from the [formal] market;
(iii) the marginal position of indigenous social classes in the structure of ownership of the means of production and in the process of capitalist reproduction; and
(iv) the distortion, lack of distinctiveness and lack of the type of "organized cohesion" of class structure present in advanced capitalist countries.

To these it must be added the lack of clearly demarcated property rights regimes and governmental ability to enforce contract that could safeguard individuals private resources from outright squander by others or make future obligations possible. The partial privatization and demarcated property right institutions and the absence of enforceable laws of contract in politically independent law courts means that land, labour, credit, and commercial opportunities have not been as yet fully liberated to fuel capitalist development. For example, the redemption of one's property such as land and credit has come to depend on private arrangements often provoking conflict, unnecessary litigations over factors of production. Existing state policies and laws in many African countries have not adequately resolved these crucial non-economic factors to liberate resources from the rigidities of pre-capitalist customary traditions to make capitalist development a smooth process.

Instead, state policies are driven mainly by neopatrimonial logic that leads political actors to use extractive state power to generate resources and to channel these resources into the maintenance of political power at the cost of genuine socioeconomic development. This political logic undermines the proper management of many African economies and the capacity of the state to intervene effectively and drive development forward. Africa's disarticulated socioeconomic structures are characterized by uncertainties, political arbitrariness and unpredictability as well as unevenness, which induce a pattern of economic development in which the growth of transaction (or trade and speculative activities) riddled with deliberate bureaucratic red-tapeism, corruption, bribery, and moonlighting) is impressive and tends to outpace the growth of production. In these economies exist the proliferation of short-term small-scale enterprises and informal sector activities characterized by low levels of productivity and investment (cf. Callaghy, 1988:79-80). This environment is not particularly conducive to modern capitalism and does not possess the facilitating factors Weber links conceptually and historically to the genesis of capitalist development.

The political conditions associated with this disarticulated socioeconomic structure can, however, be protracted and self-reinforcing because, unlike most states elsewhere in the world today and in the past, including the dynastic states of old Europe, which had to fight for their existence and thus were compelled at the point of liquidation to improve domestic organization and productive capacity, African states have the luxury of avoiding that. This is because, as already hinted, members of a new international community often treat African states as protectorates. African sovereignty was and is still guaranteed by the United Nations and rival power blocks of the cold war upheld it because each block was competing for the public appropriation of Africa. Thus, African states did not have to bother to develop the military, administrative, fiscal and economic capabilities to deter foreign intervention or confront it should it happen. With few exceptions, unlike European states and some of the modern states of today, which had to respond to external threats, African countries have not been or are not fighting each other to make nation-states. Simply stated, they were free from what has historically been an important international motivation to engage in state building and economic development (see Jackson and Rosberg, 1988, 1994).

These domestic contradictions pointed out are frustrating to African citizens themselves and those development agencies that genuinely wish to help African countries make progress. Ironically, however, the dominant development agenda in Africa is now (as it has historically been) determined mainly by external actors and agencies and their vested interests as well as their perceptions of African realities. While some of these actors and agencies might genuinely be committed to Africa's progress, their good intentions to help through development aid and programme inadvertently turns development into charity thus ruling out Africa's own initiatives to develop itself. As one critic of aid puts it, development in Africa has now turned into a self-serving international industry that sacrifices the progress of its intended beneficiaries as well as the publicly stated intentions of its practitioners in order to survive. In this regard, the international management of Africa's development process might just be another form of subjugation that aborts Africa's socioeconomic progress. And the continued failure of development in African might probably serve as a yardstick by which the developed world measures and glorifies its own civilization and human evolution (cf. Maren, 1997:1-12).

Although this view might sound overstated because the development discourse seemed to have been reformulated and revised over these years to better serve Africa's interest, there is evidence which suggests that foreign policy-makers in the developed world and officials of the International Financial Institutions as well as development agencies have not moved away from their traditional philosophical position, which was premised on the notion that development is a process of unilinear growth and progress toward those ideals and standards attained in the advanced North. Therefore, as one analyst comments, northern rather than local experts have clearer visions as to what process and projects are most relevant in Africa. The triumph of neoliberalism today and the recent revisionism of development strategy formulated in SAPs and the demands for liberal democratic politics, to a large extent, is merely a restatement in new terminology of the same philosophy that traditionally has been associated with the northern-dominated development agenda. This being the case, it is not difficult to conceive the current crisis of development as, at least, also partially originating from the continuation of the so-called development ideologies foisted on Africa with financial aid (see Maclean, 1997:132-133).

Financial aid is, however, one means of solving African state's perennial problem of revenue shortage, arising from falling world prices of primary non-oil commodities. Revenue short-fall has constrained the state's ability to pursue developmental goals and meet social welfare commitments. But this revenue problem also has another dimension. The African state's source of revenue has been traditionally assured through the export of primary commodities such as cocoa, coffee, groundnuts (which peasants produce but do not directly consume), and resources such as minerals, timber and petroleum oil, the marketing of which remains under state control. Given these stable traditional sources of revenue, the state becomes easily "content", and is thus not compelled to seek new revenue sources from, for example, an expanding private sector. While the search for a sturdy and sustainable revenue base appears to be one way of strengthening state development, improving domestic economic productive capacity and ultimately driving economic development forward, the relatively easy access to revenue appears paradoxically responsible for lack of development. Easy access to revenue and foreign aid can generate complacency and thus become a source of economic stagnation.

Furthermore, though the external environment adversely affects the African state, it is, however, benign in one respect. It supplies loans, credits and grants to the African state. This is not to suggest that such assistance is sufficient, after all, campaigns against aid and its misuse have increased in recent times in the Western world. But aid provides additional funds, the absence of which, could probably compel leaders to undertake fundamental changes in society to bring about socioeconomic development. Aid dependency, apart from the debt burden it imposes on African countries, also indirectly introduces political inertia and discourages the state from undertaking effective domestic organization necessary for increasing economic growth and investment. For these reasons, Jackson and Rosberg (1992:22-3) note that the international community and its financial institutions and aid agencies might just be part of the problem of the underdevelopment of Africa's empirical state, since they supply aid without ensuring that it is always put to "good" use.

Given the African political sociology outline above, whether the state will pursue socioeconomic development or not, and what kind of development ideology or strategy the state will embrace, are mainly a function of leaders' own choices and the domestic situations in which they find themselves. So far the state in Africa, with rare exception, has not offered much in the way of developmentalist content or intent. It has neither democratically structured polities nor developmentally-inclined authoritarian politics; neither did it have efficiently run economies. The World Bank officials who are now shaping Africa's development do not deny these internal contradictions, but the current neoliberal development agenda that seeks to "roll back" the state's arm in the economy and in its stead enthrone the market might not be in the long-term development interest of African countries.

This is because it weakens the African state, de-historicizes the development discourse and thus omits the crucial and historical role states play in the early and intermediate stages of development. What seems urgently needed in most parts of contemporary African in view of the foregoing literature review, is state reform to enable African leaders provide better governance and economic management systems and institutional arrangements that could catalyze and facilitate modern development. A reformed state with developmentally-oriented administrative machinery is necessary to set domestic priorities right, protect the welfare and needs of citizens and serve as a counterforce against the injuries of economic globalization.

Conclusion

The goal of this chapter was to explore the intimate connection between state and development, both from a theoretical and historical perspective. If there is one lesson that can be learnt here it is that socioeconomic development puts enormously complex demands on the state and their actors to put in place governing structures, rational systems of managing the public realm and institutions that ensure relative predictability, certainty and calculability in society. Without this elementary framework, long-term development cannot be sustained. In other words, sustainable development requires, above all, an enabling socioeconomic and political environment nurtured by the state. States that cannot ensure this basic framework, let alone play, a catalytic role in development process risk postponing the improvement in the material conditions of their societies. The inability or failure of any modern society to evolve a rational bureaucratic state, which is crucial for these developmental functions, is an obstacle to the promotion of development. Patrimonial political domination is anathema to long-term development because it introduces unpredictability, uncertainty and incalculability in socioeconomic and political life. It also leads to uncompromising attitude on the part of leaders towards private economics actors outside the state's purview, thus preventing such actors from undertaking long-term investment necessary for increasing societal productive capacity.

The chapter also presented an anatomical analysis of the nature and mode functioning of the modal African state which exists in many African countries today. It is the contention here that personal rule of the patrimonial kind has emerged in many African countries. This rule exhibits features reminiscent of patrimonial political domination in premodern Europe that inhibited capitalist development. Although patrimonial political systems are rational and logical from the point of view of personal rulers, and perhaps necessary or unavoidable given the present socioeconomic environment, unfortunately however, they do not always make for good logic in the administration of public affairs. From this understanding, it is hypothesized that such increasing personalization of power and praetorianism in the postcolonial period in Ghana has undermined the capacity of the state to evolve or develop into an effective rational-legal bureaucratic state vital for intervening positively in socioeconomic processes. In addition, the state has also failed to provide the minimal "contextual variables" or institutional guarantees that Weber maintains are crucial to modern capitalist development. These are the main development problems this study seeks to reveal empirically in the Ghanaian case. These problems are not easily accessible through the dominant paradigms in development studies namely modernization, dependency and the current neoliberal economic theory.

Notes

1 In this book, Weber notes that his main purpose is to demonstrate the manner in which ideas become effective force in moving history and he does not intend to present a one-sided materialist conception of history. Weber's basic methodological position is that social change is induced by many factors. But in the *Protestant Ethic*, he sets out to investigate only "one side of the causal chain" in the formation of capitalist society. The "ethic thesis" is, however, not relevant to the problem of the present study.

2 It is often said that Weber (1864-1920) spent his life having a posthumous dialogue with the ghost of Marx. Weber's "Protestant ethic" thesis (i.e. ideas) is seen by some as a critique of Marx's materialist conception of capitalism. But in recent years, scholars have suggested that there is much more agreement between Marx and Weber than previously acknowledged. See, for example, (i) Anthony Giddens, *Capitalism and Modern Social Theory* (Cambridge:1971); and (ii) Robert J. Antonio and Ronald M. Glassman (eds) *A Weber-Marx Dialogue* (Kansas:1985).

3 Weber sees the evolution of society as a rationalization process and a necessary evil, because of the encroachment of formal rationality on all facets of modern social life. This process leads to disenchantment with the modern world, since it erodes "substantive rationality" from society. Weber sees increasing bureaucratization of modern society as an answer to the problem of large scale organization and complex administration but as also as a tragedy; an "iron cage of rationality" in which modern person has to conduct his life and from which no escape seems imminent. On this pessimistic note, Weber departs from the classical sociologists Comte, Marx, Spencer, Durkheim, etc. and of late, Harbermas, who maintain in varying degrees that societal evolution and Enlightenment could represent progress for mankind (see Ira Cohen, 1981:xxvi).

4 Weber's approach is much more complex than what is outlined here. Collins, for instance, whose systematization of Weber's views is by now classic, has suggested a causal chain of capitalism that distinguishes among: (i) ultimate conditions; (ii) background conditions; (iii) intermediate conditions; and (iv) components of rationalized

capitalism. Justice cannot be done to all these conditions in this study, not because they are irrelevant but because the main interest here is on what characteristic features the rational legal state has that are conducive to capitalist development and what role the state can play in shaping its surrounding environment to promote socioeconomic development.

5 This is a condensed history and should not be taken as meaning that European capitalist development was a smooth affair or that political leaders were well-motivated to act in positive ways that favoured capitalist development and welfare of their people. This history was rather brutal and barbaric by moral and ethical standards, but, ironically, it encouraged the rise and bureaucratization of the modern state, and institutions that Weber sees as crucial in the origin of modern capitalist development.

6 The economic transformation induced by colonialism as a whole was, however, of limited importance to the world economy. This economic insignificance has continued to the present day and has even diminished. According to the World Bank (1989a), in 1970, Africa's share of world commerce was only 2.4 percent; by the mid-1980s it had fallen to 1.7 percent. When these statistics are considered against the background of rapid population growth in the last two decades, then there has been a significant drop in living standards in much of Africa. The World Bank also noted that, by 1987, the 450 million people of the Sub-Saharan region had a Gross Domestic Product of $135 billion dollars, the same as Belgium, with a population of only 10 million.

7 The states of Africa certainly differ in terms of their nature, function, ideology, etc. The patrimonial characteristics offered here should be treated as a heuristic model to which various African states would conform more or less. For further details, see, for example, Jackson and Rosberg (1982).

8 Some analysts, however, see a way out of this predicament: a "second liberation". The "second liberation" implies a liberation of the people from oppressive postcolonial state. The first being the struggle for national independence (see e.g. Davidson, 1992).

3 Historical Background of Modern Ghana

Introduction

The economic expansion of Europe and the search for state glory in the nineteenth century was not an isolated historical occurrence. It touched other peoples of the world as well, mainly through European imperialism and colonialism. Colonialism saw the agglomeration of numerous and sometimes diverse ethnic communities of Africa into colonial states with modern production and organizational systems closely linked with the economies of the industrialized countries, or what is sometimes known as the international capitalist economy. But, since the end of World War II, the indigenous peoples of these colonies began to revolt against foreign domination and took various actions that ultimately led to their political independence. Colonialism induced new socioeconomic and political changes that set the context within which the struggle for development is being waged.

The main purpose of this chapter is to review how this colonial process unfolded in the Gold Coast, now Ghana,[1] so as to shed light on Ghana's so-called privileged position specified at the outset of this study. Three main issues are focused upon: (i) colonial development policy; (ii) the socioeconomic impact of colonialism; and (iii) the rise of nationalism and the struggle for independence. This chapter is important for understanding the nature of postcolonial politics, the inter-elite conflict over ideology, and development strategy, which have become defining features of Ghana's state and society. In this chapter, the name Ghana is sometimes used interchangeably with the Gold Coast wherever appropriate, even though the period the chapter is principally concerned with is prior to independence.

But before examining the colonial intermission, it is proper to provide a brief description of the political and economic institutions of precolonial Ghanaian societies. This helps to understand the original state of these societies and to determine the extent of the transformation provoked by colonialism. Furthermore, even though precolonial economic and political institutions have not been well-treated in the development literature, assumptions about their characteristics have often been used to justify political action and the choice of development ideology in much of the postcolonial era. For example, it was not unusual for some postcolonial leaders to argue that precolonial societies were originally communal, egalitarian, and non-exploitative, and that capitalism was a by-product of colonialism, alien to and disruptive of society. In this regard, there has been a certain political ambivalence towards capitalist development and the desire to seek a new imprint for society based on foreign ideologies, usually socialism, rather than on Ghana's own peculiar realities and social experiences. Consequently, indigenous knowledge and institutions are treated as hindrances to modern development.

At the same time, some Africanist scholars maintain that Africa had long developed some time-tested political and economic institutions that could provide pillars for modern development and that part of the failure of the postcolonial development project has to do with the conventional "top-down" orientation of development strategy and the outright neglect of or assault on indigenous institutions by governing elites (cf. Ayittey, 1991:329; Ake, 1996:15-16; Davidson, 1992). What, then, is the nature of precolonial political and economic institutions in the Ghanaian context? The following section focuses mainly on describing precolonial economic and political institutions. The Ashanti society provides an illustrative case, for two main reasons: (i) on several levels, it is not different from those of other ethnic groups, such as the Ewe, the Mole-Dagbani and the Ga-Adangbe; and (ii), the Ashanti case is also the most documented mainly because of Ashanti state's complexity, belligerence and the obstacles it posed for British colonialism.

Precolonial Political and Economic Institutions of Ghanaian Societies

The West African region that became the Gold Coast under colonial rule, and subsequently Ghana upon independence, had ties with the outside world dating back before its formal colonization by the British at the end of the nineteenth century (Kay, 1972). The Ashanti (also spelt Asante) nation-state which had emerged by 1700, was one of the most powerful centralized political systems in this area. Ashanti society had all the attributes of a nation-state: a state, a given territory, a centralized system of government with police and army and a national language, law and a national council called the *Asanteman* (cf. also Wilks, 1975; Davidson, 1992). Herskovits (1964), who has studied the Ashanti state in comparative terms, argues that its complexity in the nineteenth century was superior to the European feudal state. He claims that the Ashanti state had a greater variety of land-tenure systems, better methods of exploitation of the land, more advanced form of labour utilization and more sophisticated modes of production than feudal Europe (cf. also Ergas, 1987:4). Kumasi, the capital city of the Ashanti Empire, as some call it (and now the second largest city in modern Ghana), carried on a substantial trade in gold and other exotica with the Arab world across the Sahara to the north and with the Europeans. The Carthaginians in Herodotus' historical accounts had long found a system of "silent trade" in Ghana, (cf. *Gold Coast Annual Report of 1947*) and such a trade suggests that indigenous societies have long had some links with outsiders. The Europeans mainly the Portuguese, Dutch, British, Danes, Swedes, Brandenburgers who, since the mid-1400 came and established trading forts along the coast of the country, were looking for an African Eldorado: the Ashanti gold.[2] The gold originating from this area was significant to the international economy in that, between the eleventh and the seventeenth centuries, this region was the world's leading supplier, and, by the later Middle Ages, it accounted for two-thirds of the world's gold production (cf. Hopkins, 1988:82; Howard, 1978:28).

Before the advent of colonialism, the various ethnic groups making up Ghana today existed, with the exception of the Tallensi group of northern Ghana (the only well-known example), under separate centralized political systems of varying degrees of complexity. The Tallensi had a political system that was non-hierarchical, decentralized or non-centralized. This system did not make for a state as it is understood today. In such a "segmented" system, political and social organization is based on extended families, lineages and clans that interact among themselves in a non-authoritarian fashion. A centralized political system, in contrast, is a true state, in the modern sense of the term, because it is based on a concentration of power at the top of a pyramid or has an otherwise hierarchical structure (Ergas, 1987:3).

As elsewhere in Africa, this latter political system was the dominant one in precolonial Ghana (cf. Apter, [1955]\1972; Austin, 1964). Since most precolonial societies were centralized politically and self-contained, they could be considered mini-nationalities. Partly as a result of its wealth and strong military might vis-à-vis other groups, the Ashanti nation-state became the most belligerent in the region. Like the feudal state of Europe discussed in the last chapter, the Ashanti state often sought to assert its influence in the region, as evident in the intermittent wars it conducted with other nationalities. These wars "initiated" by the Ashanti State to conquer, tax and even incorporate other groups such as the Fantis, the Brongs, the Gas and others though sometimes unsuccessful, parallel early modern European politics, where personal leaders used their state to conquer other groups. This belligerence continued so that, by 1750, the imperial frontiers of the Ashanti Empire were encompassing a significant proportion of present-day Ghana. Conceivably, Ashanti state behaviour and the efforts of other nationalities in the area to defend themselves suggest an embryonic state-making process, which was to come to an end with the imposition of colonial rule.

Such highly developed order of political achievements in these societies, however, differ markedly from political institutions in the western world. For example, there were no political parties, periodic voting and written constitutions. Nevertheless, these polities were highly democratic and those who governed were accountable and had social links with their own people. A good example of this kind of rulers is the *Asantehe*, the paramount king of the Ashanti state, also referred to by the Ashantis as *Okasapreko* meaning, "the one who speaks last" or after having heard all sides of the arguments and debates. The Ashanti political system was a Confederacy, that is, a federal grouping of Akan states under the control of the Asantehene. Although it gave the appearance of political centralization, the Ashanti state was more accurately, a decentralized political organization. The Asantehene, who sat in Kumasi, was at the helm of the hierarchy and possessed powers to make war and jurisdiction (though not in landholding). The rest of the power was actually exercised by paramount chiefs, divisional (or "wings") chiefs and village heads, with each level of hierarchy having its own counsellors through a web of subordinate authorities held by kinship ties. Devolution of power was the key to the administration of the Confederacy as a whole.

Throughout the Ashanti Confederacy, and, as in the other centralized political systems, the authority of the king/chief was tempered by the need to gain consent of the counsellors or council of the state (Asanteman in the Ashanti case) before it could be effective. The commoners, as well as the counsellors, or the "elders" had rights of their own and could also exert some measure of control over the rulers. A successor to office (to the *Stool*: an object equivalent to the British crown that symbolically enshrines the unity and identity of the state), came from the royal family, but the selection of the candidate had to be done carefully to satisfy general agreement. Among the Ashantis, for example, it was the duty of the Queen-Mother, usually the aunt or sister of the former chief, to bring forward, an eligible royal for approval or to rule until a successor is found. When the King led his people to war, it was the Queen-mother who ruled in his absence. She did this in addition to her traditional role of representing women's interest in the state. There were thus unwritten constitutional procedures for ascent to office in the Ashanti state and precolonial politics conferred much political power on women, too. Centralized political power was limited to prevent the abuses of power. The Ashanti National

Assembly would publicly admonish a newly enthroned person of authority against being greedy, refusing consultation with the representatives of the people, acting on his or her own initiative or other abuses of power (Wilks, 1975; Davidson, 1992).

The king and other chiefs of Ashanti society (and this applies to other kings and chiefs of other political communities) could be literally removed from power for exhibiting behaviour judged by their subjects as incompatible with their roles: e.g. private appropriation of state property, cowardice, insult, ruling arbitrarily and persistently ignoring the advice of the council of state or the elders. Just as there were elaborate processes for making kings, so were there institutionalized procedures for deposing them. The decision to impeach (destool) a king or chief, though a rare occurrence, was not confined to only "constitutional" authorities. Indeed, destoolment could also be initiated by the common people. A common practice in this procedure in a situation where a king refused to leave office "constitutionally" was to remove his shoes or sandals in public and let him walk barefooted, signifying that he was no longer sacred but ordinary. This was a sign that he was no longer recognized as occupying office. In the Ashanti case, King Osei Kwame (c.1771-1801), Kofi Karikari (in 1874) and his successor King Mensah-Bonsu, were dethroned for various crimes against the Ashanti state and people (cf. Austin, 1964:20). But a disposed king or chief though disgraced, could, however, re-seek office and be re-elected if he was judged to have repented from his former misdeeds. This was an indication that it was not the king, but the stool or office, which was sacred. Kings and chiefs were held in high esteem, respected and feared; nevertheless, they were not above the law. In this respect, Austin (1964:18-21) notes that, Akan kings were unlike Egyptian pharaohs: they never quite became gods. The Ashanti thus seemed to have made a Weberian distinction between office as a public role and the person occupying that role as having a private life.

The Ashanti case is not atypical, as its features were represented in other precolonial polities. It is thus tempting to conclude that, in spite of whatever limitations they might have exhibited as human institutions, precolonial politics was based on some popular conception of government and was democratic, if by democracy, we understand the existence of "check and balances", the rule of law and the right of the people to control their rulers.

As far as economic activity is concerned, precolonial Ghanaian societies can be described as subsistence economies where land cultivation or farming, fishing and animal rearing for direct consumption predominated. In this regard, they were "precapitalist" or "traditional". Economic anthropologists maintain that, in such precapitalist societies, the cultural codes for operating the economy do not typically centre on economizing or maximizing principles of the kind in modern industrial societies. In these societies, transaction rules may violate the rational choice model that is thought to be prevalent in modern western societies. Price may not be the main determinant of how goods and services are exchanged, rather, reciprocity, based on obligatory gift-giving among friends and kinsmen, and redistribution based on customary reallocation may be more important.

In classic sociological thought, we find the contrast in these sets of economic rules in "traditional" and "modern" western industrial societies, expressed in Tonnies conceptual distinction between *Gemeinschaft* and *Gesellschaft*, that is, between a community bound by kinship obligations and a community held by contract. In Weber's sociology, the distinction is between substantive (traditional) and formal (modern) rationality. Traditional societies dominated by substantive rationality are often perceived negatively as possessing institutions and values that are inimical to economic progress. An example is Marx's famous denigration of the "Asiatic mode of production" as stagnant and lacking the internal tensions that make for human progress. Equally, precolonial African economies have also been seen in substantivist terms, as based mainly on reciprocity and redistribution and possessing rigid institutions, communal property and land laws, immobile labour force, anti-capitalist values and inadequate commercial institutions, all of which limit economic progress (cf. Hopkins, 1988).

Yet, scholars from diverse disciplines (for example, Hopkins, 1988; Kay, 1972; Kimble, 1963; Hill, 1963, 1970; Hodder-Williams, 1984) have cautioned that a substantivist interpretation should not be stretched too far in the context of precolonial West Africa, particularly precolonial Ghana, lest a complex economic reality is over-simplified. This is because many of these precolonial societies had developed market and exchange institutions. Trade and commerce in these societies, as elsewhere, were as old as humankind. At the macroeconomic level, subsistence economic activities and craft manufacture, as well as trade, were engaged in variously and simultaneously. Precolonial industries included metal works, ceramics, construction, food processing, craft and textiles and the manufacture of commodities such as monkey hides, pepper for local consumption and export to many parts of the Arab and European world. According to Hopkins, all the stages involved in textile manufacture—ginning, carding, spinning, dyeing and weaving—existed in precolonial Ghana.

Again, contrary to the general view that land was communally-held and this constrained economic expansion, land ownership rights and laws varied greatly, ranging from communally-owned and worked, to land that was considered virtually as freehold. Labour, too, was not always unfree in these societies, as generally portrayed. Although mainly family-based, labour was highly mobile. For example, in her study of the fishing communities along the coast of Ghana, Hill (1970) notes that Fante and Ewe fishermen often moved periodically to the Cote d'Ivoire and even as far as Sierra Leone and the East to Nigeria to fish. Also, the legends and historical records of the origin of Ghanaian people all emphasize the importance of mobility as a means of acquiring new wealth. Mobility makes sense, because population in the precolonial period was low in relation to the size of land available. In general, underpopulation retarded Africa's economic advancement, in that, it prevented market growth and rather encouraged extensive cultivation, dispersed settlement, strong tendencies towards local self-sufficiency and also lessened pressure on resources and thus hampered technical innovations (Hopkins, 1988:76). The permeability of land laws and tenure and the high mobility of labour, however, became crucial facilitating factors in the rapid development of the cocoa industry in the colonial period.

Given the diversified nature of economic activities undertaken, it is not surprising that markets or transaction economies developed. Thus, trade and commerce were widespread in precolonial Ghana. The great markets of Salaga and Kintampo come quickly to the minds of those familiar with precolonial history. Trade, however, took two main forms: the long distance trade that involved Ghanaian societies and the larger world and that conducted domestically. The long-distance trade reached its apogee in the trans-Saharan trade that connected the Middle East and then Europe to the great empires of Africa. It was this trade that was to bring the Europeans, knowledge of the existence of gold in the Ashanti Kingdom. The other type of trade is that meant for mainly domestic exchange conducted in periodically rotating markets every fourth or fifth day. This is supplemented by itinerant peddlers and hawkers. The physical market was not only important for exchange of goods but was also an arena for social interaction, public exchange of information and gossip. In this respect, Chamlee (1993) remarks that the marketplace was the heartbeat of precolonial societies.

The social division of labour, petty commodity production and exchange were, thus, much advanced in many places in precolonial Ghana. Precolonial Ghanaian societies had developed their business acumen and introduced a monetary system based on gold dust (the most durable, portable and universally accepted commodity) and cowries nearly equivalent to modern currency (cf. Kimble, 1963:3). Institutions of money-lending and credit were also in existence before colonialism (Hopkins, 1988:68). By the standards of modern monetary systems, these currencies and institutions are undoubtedly inferior, but Hopkins implores us to regard them as "transitional" in the process of evolution into higher forms.

Kay better sums up the economic development picture of precolonial Ghana:

> in many parts of pre-colonial Ghana the social division of labour was quite advanced and petty commodity production, exchange and monetization were all firmly established. The simple uninformed view of many development economists that so-called traditional societies were more or less completely dominated by subsistence production is less applicable to southern Ghana than perhaps any other region of sub-Saharan Africa (Kay, 1972:5).

Ghanaian precolonial societies were self-sufficient economic units. They were not, however, self-contained and totally isolated from the outside world, nor were their economic systems based mainly on barter and networks of reciprocity. They exhibited the "spirit of capitalism" and their economies approached something like commercial capitalism, confirming Weber's classic argument that capitalism has existed in one form or another in both time and space.

While capitalism might have had to await the "development" of the Protestant Ethic, precolonial Ghanaian societies had long accepted commerce as a "cultural calling". Indeed, it is difficult to find traditional religions that prohibited capitalism. Furthermore, according to Hopkins, unlike patrimonial leaders of feudal Europe, rulers in precolonial Ghana did not act arbitrarily to inhibit capitalism. States supported their subjects to participate in long-distance trade, and officials sometimes transacted business on behalf of the state. This was mainly because leaders needed to raise revenue for the reproduction of their states. For example, Daaku (1970), in his study of trade and trade patterns in seventeenth and eighteenth century Ashanti society, argues that the Ashanti state was the main agency responsible for the development and innovation of the kola trade in the nineteenth century. The rulers sought the expansion of trading activities since wealth for their citizens means greater wealth for their kingdoms.

But, although states encouraged trade and commerce, the general pattern observed among the Akan states of modern Ghana was that:

> public authorities deliberately refrained from engaging in trade directly, not because they were insensitive to commercial opportunities, but because they judged it advantageous to encourage private traders and then tax them. The imposition of tolls and markets dues was undoubtedly a widespread means of raising revenue and being a relatively simple and virtually riskless exercise, it may well have been preferred, on the whole, to direct trading (Hopkins, 1988:62).

That precolonial states should behave in this manner should not come as surprise, since feudal rents that derived from land and the activities of landlords as is the case of Medieval Europe hardly existed in Africa, except perhaps in Ethiopia. Also, the "absence" of patrimonial political domination meant that indigenous capitalism did not have to suffer the obstacles that faced European capitalism in the feudal and absolutist era. According to Hopkins, prudent rulers kept close contact with local merchants and businessmen, gave monetary support to them and deliberately developed mercantilist policies and organizations to encourage commerce and trade. The interest of precolonial state and merchant class can be seen as confluent.

On the basis of this, Hopkins concludes that, the view that preindustrial states in other continents have hindered the expansion of the market by imposing bureaucratic controls on private enterprise has little relevance for precolonial Ghana and West Africa in general (1988:62). Indeed, one cannot write a good precolonial history particularly of West Africa without its trade and commerce. Given this, Kay (1972:8) suggests, perhaps with a little exaggeration, that precolonial Ghana exhibited a socioeconomic complexity whose salient feature is the emergence of an authentic Ghanaian capitalist class. But Kay is emphasizing the indigenous economic feat achieved in this period before the advent of colonialism. Recently, in her study of urban female entrepreneurs in Ghana, Chamlee (1993) reveals a cultural foundation of market and exchange institutions that date back to the precolonial period, and proposes that the Akan culture possesses some "functional equivalents" of the so-called Protestant ethic that Weber argues to have been responsible for the rise western capitalism.

The foregoing economic sociology of precolonial society suggests that, market and exchange institutions, as well the "spirit of capitalism", were integral parts of precolonial societies. In this respect, British colonial administrators did not need to "impose" a new economic imprint or economic rules on Gold Coast societies, because the so-called *laissez-faire* economic philosophy that they brought along had already found a fertile soil in which to thrive (Hodder-Williams, 1984). The cocoa industry, as well as, trade and commerce that developed in the hands of enterprising local private individuals were manifestations of this contention.

Socioeconomic Transformation of the Gold Coast: Colonial Politics and Economics

The early European contact with the Gold Coast was mainly motivated by trade in gold and other exotica that this region offered. But trade in commodities later led to the institutionalization of the trans-Atlantic slave trade whereby, many Africans were forcefully transported to the New World. The Gold Coast region actively "supplied" slaves whose labour was used cheaply to produce raw materials, such as sugar and cotton for the expanding industries of Western Europe. The slave trade lasted about five centuries and certainly perturbed the organic evolution of African societies and institutions, because not only did it introduce social anarchy in the affected societies, but it also sapped them of their most youthful, talented and innovative elements of their populations. Slaves raiders often targeted these groups because of their fitness vis-à-vis other groups in society. Historians suggest that Africa lost some 15 to 20 million of its people to the New World (some of them died during the journey, mainly because of inhumane treatment and sickness). The Gold Coast region was one of the principal sources for slave in this great forced migration in human history. But, later,

> increased nationalism, industrial competition, the need for new markets, for raw materials, and for opportunities of profitable investment... led many nations of Europe to a new interest in colonial expansion in the last decades of the nineteenth century (Bourret, 1960:18).

These needs could not be met in the context of the slave trade that was disrupting societies. Africans needed to be left in peace on the continent to provide market and supply critical raw materials for European industries. Even though European powers, particularly the British, attribute abolition to humanitarian reasons or the changing moral conscience, the main motivational reason Williams (1964) maintains was economic. The slave trade was formally abolished in 1807, albeit it took a long time before it finally ceased.

Following the abolition of the slave trade, and as a result of its growing industrial and technological power, Britain "out-competed" other European powers trading along the coast of Ghana to emerge as the main imperial power. During this period, the region supplied gold and palm (kernel) oil and received manufactured products from outside. British capital had contented itself with doing business along the coast of the southern parts of Ghana, which, before the subjugation of the Ashantis, was regarded as the Gold Coast colony. This southern part was controlled by British merchants. The people of the Gold Coast were receptive to British control of their economy and society because it offered them protection against their "bullying" neighbours: the Ashantis.

With time however, Britain sought greater control over areas outside the Gold Coast (more precisely Ashanti, and, later, the Northern Territories). In this decision, the interest of British merchants was paramount: their desire to trade on substantial scale in Ashanti state and then beyond in the northern territories. They also wished to make Kumasi, a trading centre for British imports, as well as for gold production (Kay, 1972:4). By now Britain had become the "industrial workshop" of the world and needed overseas markets for its manufacturing products because of "underconsumption" at home (cf. Hobson, [1902]/1988). But the Ashanti state and merchants had all along regarded British domination of the Gold Coast as illegitimate, giving rise to political tension. Though the first "governor" of the Gold Coast, George Maclean (actually a merchant), was an efficient and tactful administrator who succeeded in signing numerous treaties with the Ashanti State, the prospects of war with Britain, instability and economic crisis in the region loomed large (Bourret, 1960:17).

Control of Ashanti gold mines was to be a source of conflict with serious repercussions. Britain's attempts to negotiate trade links diplomatically with the Ashanti State were rejected (Kay, 1972). This uncompromising attitude on the part of the Ashanti and Britain's ambition, brought these two powers in the Gold Coast into a series of wars lasting almost a century. In the famous battle of Insamankow (1824), in which Britain was humiliated, the governor of Sierra Leone, Sir Charles MacCarthy, was killed (Apter, 1972:26). Though Britain later defeated the Ashantis at the battle of Dodowah, the expense and the disasters of many earlier campaigns were so colossal that it vacillated in its efforts to colonize the Gold Coast. British Parliament decided to abandon the Gold Coast, but British merchants saw economic value in the area and prevailed upon their government for continued colonization.

Since British capital, like all business, cannot operate profitably in an environment characterized by wars and political uncertainty, the British government decided to conquer the Ashantis once and for all. Thus, in 1874, with troops and more sophisticated weaponry from England, the colonial authorities waged their most important war, subjugated the Ashantis, burnt Kumasi city, and sent the Asantehene King Prempeh I (a mere youth), and some senior government officials of Ashanti state into exile in the Seychelles until 1924 (Bourret, 1960:21). The Peace Treaty of Fomena was signed, and this opened up the Ashanti Kingdom for British imperialism. Consequently, British capital displaced the Ashanti state, and merchants, from the ownership and production of their most important mineral: gold.

From the time of the defeat of Ashantis to the beginning of World War I, Britain attempted to consolidate colonialism: a process that began in the southern part of modern Ghana around the 1820s (Kimble, 1963:3). The various indigenous nationalities and states that currently make up Ghana were brought under the single umbrella of the colonial state. Between the 1870s and 80s, Britain extended its control over Ashanti to incorporate the territories of the North. The Volta Region, which lies along the eastern boundary of modern Ghana, was originally part of Togoland, formerly a German colony. With the defeat of Germany in the First World War, Togoland was divided: one section came under the trusteeship of Britain, and the other under France. The British Togoland section became part of the Gold Coast a year before independence through a referendum. The Gold Coast frontier has now extended to include the Ashanti, Northern and Volta regions of today's Ghana.

On the nature of colonial state, the Colony was administered by a Governor, assisted by an executive and legislative council and a staff of political and technical officials, normally representing British capital. The Governor was chief executive authority and was accountable to the Secretary of State for the colonies to whom he had to refer all important matters. The executive council, during this early period, was made up British officials consisting of the colonial secretary, the chief commissioners of Ashanti and of the Northern Territories, the attorney general, the financial secretary, director of medical services and the secretary for native affairs. This executive was an advisory body to whom the Governor had to submit all Bills before proposing them to the Legislature. The Governor had a final word in case of disagreement, but members could always report their views to the Colonial Office. When the indigenous people also disagreed with the Governor on any issue, as they did many times, they sent delegations and petitions to the Secretary of State for the colonies.

The colonial state made no formal distinction between the executive, the legislative and the judiciary organs of government, as is typically the case in liberal democracies. Such powers were fused, and those who administered the state behaved more like practical administrators than politicians. In other words, the institutional and constitutional distinctions between the state and government were blurred. The legislative council (which was gradually modified after World War II) did not begin as an elected body but as an assembly of directors of administrative departments who worked alongside representatives of British business interest. In general, the people of the Gold Coast were disenfranchised. As suggested in the last chapter, the colonial state, being an alien institution, was undemocratic, acting directly upon civil society but this society had little or no way of controlling it from below. Yet, organized opposition to the state was infrequent after the conquest of Ashanti, which helps to explain the fact although the state had developed coercive instruments, they were used sparingly, and only in extreme cases. Beside the threat of brute force, colonial rule was made easy mainly because of the kind of colonial administrative policy pursued and the economic philosophy (see next section) that informed it.

Two modes of thinking, however, dominated colonial administration. The first is that, until some time in the mid-twentieth century, British policy in the colonies was that "government is best that governs least". State intervention in the economy was not on the colonial agenda; this seems to conform to the *laissez-faire* philosophy dominant in the nineteenth century. Keynesianism, which advocates state intervention in the economy, had not yet even made its dėbut in Europe, hence, it was out of the question. But on the other hand, the British never sought to actively intervene in the colony's affairs, but to provide a Hobbesian political framework for mainly British capital to pursue its interests. The colonial state was an "imported" and alien institution; its administrators were also instinctively aware, if not fully conscious, of the precariousness of their position should there be mobilization below to challenge them. Even though, in their self-righteousness the British colonialists considered themselves more civilized, they nevertheless acknowledged that their victims had some institutions worth preserving.

Since African self-governing institutions were already in place, the British administrators were cautious not to disturb existing political orders by introducing new rules of the game. British rule was to be subtle and indirect. In this way, the impact of foreign domination and rule would be less obvious to the colonized people (Hodder-Williams, 1984:33). *Indirect Rule* then became the main instrument of administering the colony. Existing political structures and institutions were to be maintained. Kings and chiefs were to retain their positions and work with provisional and district commissioners, who were delegated from the new national political centre. To this effect, in 1883, a Native Administration Ordinance was established in the new Gold Coast to officially legitimize traditional political institutions.

The system of Indirect Rule satisfied many interests. From the viewpoint of the parsimonious British Treasury, it meant less fiscal commitment to colonial administration. Furthermore, it was a solution to the manpower problem that direct administration would have generated, given Britain's numerous colonies abroad. The system also pleased traditional rulers who could pose a further threat to colonialism because it was supposed to preserve their power base. Paradoxically, however, Indirect Rule undercut them. Two reasons account for this. First, some of the powers of the chiefs were taken over by the Colonial government at the national centre, thus they became subservient to alien rule. Second, Indirect Rule made traditional elites unpopular in the eyes of their people, and the newly educated elites, in particular, regarded chiefs as handmaidens of oppressive racist colonialist (see below).

The second mode of thinking that informed colonial administration, as in all colonial cases, derives from Social Darwinism which assumes that it is right for the "most fit" race to dominate the "less fit". All colonial powers consciously or unconsciously shared this thinking, since they believed that the "more civilized" colonialist had the responsibility to rule and civilized the "less civilized" people: hence, the self-appropriation of a "white man's burden" by the colonialist. Before colonialism, European and African traders regarded each other as self-sufficient. The issue of *civilizing* the colonized arose when questions were posed about the moral basis of colonialism. Colonialism was thus conceived as a civilizing mission through which the colonized, who had been self-sufficient in their social organization and ruling themselves since time immemorial according to their cultural traditions, were now defined as incapable of governing themselves, and must be "civilized" in the image of Europeans in order to do so.

A corollary of this thinking was that Britain would maintain a long period of control until such a time that the colony had been sufficiently tutored to deserve western type self-government. Self-government, in the western sense, appeared to be the ultimate goal of colonialism and was to be attained not by revolutions or radicalism, but through what can be called political institutionalization (Kay, 1972). Somehow or rather implicitly, as colonialism proceeded, the emergent, formally educated elites, who had skills for modern administration seemed to have subscribed to this thinking. This is evident, in the sluggish tempo with which they sought political independence for the Gold Coast and their initial antagonism to traditional rulers whose powers the colonial authority fortified from time to time. But the incorporation of these educated elites into the Executive and Legislative bodies to "prepare" them for self-rule (i.e. political institutionalization) was, however, to be an eventful process that had serious ramifications for the political evolution of the Gold Coast.

Colonial Development Policies and Economic Transformation Before World War II

If colonialism was not meant to advance upon precolonial political institutions, it was however, to mark a new expansionist phase in the evolution of the capitalist economy in the Gold Coast. The main reason behind British imperialism in the Gold Coast as noted earlier was economic: to enable British capital to gain access to resources and markets. The general belief inherited from nineteenth century economic thinking was that private enterprise and the free play of market forces would lead to international specialization based on comparative advantage of the nations concerned, and that this would promote economic growth by bringing about a more efficient allocation of factors of production (Hopkins, 1988). Development by this means was held as a positive sum game benefiting all parties. Therefore, governments were not to intervene actively in the operation of national economies. In the colony, too, governments were only to provide congenial political frameworks for private economic actors from the metropole to accumulate capital.

But, although the main purpose of colonialism was economic—to secure markets for manufacture goods and at the same time extract the colony's resources—it was thought that, if the colony remained poor, it could not consume products from the western world. This brought about a modification of the economic motive of colonialism: that it should not be crudely exploitative, but also beneficial to the colony in question. This thinking was expressed in Lord Luggard's concept of *dual mandate*: both the colonizer and the colonized were to benefit from colonialism. In concrete terms, this meant that the full benefits of the colony would not accrue to the metropole, in this case Britain, if the colony remained poor, that is, without the colonizer making some material concessions to the colonized. This view echoes Marx's famous argument about British rule in India, that colonial capitalism cannot avoid laying some material conditions for modern development, which must indirectly benefit the colonies. In tropical Africa, the colonial powers realized that, without a modern and efficient transport system, their economic goals could not be attained. In this regard, the material development of Africa was summed up by Lord Luggard in one word: "transport" (Luggard, ([1922]/1965:5).

The view that transport was the key to development emerged very early in administrative circles in the Gold Coast colony in response to the needs of British interest in mineral resources. The colonial state, in collaboration with the chiefs and the people, actively undertook the development of socioeconomic infrastructure. In 1901, a short railway, the first in the Gold Coast, was constructed from Sekondi on the coast, to Tarkwa, a gold producing centre. In 1904, this railway line was extended from Tarkwa to Kumasi, the heartland of Ashanti. By 1919, the Gold Coast had some 1,200 miles of motor roads (Bourret, 1960:23-24). The construction of a modern transport system brought about a boom in general economic activities; in particular, commerce and merchant trade expanded in the colony. Banks were established and formal currency (the Gold Coast pound) was introduced.

After the defeat of the Ashanti state, British capital got concessions from kings and chiefs to mine gold. This led to the establishment of the Ashanti Goldfields Corporation. By 1912, gold exports accounted for £1 million and became important on the list of annual exports (Bourret, 1960:23). In addition to the development of mining industries and commercial activities, other primary products such as palm oil, palm kernel, coconuts, timber, cotton and rubber caught the attention of the colonialists, and, subsequently, became significant export products in the economy.

While the colonialists were more interested in mineral extraction and trade, as that was the main reason for colonialism, the colony's economic base was soon to change in another direction. In 1857, the Basel missionaries who played an integral part of the colonialist's *civilizing mission* introduced cocoa seedlings in the colony, but its cultivation did not catch on. In 1879, however, a native named Tetteh Quarshie, returned from Fernando Po with some cocoa seedlings to cultivate. His trees grew successfully, and the potential commercial value of cocoa was quickly perceived by the colonialists and peasants. Soon, its cultivation spread among the peasants like bush fire. Thus, while between 1891 and 1895, only five tonnes of cocoa were produced, within the next two decades, the Gold Coast was producing about 106,000 tonnes. By 1910, cocoa was the number one export of the Gold Coast, making it the world's leading producer of cocoa, producing about two-thirds of the world's output (Bourret, 1960:23; Kay, 1972:6).

This remarkable economic feat was accomplished solely by the enterprise of the peasants with absolutely no European capital and very little government support (Hill, 1963). The rapidity with which cocoa production spread, coupled with the enthusiasm of peasants to cultivate it, has compelled Mikell (1989:82) to call it "the people's industry". Its impact on the lives of the people and society was colossal; it transformed many peasants into rural capitalists (Hill, 1957, 1963). Cocoa farmers exhibited capitalist values: they purchased rights to land, they left their homes and travelled long distances to settle and acquire new lands, they made long-term investment (their cocoa trees took at least three years before bearing fruit), and they hired labour. Significantly, too, the communal land ownership rights underwent change to accommodate the new economic dispensation.

While economic transformation was cataclysmic, broader social advances were held back. Since, as noted earlier, the main motive behind colonialism was economic, in the spirit of *laissez-faire* the colonial government policies de-emphasized education, mainly because of the consciousness it could possibly raise among the local people. As alien rulers, and few, the colonial administrators were conscious of their precarious position in the colony and the backlash they could suffer should the local people be educated to question European rule.

This was not surprising, since right from the beginning British colonialism was highly contested by the Ashanti state, and the early attempts by the colonialists to seize all "unoccupied" lands for the British Crown was also fiercely resisted by the people along the coast. But social reality is always complex and events have a way of upsetting the world we seek to construct. In spite of such carefully guarded colonial objectives, progress was made in formal education, initially, mainly through the work of Christian missionaries who were zealous to Christianize and train people to read the Bible and write. The colonial administration also required manpower for its civil service. By 1919, the colonial government had some 19 schools and the missionaries 204, which were assisted by government funds. There were also about 250 unassisted schools, mainly in the hands of the indigenous people whose enthusiasm and insatiable appetite for education was often commented upon and praised by colonial officials and visitors to the colony (Kay, 1972:32). Such was the material development picture of the colony, at least before World War I. The war, however, brought a temporary halt to the Gold Coast's "economic renaissance".

With the end of World War I, the development of infrastructure gained a new impetus. A new colonial governor, Sir Gordon Guggisberg, had arrived full of ambition to develop the potential of the colony. Guggisberg's rule was to be the most cherished in the Ghanaian colonial history, mainly because of the unparalleled achievements of his administration and his unusually strong emotional attachment to the colony and the people. This affection is obvious from all his speeches in the Legislative Council. He normally referred to the Gold Coast as *our country* (see Colonial legislative documents in Kay, 1972). His rule was to lay the main socioeconomic foundation of modern Ghana and was to have repercussions for later political development.

Guggisberg had previously worked as head of the Survey Department in Nigeria in 1914, and probably came across Lord Luggard's thinking about transport, since Luggard was a Governor of Nigeria but did have his own development philosophy. He shared Luggard's view that a sound economy depended on the development of a good transport and communication network. In the Gold Coast, such a network would lead to the opening up of new areas and the development of new productive activities for markets. In addition, he believed that education must be the keystone of Gold Coast's progress towards a higher state of civilization and social progress (see Governor's Address to the Legislative Council in *Gold Coast Gazette*, 17 March, 1923). But Guggisberg was also aware that, without a sound economy with a strong revenue base, social services such as education and health could not be sustained. Guggisberg sought to create a "developmental state" out of the colonial state to transform the colonial economy and society.

Consequently, the Governor and his administration formulated a Ten-Year Development Plan (which owed less to the deliberations of the Colonial Office) covering the period 1919-1928 and involving an expenditure of £25 million (Hopkins, 1988:190). Since British colonial policy was that colonies should be self-financing, this meant that the Gold Coast had to shoulder all the cost of its own development. During this time, however, the Gold Coast colony already had a trade surplus of £1,250,000;

thus the rest of the expenditure was to be in the form of loans from abroad augmented with revenue from domestic economic activities. In order to put the economy on a sound financial footing, Guggisberg set down definite principles by which all future estimate committees were to be guided. He also set up a General Reserve Fund of £500,000 for the colony which was to be built upon annually by its own interest and not to be used until it reached an amount proportional to the capital value of the country. This reserve fund was also maintained by successive colonial governments so that, by 1946, it had reached £1.5 million (see *Gold Coast Gazette* op. cit.). The domestic revenue came mainly from taxes on cocoa and duties on kola and timber.

One notable aspect of the colonial tax policy, particularly that on cocoa, was that, it should be high but not too high to serve as a disincentive to production (Frimpong-Ansah, 1991:20). This decision was based on economic empiricism of the understanding of peasant mode of production that combines subsistence food and cash crop production resulting in their partial proletariat status: a situation Hyden (1980) refers to as "uncaptured peasantry" trapped in the "economy of affection". The logic of Guggisberg's argument was that, if the farmer was "overtaxed", he or she could abandon cocoa cultivation, which was the backbone of the economy. This wisdom could easily be observed by a colonial government that took upon itself fewer developmental tasks. But, as the quest for socioeconomic development and consumption became greater in the postcolonial period, later governments were unable to strictly adhere to this "golden rule".

The immediate post-World War I economic climate was favourable to the Gold Coast economy. The price for cocoa soared to £80 per tonne, and even in some years, to £120 (Bourret, 1960:27). But this economic fortune was short-lived, as the Depression of 1920-23 and other unforeseen circumstances set in and the Government had to reduce expenditures. This depression, and the accompanying declined terms of trade, not only limited government performance but also heightened the problems associated with monoculture economy that had now emerged in the Gold Coast. Yet, much was achieved under the Ten-Year Development Plan.

The most notable accomplishment of the Guggisberg development plan was in infrastructure. The railway line between Accra and Kumasi was completed in 1923. A new line from Tarkwa to Prestea, the heart of gold mining district, was also constructed. Between 1923 and 1927, another railway line from Huni Valley to Kade was built to open up new cocoa and timber producing areas. Ghana's first harbour at Takoradi (the second is at Tema built after independence) was built to enable the shipment of minerals and cocoa to the outside world. In all, 233 miles of railways were built under the Ten-Year Development Plan. Surveys were also made of 250 miles of prospective railway lines (Bourret, 1960:29). The Governor's intention was to extend the railway line to the north of the country, which was lagging behind in the Gold Coast's "economic revolution". He was, however, unable to realize this project before his term of office came to an end in 1927.

By the end of World War I, the Gold Coast had only 1,200 miles of motor roads, many of them built by the colonial state, the indigenous people and their chiefs eager to transport cocoa to the ports for the outside world. The development of the transport system had a multiplying effect on the economy; cocoa production boomed, likewise, the production of minerals such as gold, diamonds and timber. Communication facilities such as telephones and telegraphs were also expanded. These infrastructure developments were however, skewed in favour of regions that produced the resources (cocoa, gold, diamond, timber etc.) that interested the British interest. Other regions such as the Northern Territory that was not favourably endowed with these resources were left out in this socioeconomic transformation.

The Gold Coast was held to be an economic success, but this needs to be seen as only in relation to the performance of other African countries. Gold Coast prosperity was mainly due to the enterprise of the peasants, particularly cocoa farmers who had responded to the opportunities brought by the insertion of the Gold Coast in the international economic system. Cocoa has remained to date the mainstay of the Ghanaian economy. But in this colonial period, Ghana's neighbours, Togo and Cote d'Ivoire, were also becoming producers of cocoa, and the price of cocoa paid to farmers in Ghana vis-à-vis prices in these countries would come to have an effect on Ghana's cocoa industry.

In the social field, too, the Ten-Year Development Plan had much to its credit. Before the end of World War I, education was mainly in the hands of missionaries and leading indigenous people who were anxious to expand education in the colony. Shortly after his arrival in the Gold Coast, Guggisberg appointed a committee to study the education situation and to establish principles that could guide schools, which were mushrooming all over the colony. The colonial administration was particularly worried about ill-equipped schools being rapidly set up by the indigenous people. The recommendation of this committee led to the introduction of an educational ordinance in 1925 which aimed inter alia at high standards, equal opportunity for boys and girls, teacher training, equipping schools with the state of art equipment, leadership, and training in agriculture and handicrafts, rather than solely literary subjects. The demands for higher standards led to the closure of many schools, but efforts were made by the government and missionaries to put up better schools and train personnel. The main result of government intervention in this sector was the Prince of Wales College at Achimota, formally opened on January 28, 1927 by

Governor Guggisberg, which had a Ghanaian, Dr. James Kwegyir Aggrey, as vice-principal. Boarding-schools were expanded in the Cape Coast community, some of whose graduates went to England and the US for further studies to become civil servants, lawyers, doctors and architects. A large modern hospital, acclaimed to be the best in Africa at the time, was built at Korle-Bu in Accra. Other health services were expanded.

But, while infrastructure development received a deliberate boost from the colonial government, general economic policy was that which emphasized private choice and market principles. This orientation derived not only from the *laissez-faireism*, the basic economic philosophy dominant in the nineteenth century, but also from the economic motives behind the colonial project. Interestingly, however, given the economic characteristics of precolonial societies noted above, this philosophy was receptive in the colony. This also helps to explain why the colony was generally "receptive" to colonialism and its economic policies. The economic rules of the game were not altered radically enough to upset the prevailing economic order, which could possibly engender internal tensions and revolt.

The policy of minimal government intervention in the economy favoured British commercial interest, since British capital actively lobbied for direct subjugation of the Gold Coast. Thus, with regard to commerce, the colonial state allowed a free market economy to operate with little or no restraint. Since many indigenous merchants lacked huge capital and the modern skills, most might have lost out in the near-perfect competitive colonial economy. Long-established British and other foreign/western expatriate firms came to dominate the mining and commercial sector of the Gold Coast economy and thus came to control the international trade. They specialized in selling manufactured goods, and some of them even entered the marketing of cocoa and other primary commodities such as coffee and palm oil. Prominent among these firms were the United African Company, (UAC), Lever Brothers, SCOA (Societe Commerciale Ouest Africaine), Cadbury Brothers and John Holt & Co. Many of these firms were successful because of the government's obliging attitude towards them, and mergers and concentration. In addition, came Levantine companies (a term loosely used to refer to the nationals from the countries bordering the eastern Mediterranean, particularly the Lebanese). Many of these Levantine migrants did business on a small scale, operating motor transport taxis, restaurants, and commercial shops (Hopkins, 1988).

Given the significant intrusion of foreign capital into the economy and the colonial administration's indulgence of them, indigenous business is often perceived to have suffered unduly. The view is often formulated in more general terms, that, at independence, Africa lacked the indigenous bourgeoisie to fulfil their historic mission of revolutionizing the productive forces of society and this tends to provide justification for a greater role for the state in the economy. Often, in Marxist-oriented development literature, however, the economic activities of the national bourgeoisie are seen as tied to the interests of foreign capital and/or forces of imperialism, and thus considered detrimental to the economic advancement of the larger society (see Chapter 1). There is, therefore, a proclivity among some leaders—particularly those with socialist inclinations—to see the defence of "national interest" in terms of how much they expand the state sector and simultaneously "discourage" the advancement of their national bourgeoisie. But the defence of national interest need not always be so narrowly conceived or seen in such doctrinaire terms, because the economic activities of the national bourgeoisie have been key ingredients in economic modernization of some societies. Thus, in frowning upon the progress of their bourgeoisie, these leaders might inadvertently be hindering the economic development of their own countries.

The position of the indigenous bourgeoisie vis-à-vis expatriate firms deserves a further analysis in the Ghanaian context, because as with all generalizations, the attempt to see the forest overlooks the rich variety the grasses offer. Given the economic order created by the colonialists, there is reason to believe that indigenous merchants suffered a decline in fortune in the colonial period and some even withered away. Yet, a number of historical studies on Africa, suggest that they were not totally liquidated and that some might have survived and even prospered. This is because some of them were astute enough to supplement the activities of foreign capital. Most foreign trading firms confined their activities to cities and a few big towns. Indigenous merchants then filled the economic void left in other towns or the hinterlands.

Furthermore, some of them acted as brokers or apprentices for the big European companies, acquiring investible capital and managerial skills and later moving on to develop their own businesses. Others diverted into other activities, e.g. motor transport (cf. Boahen, 1975:102-107; Markotviz, 1977:235; see also Hopkins 1988:205). In this process, some wealthy businessmen emerged during the colonial era. For example, Kimble (1963:5) mentions that one Mr. Robert Hutchison (a Ghanaian) claimed to be worth £60,000 in 1855. Other prosperous indigenous merchants included the Akrosa brothers of Fetu, the Asomani of Akwamu, John Kabes of Komenda, John Konny of Ahanta and the wealthy timber merchant George Grant of Sekondi. In 1656, the merchant prince John Claessen, one of the most prosperous Gold Coast businessman, turned down an opportunity to become the king of the important regional state of Fetu because according to him, kingly duties would deprive him of time for his business. Also, the emergence and importance of an indigenous merchant class in the colony made the editor of the London-based *African*

Times to write enthusiastically that "we want a race of native capitalists in West Africa... and we will do our best to help in the creation" (cited in Markotvitz, 1977:241-2).

This sociological picture indicates that indigenous businessmen began emerging in Ghana. Although their number was undoubtedly small, and it is doubtful that they could have revolutionized postcolonial productive forces, since their activities were mainly commercial in nature, these indigenous entrepreneurs nevertheless constitute a significant category in the evolution of an authentic, productive bourgeoisie in the Gold Coast. But as an incipient bourgeoisie, its advancement was constrained by the activities of expatriate firms that dominated the economy and also the colonial administration, which benignly neglected their interest. It is little wonder that these actors closed ranks with the intelligentsia to demand independence. Apparently, they saw independence as an opportunity to advance their interest, and, equally to expect the independent state to protect and nurture their reproduction.

Yet, this group is just the tip of the iceberg of an emergent capitalist class; there were others. Polly Hill (1970) in her economic anthropological studies, also notes that, a rural capitalist class of cocoa farmers, fishermen and cattle owners were also accumulating wealth like economic actors in the advanced western nations (cf. also Hodder-Williams, 1984:56). In addition, commercial activities, particularly petty commodity trading, which had long been associated with women, particularly those in southern Ghana, also boomed (Rimmer, 1992). What these analyses point to is that the Gold Coast economy was undergoing capitalist development with the concomitant emergence of socioeconomic groups (a minority, but potentially significant politically), interested in the preservation of the colonial economic order. In this regard, any attempt to impose a new set of economic rules was likely to be conflicting and contested. Such then, was the early "success" of Ghana's economy, which attracted economic migrants from the "less successful" neighbouring economies, such as Mali, Upper Volta (now Burkina Faso), Togo and Nigeria at this time. On the whole, it can be argued that the structure of Ghana's postcolonial economy, which is geared mainly towards the production of what it does not directly consume and consumption of what it does not produce, was laid in the colonial period.

Things Fell Apart: Socioeconomic Transformation and the Labyrinth of New Conflicts

Social change or modernization can be a disintegrative force in society and disruptive of the way people behave and relate to each other. Cocoa production has become the mainstay of the colonial economy, bringing wealth to a large mass of common people, at least in the cocoa growing communities. As wealth permeated all social strata, it consequently accelerated status mobility, individual choice and independence. Colonialism also accelerated the advancement of the capitalist mode of production and exchange, monetized the economy, introduced wage-and salaried labour, and commodified social relationships. This no doubt would have generated problems such as alienation, conflict and exploitation. The transformation of the economic base of society, however, brought two main but related new conflicts that affected the structure of existing society and the evolution of politics in the colony.

The first major conflict was between the common people and the traditional aristocracy. Cocoa cultivation, as we previously noted, spread quickly among the common people. New economic occupations such as commerce and salaried and wage work in the public and mining sectors, also became significant. This means that an increasing number of people now had money directly in their hands. Money brought a sense of independence to the holder, as such respect for traditional rulers and elders began to wane. As individuals and groups became empowered economically, they demanded to be recognized as independent actors, and, consequently, they came to rock the boat of traditional social stratification and political institutions based on ascription and hereditary status (Mikell, 1989:xv).

It was, however, not only the commoners who challenged the power base of the aristocracy but also the modern educated class. Socioeconomic progress in the colony created a new middle class, e.g. teachers, clerks, clergymen, merchants, doctors and lawyers, in the towns. These groups became conscious of their modern skills and the resources they control. Consequently, they also threatened the powers of traditional elites. Even long before a fully-fledged, educated African emerged, traditional rulers were losing much of their respect, since the colonial state, through its policy of Indirect Rule, was calling upon chiefs to implement "unpopular" policies. Chiefs were being seen by their people as puppets serving the interest of the Colonial state and foreign capital. They were criticized for selling or leasing lands without securing adequate provisions for guarantees. Paradoxically, the Colonial State had also introduced into the country its own agents: provisional and district commissioners whose powers undermined traditional authority.

The chiefs' declining legitimacy was thus aggravated by the rise of new educated elites and businessmen who possessed modern skills that could enable them to participate in the emerging modern economy and political system. These skills were, however, lacking among the old elites and rulers many of whom were illiterate. The new elites even resented the colonial state's attempt to incorporate chiefs in government. There was thus a conflict between old elites, on the one hand, and new elites and their "bourgeois" counterparts, on the other. The most notable of this conflict was the one that involved the less educated elements, the Standard IV graduates who had only clerical skills that were usually not enough for civil service employment, except at levels such as messengers. This group's aspiration could not be contained in the old socioeconomic and political order. Their frustration was vented at the chiefs, who were now unpopular, and the colonialists, who were perceived as failing to create jobs for them. The *ancien regimes* of the "ashantehene", the "nanas" and "togbuis" (the kings and chiefs), came under severe strains because the cocoa farmer, cloth merchant, and civil servant, who now had economic power, desired freedom (see Boahen, 1975:102-107).

A consequence of their frustration was expressed in "anti-chief" social movements. For example, according to Austin, who has closely followed the Gold Coast's political development, in the Akan State of Kwahu, the *asafo* company of young men was reported by the district commissioner in 1919 as having become "a dangerous and most undesirable element in the politics of Kwahu". The elders also complained about "the growing power of the young men which was breaking the power of the chiefs". In Ashanti Bekwai, young men complained that they were not consulted in the choice of Headchief and they would not serve him. Chiefs themselves were often caught up in these political bickerings, as they sometimes took sides in the conflicts. In some cases, the situation was almost anarchical. Such challenges to tear asunder the old establishment were reported in many parts of the colony (Austin, 1964:22).

The colonial state's response to these anti-establishment movements was to improve local administration by strengthening the position of the chiefs. The colonial government took the "commoner" challenge very seriously. In May 1925, it promulgated a new constitution that gave more positions to chiefs in the legislative council than the educated elites. The educated elites, however, complained bitterly because they felt they were not well represented, and that the chiefs might become tools in the hands of the colonial state, which could delay independence. The colonial administrators, perhaps out of selfish motives to maintain their power, did not want African's traditional political institutions, which have worked effectively for centuries, to be brushed aside by what they saw as small vociferous groups lacking a social base in society.

In 1927, the existing Native Administrative Ordinance was modified, giving the paramount and divisional chiefs, powers to make by-laws covering a wide range of issues defined as best left in the domain of traditional authorities, e.g. the election and destoolment of chiefs, the administration of public lands and forests and the maintenance of roads. But in judicial matters, only minor civil and criminal cases were given to the traditional authorities. Traditional courts were concurrent with but subordinate to the British courts in matters that could not be "adequately" dealt with by them, e.g. the imposition of the death penalty. Because chiefs were to play important roles in local administration, the colonial authority reasoned that their incorporation into the Legislature would preserve local institutions and make them more responsive to government policies. While the colonial state recognized the importance of traditional political institutions in society and sought to preserve them, the new educated elites considered them anachronistic.

The second major conflict was between the educated elites and the colonial state. Increased wealth in the colony also led to the quest for more education. The expanding educational system was producing graduates whose interests and aspirations could not be satisfied by the old traditional set up. At the same time, there was little of what Huntington calls "political institutionalization" taking place at the national centre to accommodate the rising demands from the new educated elites. Furthermore, the formal sector was not expanding enough to provide jobs. This created frustration and a "cultural awakening" among the educated elements that the status quo was unjust and should be reformed. Modernization of the colony had generated a momentum that seemed to be getting out of control. As one analyst succinctly puts it, along with the market, came "western" education, culture, technology, and religion, spreading unevenly, generating new interests, and conflicting and sometimes resurrecting and reinvigorating ancient commitments and alliances (Markovitz 1977:7). Civil society had emerged on the scene, attempting to secure its historic space and role in a developing market economy.

The conflict between educated elites and the colonial state revolved around many factors. Principal among these were: discrimination against the educated elites in the labour market particularly the civil service employment; the issue of alien rule and racism; lack of political representation; and more, importantly the "refusal" of the colonial government to undertake a more comprehensive development and industrialization of the colony. During one of Legislative Council debates in 1935, for example, it was noted:

> ...Your Excellency, as industries are the mainstay of any country, I desire to emphasize the necessity of fostering native industries such as soap, salt, tobacco, sugar, rice, pottery, cloth-making, fishing, and palm-oil with a view to establishing good internal trade in this country. Why should crafts that have held the people from time immemorial be made to give way for foreign cargo? Home weaving with all its beauty, for example, make way for Manchester and Japanese goods? Now the Department of Agriculture has often been severely criticized because of the failure of that Department to help the country in creating new industries...are we to understand, Sir, that our Agricultural

> Department is incapable of helping us to produce locally the articles mentioned in the notice? Surely they are capable, unless, in order to protect the European trade, it is not intended to help us develop economically. It may be contested that this is misconception, but I say, Sir, that this is what really exist in the minds of the people of this country... (*Legislative Council Debates* of 1935; 115 cited in Bourret, 1960:129-30).

Thus, the kiss of death given to the indigenous industries by the products originating from the metropole and the lack of industrialization in the colony was quickly perceived and the possibility of developing modern industries based on Africa's precolonial industries was also suggested. The modernization of agriculture, particularly food production was also ignored by the colonial state. The farmer continued to till the land with traditional tools and agricultural productivity was low.

There was also a feeling that the colony was not benefiting fully from its mineral wealth. This complaint, according to Bourret, had some ring of truth. By 1937, the Gold Coast ranked as the third largest producer in the world for diamonds and manganese. Bauxite deposits were estimated at some 229 million tonnes, and was among the largest known to exist. The Ashanti Goldfields Corporation at Obuasi was producing about 10 percent of the world's gold and was a profitable venture. In 1937 alone, it exported minerals worth £5,583,905, but British mining companies did not pay any tax on their exports besides the insignificant rents to chiefs who leased the land for mining and paid a token tax to the colonial government. The Corporation made a windfall profit because the cost of African labour in the mining industries at this time did not exceed £1 million and the local taxation did not exceed £400,000.

Indeed, until the end of World War II, British companies never paid any significant tax on mineral exports. Civil associations, such as the Achimota Discussion Group which had been reflecting over colonialism and its future in the country, complained on many occasions that the Gold Coast had too favourable a tax system compared to rival gold producers, such as South Africa and Australia (cf. Bourret, 1960:127-8). The bulk of the mining profits thus went to Europe rather than remaining in the Gold Coast for reinvestment. The accusations (and hence conflict between the younger generation and chiefs) that chiefs gave out lands and mines without a sufficient provision for retaining much of the mineral wealth were not without substance. The colony was, thus, not benefiting from its mineral wealth, indicating that the overriding objective of colonialism was as Baran ([1957]\1964) maintains, to take as much wealth and resources as possible from the colony and to use the colony as market.

However, the above quotation also suggests that industrialization was conceived of very early by the educated elites who were to become members of the first political party in the Gold Coast. This demand popularly expressed throughout the country was also to exert pressure on Nkrumah and his CPP regime to industrialize and modernize the country at a quicker pace than perhaps feasible. The emerging elites thus understood the colony's underdevelopment as due primarily to foreign domination and lack of industrialization. Apparently, they were dissatisfied with the colonial policies that held back the economic and political progress of the Gold Coast.

One major consequence of these complex conflicts was a general feeling among the elites that the time had come to exercise some control over their country. Indeed, by this time a number of civil associations had emerged in the Gold Coast with diverse interests but also gradually becoming politically significant. Notable among them was the Aborigines' Rights Protection Society, formed in 1897 by the coastal lawyers of Fanteland mainly to protect native lands from being sold to the Europeans by the chiefs. This group was already well known in the Colonial Office in London for the delegations and petitions it had sent to London to protest against the colonial government's attempt to confiscate land for the British Crown (Rimmer, 1992:21). Other associations included the Achimota Discussion Group, the Gold Coast Youth Conference organized by Dr. Joseph Boakye Danquah and other professionals in law, medicine, architecture and business. Conceivably, an emerging civil society and bourgeoisie were colluding to demand the democratization of the colonial state and the Africanization of the civil service.

Having developed an export-oriented primary economy, the colony was exposed to the vicissitudes of the international economy. The danger of a country depending a single primary commodity was again brought home by the Great Depression, and this had repercussions for political evolution. By the early part of the depression, world price for cocoa tumbled from £50 to £41 per tonne in 1929 and in 1930 to £20. In 1930, export earnings and revenue declined precipitously, leading to a budget deficit. The colonial administration decreased development and social services. The Africanization of the government and civil service was also halted, some 200 African and European officials were compulsorily retired, and income tax imposed throughout the colony (see Bourret, 1960). This meant that some of Governor Guggisberg's ambitions of greater material development and political representation were not being realized. Guggisberg had planned that, by 1936, Africans occupying European posts should increase to about 151, but, by 1938, only 41 Africans were in such posts. While much of these changes and unfulfilled goals could be explained by changed circumstances, the colonial government at this time was also not interested in further development of the colony. This situation brought frustration to the colony's graduates and even to the numerous educational establishments that were producing them.

Consequently, the colonial government's decision to increase taxes were resented on the grounds that the depression had contracted economic activities and generated unemployment problems. The African opinion was that the reserve funds created by Governor Guggisberg should be used for the development of the country. Furthermore, it was suggested that taxes could only be imposed if there was to be expansion in political representation in the Legislative Council and a permanent representation on the Governor's Executive council. The popular adage of "no taxation without representation" was echoed in the Gold Coast, but these requests were rejected because the colonial regime saw the educated elites as not representing the general will of the people. Indeed, in the post-Guggisberg era, there was less enthusiasm to advance the colony's economic and political progress. The protest and discontent continued, until 1934 when the economy recovered as cocoa prices soared again, reflecting a recovery of the world economy from the depression, and the revenue base of the colony improved. This put a damper on the conflicts, which were to escalate in the coming years.

By 1936/37, the price of the colony's main foreign exchange earner, cocoa, rose from about £21 per tonne in 1935 to £44. But, in the following year, the major European buying agents, who also controlled much of the import and commercial trade, lowered the domestic price of cocoa paid to local farmers. This measure infuriated the people, leading to chiefs, intelligentsia and peasants uniting successfully to withhold cocoa in 1938. The "Cocoa Boycott" of 1938 was also a boycott of European goods and stores. This action led some farmers to abandon cocoa production in favour of food crops. In order to protect the cocoa industry and farmers from the vagaries of international market prices and to forestall such domestic conflicts, the Cocoa Marketing Board (CMB) was established in 1947 with its main *raison d'etre*: the stabilization of domestic cocoa prices.

The CMB was mandated to lower prices paid to local farmers so that the difference between the world and the domestic prices could be kept as reserve to guarantee prices to farmers whenever the world prices fell to an extent that could undermine the industry. This forced collective savings scheme was, however, hardly used for this purpose. Instead, it generated foreign reserves, which were held in London. For example, in its three years of operation, the CMB secured over £40 million, and, in 1950/51 alone, it earned £19 million. By 1955, Ghana's foreign reserves totalled about £211 million (cf. Rimmer, 1992:56-7), but much of this was neither used for the development Gold Coast nor in ways that could benefit the farmer. Rather, it was invested in British government securities, contributing to the post-war reconstruction of the metropolitan economy (cf. Beckman, 1976:184). The colonial state was thus predatory, in that its *modus operandi* was to ensure that the colony's resources were siphoned off as early as possible without caring to develop it. This helps to explain why infrastructure development was skewed in favour of those areas that had resources and access to the sea. It also helps to explain why no cocoa processing industry was established in a country that was the leading producer of cocoa, as well as why the colonial government did not bother to scientifically revolutionized the food production system.

But, as events in the Gold Coast turned very conflictive, the colonial authority was forced to make some political concessions. A number of educated Africans were incorporated into Legislature, and their number was increasingly expanded as the political tension unfolded. The colonial government also gradually Africanized the civil service, particularly its lower levels.

Though the civil associations and discussion forums were demanding political reform and even complete autonomy, they sought to achieve their demands through constitutional channels. Indeed, the older men and the intelligentsia of these groups emphasized a gradual evolution of the colony towards independence. A very respectful nationalism had emerged, which was much praised in the colonial establishment. Compliments littered annual reports of the postwar period that the Gold Coast was a peace loving country exhibiting an unusually high political competence in Africa (Austin, 1964). It was also referred to as Britain's "model colony", full of educated and gentle people with respect for civil values and British institutions. As Casley Hayford, one famous indigenous scholar, put it:

> ...every son and daughter of the whole territory is well and rightly persuaded that our destiny is linked with Great Britain, and that under her tutelage, we shall succeed in achieving the best of every phase of life. Our quarrels and bickerings are domestic affairs... on the larger issues we are agreed... (quoted in Bourret, 1960:70).

Within this era, the colonial authority could pride itself that "responsible" educated Africans were emerging who would be groomed for independence. But, the effects of World War II were to dramatically accelerate the events in the colony towards independence, to the bewilderment of both the colonial administrators and these so-called "respectable" nationalists.

The Impact of World War II and the Rise of Gold Coast Nationalism

Although World War II, which broke out in the autumn of 1939, did not affect the Gold Coast directly, the colony gained strategic value, and this had profound effect on its material welfare and political development. During the war, as Bourret (1960) explains, Britain had counted on using the Mediterranean route to the Far East, but when Italy joined the Nazi camp, this route was blocked. The defeat of France after the fall of Dunkirk not only prevented the use of air bases of North America, but also meant that British colonies were flanked with potential enemy territory. Almost overnight, West African, and particularly Gold Coast airfields became of strategic importance to the Allied Forces.

Before this war, the Gold Coast had only a Volunteer Naval Force that was designed to defend the home waters. The war period, however, saw new developments in the Gold Coast: harbours and airfields were expanded, and internal communication systems were developed in response to the war needs. The people of the Gold Coast supported the British cause both materially and financially. They contributed as much as £340,175 as voluntary gifts, subscribed to war certificates worth £205,000, and offered £815,000 interest-free loan to Britain. They also volunteered for service abroad. All in all, some 70,000 personnel were drafted (cf. Bourret, 1960). The war effort also demanded raw materials such as minerals, rubber and vegetable oil which the Gold Coast populations supplied eagerly.

The demands of the war, however, were to put a number of strains on the colonial economy. First, the colonial Government curtailed developed programmes in the colony, and this was resented by the new elites. Second, the drafting of the people undermined food production at the same time that food imports declined. Ironically, however, the war efforts brought a concentration of European and African population in the urban areas. To make matters worse, the development of war services with wage incomes increased the money supply, which induced inflation and illegal informal sector economic activities. The cost of living in this period rose from 50 percent to 75 percent in some coastal towns (cf. Bourret, 1960:154). The government introduced some fiscal measures such as high wages and controlled prices, but these were too feeble, given the magnitude of the problems. Yet, these socioeconomic ills could be justified and tolerated so long as war could be used as an excuse. After the war, however, their existence was difficult to justify.

The end of the war raised expectations for early solutions to these socioeconomic problems. But these solutions did not come as expected, since the colonial masters themselves were demoralized and attention was turned to reconstruction at home. The conflict between the new elites and chiefs did not subside either, but escalated. Chiefs were threatened with destoolment. To diffuse the tension between the chiefs and the new elites, in 1944, Governor Alan Burns introduced a comprehensive Native Authority Ordinance, which allowed chiefs to impose fines and taxes for the native treasury. They were also given more power, making it, in Burns' own words, "no longer possible for a handful of malcontents to paralyze the activities of a State" by "fishing in muddy waters of Stool intrigue" (cf. Austin, 1964:24).

Pressure from civil society also mounted for further political representation, not only because of internal contradictions, but also of the psychological "benefits" of the war. The war broadened the horizons of many; those who were left behind, wondered what was happening to their sons, fathers, husbands and relatives who had been drafted. For those who directly participated in the war, the war brought out the "fears", "evils" and "barbarism" of the white man's "civilization". In a manner, the war was a manifestation of the probable havoc that beliefs of racial superiority could wreck on humankind when carried to the extreme. At the same time, however, the coalition against Hitler could be conceived of as a rejection of racialism. The new "cosmopolitanism" that the soldiers and colonized imbued, coupled with a change in international opinion towards colonialism, undermined white rule in the colony.

The postwar international politics was, in one respect, a rejection of colonialism. The end of the war in 1945 saw the establishment of the United Nations, with a Charter that had a section imploring colonial powers to promote the advancement and eventual self-government of their colonies. The US had emerged as a superpower, and anxious, to expand its hegemony, made it known to European colonial powers that NATO would not be used to maintain colonies. The independence struggle of India also had an osmosis effect on nationalist aspirations in the colony (see Nkrumah's "Positive Action", below). All these factors contributed in diverse ways to make the conflict between the educated elites and the colonial state highly inflammatory.

To control the conflict before it reached uncontrollable dimensions, Governor Burns introduced a new constitution (the Burns Constitution) in 1946, which provided for a clear African elected majority in the Legislature, the first ever in colonial Africa. Although the Legislature had little authority over policy, it controlled the purse, and thus represents another major step in the political development of the Gold Coast. Furthermore, he appointed Nana Ofori Attah, one of the most educated Africans, and Mr Arku Korsah as members of his Executive Council. These men were welcomed by the intelligentsia. The Legislature was inaugurated on 23rd July 1946, and the occasion celebrated accordingly. The colony was evolving constitutionally towards representative government as envisaged by the colonial authority albeit, perhaps with a faster tempo than they might have wished. The elites also seemed somehow satisfied and were moderating in their demands for change.

The success of representative government, liberal democratic theorists maintain, is mainly premised upon strong material development. The Gold Coast as a whole had experienced much socioeconomic transformation, possessing wealth, educated elites and administrators. Representative government was being institutionalized, which could enable it to achieve a smooth transition to self-government. But the postwar period was one of socioeconomic crisis. Thus, beneath the euphoria surrounding the political advance made by the Burns Constitution rested deep undercurrents of dissatisfaction.

With the end of the war came the de-mobilization of some 60,000 soldiers, and most expected the colonial government to remunerate them for their contribution to Britain's cause; a situation that never materialized as the economy was in dire straits. Goods were in short supply, inflation was high, jobs were unavailable and social services were inadequate. Much sociological change had taken place within this short period and much rapid socioeconomic progress was expected by all. Dissatisfaction against the colonial government increased. Given these socioeconomic setbacks and the "failure" of government response, in 1946, the educated elites, the emerging middle class and a few chiefs (not the majority), established the colony's first political party, the United Gold Coast Convention (UGCC) under the direction of the Mr George Grant, one of the wealthiest businessmen, and Dr. J. B. Danquah, to work for self-government within the earliest possible time.

It is important to stress the class origin of the members of this first political party, for it helps in comprehending the dynamics of politics in the later years. Most members of this party were the educated elites and chiefs who constantly insisted on self-government at the earliest possible opportunity and through a non-radical constitutional means. There is some element of self-interest here (which was also to define CPP politics). After all, they perceived themselves as the only heirs to the colonial throne, hence there was no need to be radical. By this token, theirs became bourgeois nationalism, in contrast to a popular and radical type that was to develop later from within their party. Their approach to the national question was one of accommodation and moderation. This "respectful" approach was endorsed by the colonial government, since the colonial authority had long held that self-government—liberal democratic style—was the ultimate goal colonialism and should be attained through tutelage. The UGCC was, however, to make a fateful decision that was to reverse its political fortunes and undermine its claim to power.

In 1946, upon the advice of its leader Dr. J. B. Danquah, the UGCC party invited Dr. Kwame Nkrumah, to become the party's permanent secretary. Nkrumah had earlier studied in Achimota College in Ghana, Lincoln University and the University of Pennsylvania. During his sojourn in the US, Nkrumah actively participated in many political organizations and associations that purported to advance the interest of Black people in general and self-government for Africa. Nkrumah also came to Britain to organize a series of conferences in Manchester. By electing Nkrumah, the UGCC leadership hoped to tap his rich organizational experience for its activities.

Nkrumah arrived in the Gold Coast in 1947, amid the economic problems and political tensions specified above. More tension arose because the cocoa industry was afflicted by the "swollen shoot" disease, which compelled the Colonial State to establish a Cocoa Research Centre at Tafo. Being a new problem in nature, there was no immediate scientific solution, indicating that, in spite of its contribution to the economy, the cocoa industry was not put on serious scientific footing. The government advised farmers and also sent out labourers to cut down trees, and this precipitated more social unrest in the cocoa growing areas.

On February 28, 1948, a group of ex-servicemen obtained permission to hold a parade and march to present their grievances to the government at Christianborg [the Castle], but these turned into social disturbances, government response to which left two people dead and five wounded. On the same day, disorder broke out in the capital city of Accra to protest against high prices. People were disenchanted by government policies and its inability to lower prices. The popular opinion saw government's "ineffectiveness" as a conspiracy with European merchant houses. They rioted, looted shops, and went to release prisoners in Accra's main jails. The disturbances lasted three days. In all, 29 people died, 237 were injured, and about £2 million worth of property was looted or destroyed (Bourret, 1960). The riots also spread to Kumasi. The Colonial government blamed the UGCC for the riot and detained the six leaders (who became known as the Big Six) of the UGCC, including Nkrumah and J. B. Danquah.[3]

The Watson Commission (from England and named after its head) was established with wide terms of references to investigate the immediate causes of the riot and the general problems of the colony, and also make recommendations. The Commission found no evidence that the UGCC leaders were behind the rioting or wanted to topple the regime. Its report, however, said, among other things, that the greatest problem in the colony was suspicion, which surrounded every government activity and policy; that the causes of the riots were due mainly to the socioeconomic straits of the colony; that political underdevelopment and insufficient jobs had created frustration for the educated class and that the people had wanted industrialization. As the Commission reported: "at every turn we were pressed with the cry of industrialization" (cf. Kay, 1972:77). With regard to chiefs, the Commission claimed the general view was that chiefs were instruments of British policy and popular opinion held that they should confine themselves to traditional rather than modern politics. The Burns constitution, which was earlier hailed as a great political advance, was now described as "outmoded at birth" because it did not take into account postwar developments. The Watson Commission recommended the enlargement of the Legislative Assembly and the development of African responsible government and independence for the Gold Coast within 10 years.[4]

The Commission's recommendations, however, did not please the conservative elements of British government and opinion. It was criticized for making radical recommendations that would accelerate Gold Coast's autonomy (Bourret, 1960:170). Sensing that the UGCC members had been respectful agitators, the colonial regime calculated that, if the choice was thrown to them, they would advocate a gradual approach to independence. Thus, in 1949, the Coussey Commission was set up, headed by an African judge, Sir Henley Coussey, to suggest how the colony's problems could best be resolved. The Coussey Commission called for gradual reforms in government at the central, regional and local levels. It recommended the adoption of British-type parliamentary system of governance but with a unicameral legislature in the colony. It called for the establishment of regional and local government institutions and advised that the members of the legislature should be elected directly or indirectly and the Executive Council should have African majority. The Executive Council was to be responsible to both the Governor and Legislature, rather than only the Legislature. The tone of the Coussey Commission's recommendation was less radical than that of Watson. Thus, its "wisdom" was praised in England and its recommendations were to form the basis of the next election and the constitution of 1951, which was to move the colony towards self-government.

After the 1948 events, the leadership of the UGCC began to experience internal division, mainly because Nkrumah wanted independence earlier than envisaged by other party members. This division was to change the character and pace of the independence struggle. Nkrumah's emerging radicalism was not shared by other UGCC members, hence the beginning of inter-elite rivalry that was to haunt Ghana throughout the postcolonial era. At this historic moment, had Nkrumah gone along with the others, or vice versa, the whole history of Ghana would probably have been different. But Nkrumah insisted on his cause and broke away from the UGCC which he perceived as too conservative, not addressing the issues that concerned the ordinary people. He converted the youth wing of the UGCC and founded his own Convention Peoples' Party (the CPP) which was more radical than the UGCC.

The CPP sought to mobilize the aspirations of the masses and attain early independence. It is important to note that the educational revolution in the colony had created two analytically distinct groups: (1) a well-educated minority coming from Achimota College and schools from Cape Coast; and (2) a relatively large group of adolescent primary and middle school leavers who had ten years of elementary education after the age of six (often referred to as Standard IV graduates). Those in this latter group who completed their school had limited clerical qualification and sometimes no skills, and thus were scarcely employable. The more successful ones were primary school teachers, petty traders and storekeepers. Others found more subordinate jobs in Cape Coast, Kumasi and Accra. The less successful ones become market-stall assistants, messengers, divers' mates or apprentices to master carpenters or motor-fitters in the urban centre.

When Nkrumah broke from the UGCC, which was dominated by the intellectual cream of society, it was mainly to this Standard IV group that he turned for support, and the CPP manifestos mostly reflected the demands of this group. In the prevailing political climate, the CPP made promises for quick socioeconomic development: industrialization, jobs for all, free education and national health service and equal opportunities. "If we get self-government, we'll transform the Gold Coast into paradise in ten years", Nkrumah maintained (cf. Rimmer, 1991:4). "We have the right to live as men"; we want "Self-Government Now": these were familiar sayings of the period. Nkrumah and his youthful supporters burst upon the political scene. Like the UGCC before it, the CPP attributed societal grievances and the country's underdevelopment to the colonial establishment and imperialism. Being more popular and in tune with the "realities", the CPP swept aside the UGCC. Nkrumah understood politics to be the key to Ghana's development. "Seek Ye First the Political Kingdom, and All Things Will be added unto it", he declared. A secular aspiration was to be enacted with a biblical injunction. Nkrumah and the CPP rejected the Coussey Commission's proposals and the 1951 Constitution because they sought to delay independence for the Gold Coast (Bourret, 1960:175).

Nkrumah had now injected a new radicalism into the Gold Coast nationalism, but Austin (1964:12-28) suggests that elements of his party were already a familiar phenomenon in the Colony and Ashanti Kingdoms where they had already articulated "anti-chief" sentiments. Nkrumah, and not the UGCC, voiced the aspirations of this mass and had concern for the less privileged rank and file of society. The UGCC, however, lacked a

CPP-type of popular base and also failed to articulate a "credible" party programme. Its members saw Nkrumah as one who had betrayed the trust they reposed in him and never forgave him and thus came to oppose most of Nkrumah's ideas and development vision. This is evident in the character of their political opposition to the CPP regime and the systematic efforts they made after its fall to rid Ghana of Nkrumah's ideas and development projects (see later chapters).

In his newspaper, the *Accra Evening News* of 17th December 1949, Nkrumah expressed his party's political ambition:

> As for us, we have already declared our stand with regard to the new constitution that only full Dominion status... will satisfy us and by this demand we shall stand and face all storms however strong they are. *Ghana must be free and free now* (emphasis added) (quoted in Bourret, 1960:172).

The unified nationalist struggle had decomposed and split into two main groups. A conflict had emerged among the elites who were to lead the Gold Coast into independence and beyond. With his charisma and considerable oratory gift, Nkrumah and his party re-organized the unemployed and ordinary people and took civil action to force the Colonial Government to concede independence. The "Positive Action", as this action was called, was a non-cooperative and non-violent strike, akin to Mahatma Gandhi's "Positive Resistance". The "Positive Action" was successful, for, it paralyzed the economy, and the Colonial Government was compelled to arrest and put Nkrumah in gaol, but this only made Nkrumah into a national hero. Consequently, he and his party became popular throughout Ghana.

Nkrumah was thus in gaol during the main campaigning period for election in 1951. But the CPP political programme was kept alive mainly because of the organizing ability of Nkrumah's lieutenant K. A. Gbedemah. In the 1951 election, the CPP won 34 out of the 38 seats. Only three seats were won by the UGCC, and one by an independent candidate. Given the victorious outcome of the elections for the CPP, the Colonial Governor, Sir Arden Clarke, had no choice but to release Nkrumah from prison and ask him to be the leader of the new African government. Later, at a press conference, Nkrumah declared that he was a Marxist socialist and a non-denominational Christian, that he was not against Britain, and he did not stand for racialism nor for discrimination against any race or individual, but was opposed to all forms of imperialism (cf. Bourret, 1960;176).

Under the 1951 constitution, Nkrumah became leader of Government Business which was a diarchical political arrangement through which he was to be answerable to the Governor. A transitional period was to expire to prepare the state and society for full independence. The CPP became highly popular at this time because of its organizational ability that the UGCC never rivalled. Yet, there were indications, even if not directly stated, that the colonial government would have preferred the UGCC to win over the CPP. They were, however, overtaken by events. A clear majority for the CPP made the party feel confident enough to seek its radical objectives. Shortly after becoming part of the government, the CPP put forward a proposal for early independence.

By 1953, a new constitutional amendment was made, calling for the withdrawal of European ministers and the creation of a legislative assembly with 104 seats. A new election was held in 1954 to test the political barometer. The CPP again won this election but had to face not only a challenge from the old enemies—the UGCC, now renamed Ghana Congress Party (GCP)—but also minor opposition parties based on regions and ethnicity. Notable among these was the Northern People's Party, organized to counteract what was being seen as a monopolizing CPP rule.

There were 104 seats at stake in the 1954 election and the result was as follows: the CPP won 71 seats and the rest of the opposition parties won the remaining 33 seats. Among the opposition parties, surprisingly, the Northern People's Party won as many as 12 seats and the Ghana Congress Party won only one. Nkrumah was able to mobilize the common people, mostly members of the "anti-chief movements" and cocoa farmers, even in the Ashanti region, to win the election. The CPP victory was not only a matter of political slogans. There were tangible socioeconomic achievements of the regime that sharply contrasted sharply with the poor development record of the colonial period (see Chapter 4). This victory was repeated in the July 1956 election which was to lead the Gold Coast to independence. CPP's victory was so seemingly convincing that Britain had no choice but to grant the Gold Coast self-government. Independence came on 6 March 1957, thus making the Gold Coast the first nation in Sub-Saharan Africa to do away with direct foreign domination.

In general, interest in socioeconomic change has been characteristic of all African colonies but as Bourret (1960:179) maintains, the inhabitants of the Gold Coast have shown themselves to be the most alert on the continent and have made most of their education opportunities. The socioeconomic progress already attained in the Gold Coast and the human skills developed made for the prognosis outlined in Chapter 1: that socioeconomic development would be easy and Ghana would develop into a coherent national unit.

Conclusion

The above historical survey indicates that development in Ghana along modern lines, for better or for worse, began in earnest with British colonialism. Precolonial Ghana possessed democratic political and market institutions. No doubt, institutions that have evolved over centuries and have worked effectively can be destroyed if a foreign structure is imposed from without. In terms of political development, colonialism halted indigenous state-making process and disturbed the organic development of Ghanaian societies. But at the same time, it integrated the numerous states and ethnic groups into something like a national community with a bureaucratic state out of which the postcolonial state would grow.

Colonialism gave an impetus to an economic process that was already under way, but was hindered by lack of markets and technological innovation. Its contribution was the removal of some of these constraints by the development of transport and communication systems, education and social services, the imposition of a bureaucratic-legal systems, modern laws and monetary system and also integrating the local economy into international economy. The *laissez-faire* economic philosophy and physical infrastructure reinforced existing capitalist values and accelerated capitalist development. But above all, the most potent force of change was the indigenous people's institutions and their energy and ability to seize the opportunities created by the insertion of their economy in the emerging global capitalist system.

Though the "spirit of capitalism" was prescient in precolonial Ghana and capitalist transformation had advanced during the colonial period, it would be an exaggeration to describe Ghanaians as individualistic, in the western sense. They were strongly communal in orientation, but beneath this lurked a deep-seated desire for material gain for themselves and their kin based on private choice. By the early 1950s, many people have come to share the view that the colonial political order (and not necessarily the economic order) was frustrating the realization of their aspirations. Consequently, they came to believe that independence would be beneficial. Also, a relatively buoyant capitalist economy emerged with social groups interested in its maintenance. But the radical elites have marginalized those liberal elites that could ensure a smooth transition of the colonial economic and political order into the postcolonial era. Thus, beneath the positive prognosis that development would be easy for Ghana, there was something crucial many analysts missed or did not give significant attention to, at the time: Ghana was at birth a fractured and ideologically divided society. This phenomenon, as it will be argued in the study, contributed in no insignificant way to the failure of development in the postcolonial period in that governing elites were unable to articulate a unified long-term development strategy and goal for Ghana.

Though one can argue that colonialism did lay much of the foundation for modern economic and political development, when a broader criterion of self-reliance economic development benefiting a country's population is applied, it fails in its impact. Colonial rule oriented major productive economic activity outwardly, in the form of cocoa production. Ghanaians were divorced from the ownership of mineral production. Much of the profit generated by the new mining companies was repatriated abroad, and foreign capital dominated the banking and trade sectors. Industrial development was shunned by the colonial powers and the human resource development was limited. Ghana was economically poor and "fragile" country, even though it appeared to possess some advantages over many other African countries. The Ghanaian economy has, however, become an integral part of the international capitalist economy. External rulers have different motives for colonizing other people, and, thus, even in their most benevolent moments cannot come to act in the interest of long-term economic development, as indigenous rulers might do.

The end of colonialism, however, brought an opportunity for Ghanaians to redress the development and welfare balance, to exercise some control over their resources, to devise strategies to achieve developmental goals, and to re-assert their human dignity, which was compromised as a result of slavery and foreign domination. As indicated in Chapter 1, some analysts emphasize that previous colonial relationships of centre-periphery survive in the postcolonial period and continue to militate against socioeconomic development in many African countries. This situation is dubbed "neocolonialism". Surely, "structuralism"—the colonial inheritance and the influence of the external economic environment—did not vanish with the attainment of political independence. Yet, the potential value of independence as a crucial step towards the attainment of some degree of economic development cannot be overemphasized. To rule out this, is to underestimate the capacity of agency or local initiative as crucial component in the process of socioeconomic transformation. Thus, Nkrumah's argument that Ghanaians should seek first the political kingdom cannot be out of place. If this was not important, then it would be difficult to justify the quest for independence in the first place. How, then, has the political kingdom been used for developmental goals? This question is addressed in the subsequent chapters.

Notes

1 The name Gold Coast signified this colony's wealth of gold which attracted Europeans. But as Huq (1989) notes, the name was changed mainly on account of the belief that the Akan ethnic group, which constitute about half of Ghana's population migrated to its present place from the old Ghana Empire which collapsed in the 13th century. The name Ghana was to serve as a symbol of the "re-birth" of the nation. Mazrui and Tidy (1984) quoting Nkrumah's speech to the National Assembly in November 1956, add that, Nkrumah, felt that "The Gold Coast" was a description with as many translatable meanings as there are many languages in the United Nations. Nkrumah was dissatisfied with the situation where countries did not adopt the English title "The Gold Coast" but rather translated it into their own languages. Hence the change.

2 The Portuguese (the first Europeans) arrived in 1471 and they found their "Africa Eldorado". In 1482, they built the first fort and significantly named it, Sao Jorge da Mina (i.e. Saint George of the Mine, now known as, Elmina Castle). See *The Gold Coast Annual Report* of 1947.

3 The others were William Ofori Attah, Akufo Addo, Ako Adjei and Obetsebi Lamptey. These men became political heroes and were later known as "The Big Six".

4 Sir Alan Burns' term as governor ended in August 1947. Sir Gerald Creasy arrived in the Gold Coast in January 1948 at the time when the colony was in much economic and political predicaments. Sir Charles Arden-Clarke (the Gold Coast's last British Governor) replaced Governor Creasy in 1949.

4 Nkrumah and the Great Transformation (1951-1966)

Introduction

Although independence seemed to have come to Ghana earlier than anticipated, there were at least four reasons why it was thought proper: (i) an indigenous political experience that went farther than in any other African colony; (ii) an administrative capacity and efficiency of the civil service; (iii) a national unity unburdened by strong ethnically-based politics; and (iv) an economic viability predicated upon previous achievements (cf. Rimmer, 1991:3-6). Sir Arden Clarke, the last colonial governor, also observed that, given the significant progress already made economically and the relatively large pool of trained manpower, if Ghana could not successfully carry out its development programme, then few African states would be likely to succeed (cf. Seidman, 1978:3). But, as indicated at the outset of this study, Ghana's postcolonial development has suffered in several respects and Ghana has become more dependent on the "aid industry" in recent years.

This chapter begins the story of how all this happened. It focuses on the early part of the postcolonial period when Nkrumah and his CPP ruled Ghana and made radical attempts to transform the society. It explores the nature of the state and relationship between it and development, the socioeconomic achievements of the CPP regime as well as the tensions, paradoxes and conflicts generated by its development strategy. It also analyzes the causes of centralization of political power and the patrimonialization of the state as well as the resultant political conflicts and the weakening of the state.

The chapter locates the early depletion of foreign reserves as a result of the search for rapid socioeconomic development, deterioration of the terms of trade for Ghana's cocoa, economic mismanagement, the failure of agriculture, public waste and corruption as some of the factors responsible for the economic crisis of 1966. At the same time, it argues that although the CPP regime was able to accelerate socioeconomic development of the nation, it was unable to evolve a development strategy that could be more sustainable and compatible with the resources of the so-called fragile economy. Furthermore, its development ideology was not sanctioned by the elite group (the UGCC) from which the CPP emerged during the independence struggle. Consequently, there was inter-elite conflict over development ideology. The chapter sees the 1966 military coup that ousted the CPP regime as the culmination point of the conflict between the regime and those with vested interest in the continuation of the colonial economic and political legacy. This conflict thus resulted mainly from structural dislocation with existing socioeconomic processes and the attempts to radically evolve new set of rules and institutions for society. The period covered in this chapter is from 1951 to 1966.

The State and Development: From Gradualism to Structural Transformation

Soon after independence, Prime Minister Kwame Nkrumah stated what he wanted to do for Ghanaians in this way:

> my first objective is to abolish from Ghana poverty, ignorance and disease. We shall measure our progress by the improvement in the health of our people; by the number of children in school and by the quality of their education; by the availability of water and electricity in our towns and villages; and by the happiness which our people take in being able to manage their own affairs. The welfare of our people is our chief pride, and it is by this that my government will ask to be judged (see *The Ten Great Years 1951-1960*, 1960).

In 1951, when Nkrumah became the leader of Government Business, there was an improvement the colony's financial position, mainly as a result of favourable international economic environment for exports. This made it possible for Nkrumah to implement the Ten-Year Development Programme launched by the colonial government in 1946 within a five-year period. This colonial plan did not aim at any structural transformation of society, trapped as it was in the British economic philosophical tradition of less government intervention in economy. It did, however, establish an Industrial Development Board (IDB) with modest ambition: to develop African manufacturing industries with a view to handling them over to local private enterprises (Asante, 1993).

Feeling confident about the future, the government invited Professor Arthur Lewis, a development economist from Manchester University, in 1953, to study the country and submit a report "On the Problems of Industrialization and Economic Policy". The Lewis Report noted that a major industrialization programme was unjustified at that moment because of the following: (i) the domestic market was limited; (ii) a big industrialization programme would mean the transfer of rural resources to the modern sector, which would reduce the living standard of the farmers; and (iii) there was a shortage of labour that industrialization would aggravate. This last factor appears to be a bit exaggerated since there was no shortage of labour in Ghana at this time.

The Lewis recommendation, however, emphasized three important areas in the economy that needed government attention: (i) food production, (ii) public services, and (iii) the development of physical infrastructure. Lewis stressed the development of infrastructure so as to lay a basis for private foreign investment without the government having to bother offering special investment favours. In his opinion, the number of industries that Ghana could set up was small because Ghana's industrial raw material base was limited. He further noted that the government was so short of money that it should be reluctant to take on ownership and operation of industrial undertakings, except in the case of utilities and for the purposes of pioneering. Lewis argued that it takes time to lay the foundations of industrialization.

Given these constraints, Lewis urged the government to set up the Industrial Division in the Department of Commerce to announce its willingness to give limited aid, by way of temporary protection or subsidy, to newly established factories, particularly those that were willing to employ local people in senior posts. He also advised the government to make known its positive attitude to foreign enterprises, welcoming such enterprises, especially if they entered into partnership with the government or Ghanaian enterprises. Furthermore, Lewis suggested that the government should guarantee the free transfer of profits and dividends and a fair compensation, independently determined, if nationalization ever took place. A small industrialization programme was, however, justified in Lewis' view (see Lewis Report in Kay, 1972:88-89).

The Nkrumah regime accepted some of Lewis' recommendations, but not his cautious approach to industrial development, because rapid industrialization was the regime's ultimate development goal. In a broader perspective, Nkrumah's view about development belongs to the tradition of "economic nationalism" that according to Gilpin (1987) is often based on the belief that politics drives the economy and the outside world is hostile. This philosophy laid behind the mercantilist ideology dominant several centuries back in Europe and has undergone modification to become Keynesianism in contemporary times. Nkrumah believed (see below) that, unless Ghana industrialized and lessened its dependence, it would lose all that it fought for through independence. Nkrumah's views were also congruent with a number of development ideas hegemonic in the postwar period: (i) mainstream development economics which by now was positing a "big push" strategy to be adopted by the state to overcome underdevelopment; (ii) socialist ideology which also conceived the state as the main motor of development; and (iii) Keynesianism, which also advocated a strong role for the state in market economy.

Beside the ideas emanating from the outside world and the discipline of development economics, the local context within which Nkrumah was operating also favoured economic nationalism that could deliver rapid socioeconomic development. As noted in the last chapter, lack of industrialization, employment opportunities and social development were some of the grievances of the nationalists who challenged colonial rule. These grievances did not diminish with the attainment of independence. Indeed Nkrumah and his CPP had responded to the demands of mostly the youth and the urban unemployed at the time and had promised them employment as well as welfare as soon as independence was achieved. The promise to seek first the political kingdom and everything else would be added and to develop Ghana into a paradise within a decade "from above", was consecrated by rising social expectations "from below". Indeed, the political utterances by the CPP to outbid the claims of the UGCC party during the independence struggle excited the people's demands from their new leaders. There was thus societal pressure on Nkrumah and the CPP to deliver more goods than it seemed feasible at the time.

Another factor favouring economic nationalism was leadership ambition to secure continental unity for Africa. Nkrumah had conceived Ghana's independence in much broader terms (than the UGCC members) as a prelude to the total liberation of the African continent from colonial rule. Ghana's independence, he maintained would be meaningless, unless it was linked with the liberation of other territories in Africa (Nkrumah, 1957:x; 1961). Rapid socioeconomic development and industrialization would serve as a testament of the tangible achievements that independence could bring. This, in turn, would act as a fillip to those countries still struggling for independence. Rapid socioeconomic development would also boost Nkrumah's Pan-African objectives towards the realization of African unity that Ghana was to champion. Nkrumah's vision was that Ghana would lead African countries into African Unity (or United States of Africa). As he noted, "across the parapet I see the mother of ... African Unity and Independence, her body smeared with the blood of her sons and daughters in their struggle to set her free from the shackles of imperialism. And I hear and see springing up cities of Ghana becoming the metropolis of science, art, industry, philosophy and learning" (cited in Asante, 1993). Thus, leadership ambition also exerted pressure on the regime to accelerate the pace of development in the country to justify a leading role for Ghana in African affairs.

Lewis' approach to development was, however, gradualist in orientation and emphasized the development of infrastructure and agriculture over industrialization. Working under the colonial Governor, the Nkrumah regime did not have much room to initiate a radical transformation of society. In this period, Nkrumah accepted the Lewis development strategy and also held the view that the private sector should be supported to contribute to Ghana's development, because "of the fact that it will be many years before the Gold Coast will be in a position to find from its own resources people who can combine capital with experience required in the development and management of new industries...." (cf. *Prime Minister's Statement on Foreign Investment*, March 1953, in Kay, 1972:89-90). To this extent, Nkrumah made it clear that the Gold Coast was prepared not only to encourage the entry of foreign capital but also to ensure that the interests of investors were adequately safeguarded.

Yet, it was not only foreign capital that was to be encouraged, but also indigenous investors. In 1955, the government established the Industrial Development Corporation (IDC), which was, among other things, to investigate the possibilities of new industries, that it would itself establish either as subsidiary companies or in association with reputable firms. It would also offer loans to small businesses and help them acquire machinery. Some of the areas identified for industrialization were the production of matches, brick-and tiles, liquor, vegetable oils, paint, soap and fishing boats. Thus, there was an intention and even attempt by the government to harness the resources and activities of local entrepreneurs to supplement its development efforts.

The IDC policy remained fundamentally unchanged until it was disbanded in 1962 when the regime definitively broke from gradualist approach to development (cf. Kay, 1972:47) to implement the so-called socialist development strategy. The development strategy at this time was supposed to be based on the alliance between the state and the private sector (both foreign and local). This was in harmony with Nkrumah's own observation that history has drawn Ghana and the West together, a fact that could not be erased overnight (cf. Etsiah, 1989/90:38). Before proceeding further, it will be helpful to examine what socioeconomic progress the government achieved before it opted for a more ambitious development strategy.

Socioeconomic Achievements of Gradualist Development Strategy

The Nkrumah Government implemented two development plans between 1951 and 1962: the Ten-Year Development Plan (1950-1960), which, as already noted, was compressed into a shorter period, and the Five-Year Development Plan (1959-64), which was abandoned after Nkrumah's tour in 1961 of the former Soviet Union, Eastern Europe and China in favour of the third, the Seven-Year Development Plan 7YDP (1963-1970), the so-called socialist blueprint.

From 1950 to 1962, when the government implemented the first two plans, it concerned itself mainly with laying socioeconomic infrastructure. The government created a nationwide road system that made it possible to reach all regions on macadam all-weather roads. Feeder roads were also extended and bridges built to span many rivers. In 1951, Ghana had 3,500 miles of trunk and town roads out of which only 939 miles were tarred. By 1960, however, there were 4,420 miles and over 1,900 miles of them were tarred to the highest standard at the time. The most significant bridge, the 800-foot long Adomi Bridge spanning the Volta River, was completed in 1956. The railroad system was also extended to new cocoa growing areas. The government also constructed a £2 million modern four-lane motorway between Tema and Accra. So spectacular was this development that, by 1965, Ghana was judged to possess a physical infrastructure capacity that could last for years to come, and little additional investment was necessary for feeder roads development (Seidman, 1978:140). This conclusion is not without substance, because until the late 1980s and early 1990s when the PNDC-NDC came to power, no new major roads were added to what the Nkrumah Government had provided, although much of this infrastructure decayed in the post-Nkrumah era.

In addition, the government undertook a major hydroelectric project on River Volta—the Akosombo Dam—and linked it with the development of a modern harbour at Tema to provide a firm foundation for Ghana's industrialization programme. The Dam was the largest single infrastructure project at the time, and one for which the Nkrumah government did not have enough money to finance. But after an elaborate negotiation by the government, coupled with the intervention of President Eisenhower of the US, the US firm Kaiser Aluminum Company agreed to underwrite part of the Volta project on condition that it form a consortium with Reynolds Aluminum to utilize about two-thirds of the electrical power to be generated for smelting aluminum in Ghana.

The aluminum smelting project: Volta Aluminum Company, which is locally known as VALCO, was established but its agreement did not favour Ghana because it secured a number of generous concessions from the Nkrumah government. Beside having access to very cheap power by world standards (which means low tariffs to the government), Kaiser also got a tax moratorium of 30 years, whereby VALCO was not to pay taxes on imports of construction equipment or alumina, and no taxes on exported aluminum ingots or profits for 10 years. After this period, it was to pay 40 percent on retained profits in addition to 2 percent on those remitted. VALCO was also not to use Ghana's large deposit of bauxite as raw material in its activity. Being a capital intensive project, VALCO was using only about 0.03 percent of Ghana's labour force at that time (Seidman, 1978:138).[1] Given its limited benefits to Ghana, the VALCO agreement was to become a source of conflict between Ghana and VALCO, particularly in the early 1980s when the Rawlings regime sought greater control over the economy. The total cost of the artificial harbour said to be the biggest in Africa at the time, was, however, financed by the government.

One other subsidiary benefit of the Volta Project was the lake created by the construction of the dam (renowned to be the largest man-made lake at the time). This lake was to serve as a transport system linking southern and northern Ghana. Furthermore, its waters were to be used for irrigation agriculture in the Accra Plains and the savannah regions in the north. Moreover, the fishery resources of the lake were to enable Ghana achieve self-sufficiency in fish production (cf. Nkrumah in Obeng, 1997).

Ghana also made progress in air and maritime services, communication and public utilities. Ghana Airways was established and was employing Ghanaian pilots. The national maritime company, the Black Star Line, and the Nautical College were established to train Ghanaians as merchant seamen. Also established was a military academy to train soldiers. Post and telephone services were expanded throughout the major cities and towns. In 1950, there were 6,000 telephones in service throughout Ghana, but by 1960, they increased five-fold to 31,100. The government also expanded the State's broadcasting system. Radio receiving stations and technical facilities were established in many towns in the country and fleets of mobile cinema vans undertook countrywide civic education programmes on subjects such as health and education. While in 1951, only 157,400 people in the rural areas had water supply, by 1960 this figure had tripled. The government also began nuclear research with the help of Russian scientists.

Primary, middle and secondary schools, as well as technical and teacher training colleges were expanded rapidly. Primary education was compulsory and free. Between 1951 and 1956, enrollment in primary and middle schools increased by about 150 percent. Within the decade of 1950-1959, enrolment in government-approved secondary schools increased five-fold likewise in government technical and vocational institutions. The University of Ghana was expanded and began employing Ghanaians as lecturers. In 1951, the Kwame Nkrumah University of Science and Technology was also established in Kumasi to provide training in agriculture, estate management, engineering, pharmacy, science, architecture and building. So much development in education occurred beyond what the pre-independence Commission on Education of 1948 believed could be possible within twenty years. The cost of university education including tuition and housing was totally free.

Health services were also improved. New hospitals and clinics were established throughout the country. Korle-Bu Hospital in Accra was extended to become the Teaching Hospital for the Medical School of the University of Ghana. In 1955, Kumasi Teaching Hospital for the Medical School of the University of Science and Technology was also established. Social welfare and community development programmes were also expanded. Hotels were built in the major cities and regional capitals notable among these are Meridian and Ambassador Hotels and Hotel Continental. Cultural development was not ignored. The Ghana National museum in Accra was opened, the Asante Cultural Centre was also established in Kumasi, and the Arts Council of Ghana in 1958 to foster the arts and crafts. The Ghana Academy of the Arts and Sciences was also established. Sports stadiums were built in Accra and Kumasi as well as in some other regional capitals.

In the tradition of modern states having control over their financial policies and institutions, the government moved quickly to establish monetary and fiscal sovereignty for Ghana. Ghana withdrew from the West African Currency Board (WACB) established by Britain, to establish its central bank, the Bank of Ghana, in 1958, with powers to issue its own currency and control reserves. Other banks established included the Ghana Commercial Bank (1952), National Investment Bank (1963), the Agricultural Credit Bank (1965) and State Insurance Corporation (1962). Some of these banks had branches in the large towns. They were to encourage savings, give out credit and engage in investment, but they were also meant to challenge the dominance of foreign-owned banks, such as Barclays Bank and Standard Chartered Bank operating in Ghana. This rapid socioeconomic development and modernity certainly raised the quality of life at this moment.

Yet, by 1961, the government became dissatisfied with the gradualist pace of development unfolding in the country. The Lewis report urged the government to lay socioeconomic foundations to attract foreign investors as a means to industrialization. But foreign investors were not coming to Ghana as anticipated. This situation, as will be shown in the subsequent chapters, has changed little in the post-Nkrumah years. In the absence of foreign investment and a strong domestic bourgeoisie to initiate Ghana's industrial development, the Nkrumah government confirmed a state-led industrialization programme. Gradualism gave way to a more radical development strategy aimed at the structural transformation of Ghanaian society.

In 1964, the government introduced its Seven-Year Development Plan (7YDP): the socialist plan. But before this plan, the Nkrumah regime was already according a leading role to the public sector in the development of the economy and society. While this approach is consistent with the recommendations of mainstream development economics at the time (cf. Killick, 1978), some analysts see socialist ideology at work. Petchenkine (1993:16), for example, argues that the second plan was actually borrowed from Soviet economic history and the choice of time limit was consistent with Kruscshchev's seven-year plan started much earlier (see also Dowse, 1969).

The 7YDP was meant to radically advance previous achievements. It was drawn up within the framework of a 20-year perspective to restructure the production base of the economy. It proposed that the state should enter directly into productive activities in both agriculture and manufacturing industries, with the ultimate goal of developing a socialist society. Rapid integrated industrialization was the main core of the plan that aimed at a shift from import-substitution to export processing to intermediate, and, finally, to the production of heavy industrial goods (Seidman, 1978:89). The government's main objective in the plan was to bring about quickly the economic and social reconstruction of the country to provide higher standards of living for all, to create a welfare state based upon socialist pattern of society adopted to suit Ghanaian conditions, and to give active support to the conception of a Union of African States or Republics (cf. *The Ten Great Years: 1951-1960*). According to Petchenkine, this plan also followed a new programme adopted by the Communist Party in the USSR and shared similarities with other programmes in the Eastern European countries.

In launching this plan, the regime proclaimed its new party programme, called *For Work and Happiness* for the attainment of a socialist state. That Nkrumah was to gear the country's development in a socialist direction should come as no surprise, because in his autobiography published significantly in 1957 (the same year Ghana got independence) he espoused a socialist ideology or philosophy:

> ...no race, no people, no nation can exist freely and be respected at home and abroad without political freedom. Once this freedom is gained, a greater task comes into view. All dependent territories are backward in education, in agriculture, and in industry. The economic independence that should follow and maintain political independence demands every effort from the people, a total mobilisation of brain and manpower resources. What other countries have taken three hundred years or more to achieve, a once dependent territory must try to accomplish in a generation if it is to survive. Unless it is, as it were, 'jet-propelled', it will lag behind and thus risk everything for which it has fought. *Capitalism is too complicated a system for a newly independent state. Hence the need for a socialistic society* (Nkrumah, 1957:x; emphasis added).

The philosophy emphasizes the importance of securing political freedom for development and prosperity, which, in turn, would lead to respect and dignity for colonized people. It sees progress and development as advances in education, agriculture, and industry. It also recognizes that to overcome underdevelopment requires hard work from everybody and it also stresses the need for "catching up" with the developed world within the possible shortest time. The sense of urgency cannot be overemphasized. It echoes Lenin's advice to the Soviet people that they were fifty or hundred years behind the advanced capitalist countries; either they made good this distance in ten years or they would be crushed (cf. Hettne, 1990:38). Above all, it spells out Nkrumah's aversion to capitalism. Killick (1978) has suggested that Ghana's immediate postcolonial development strategy can be explained in terms of *Development Economics*, but it seems that the "success" of state socialism in the former Soviet Union and Eastern Europe also influenced the Nkrumah regime and strengthened its belief that socialism represented the most efficacious means of attacking underdevelopment in Ghana (see, Asante, 1993). The so-called socialist development strategy was expressed in two main areas: agriculture and industry.

State-Propelled Agricultural and Industrial Development

Before launching the 7YDP, the CPP regime had relied on the Agricultural Development Corporation (ADC), established earlier to develop not only alternatives exports to cocoa but also food production and marketing facilities. The ADC organized numerous co-operatives for food production and marketing across the country, but was not a profitable organization. It accumulated debts and financial losses while its main offices were over-staffed and its officials enjoyed considerable fringe benefits. A study done by the National Investment Bank at the request of the Nkrumah Government noted that the co-operatives developed by the ADC did not operate as true co-operatives. Those who administered them knew little of co-operative practice. In some places, they were administered by chiefs who assigned work in a semblance of old communal patterns. No proper auditing and accounting were done, and they lacked managerial and technical skills (cf. Seidman, 1978).

As a result of these bottlenecks, the government in 1963 re-organized the activities of the ADC and sought in the 7YDP to create a socialized agricultural production system. Dadson (1993:310) has summarized the main reasons for this new approach to agricultural development as follows: (i) to raise agricultural output and efficiency by introducing technology into agriculture, and by expanding the area under cultivation; (ii) to alter the structure of commodity production in favour of industrial and export crops, and livestock; (iii) to supply the urban population with low-cost foodstuffs; and (iv) to attract the rural educated youth into agriculture and thus reduce unemployment and improve the quality of human resources in agriculture.

With the socialization of the agricultural sector, four agencies—the State Farms Corporation (SFC), the Workers' Brigades (WB), the United Ghana Farmers' Council (UGFC), and the Young Farmers League (YFL)—were now in charge of agricultural production. With the exception of the SFC, all these had been established for various non-agricultural functions: the Workers Brigade had para-military training programmes for the unemployed; the UGFC was a farmers' political organization and cocoa buyer; and the YFL was a farm training organization for the educated youth. All these organizations were integral parts of the CPP party, which had been seeking to bring under its umbrella existing and emergent civil societies. The State Farms Corporation was to diversify the economy by producing, for example, maize, rice, millet, vegetables, fruit, oil palm, sugar cane, tobacco, rubber, yams, banana, sorghum, cashew and kenaf, so as to lessen Ghana's over-dependence on cocoa and also to provide raw materials for Ghana's agro-based industries. By 1965, there were some 123 farms with a labour force of 20,000 cultivating 64,300 acres of land. But this acreage represented about 1 percent of the total area under cultivation in Ghana (cf. Hutchful, 1987).

As far as industrial development was concerned, it will be recalled that the Industrial Development Corporation (IDC) was to establish small-scale industries to be privatized. This was in harmony with the views of the development experts who visited Ghana, although mainstream development economics was now advocating state-led Import Substitution Industrialization (ISI) programme for the so-called developing countries. The IDC was able to enter into partnerships with some foreign companies to build a storage plant; an engineering firm and a construction firm (Seidman, 1978:205-6). But these achievements were too feeble to be considered a serious industrialization programme.

Given these modest industrial achievements, the government liquidated the IDC and sought to implement an ISI programme and, later, an export-oriented programme. This involved the widening of Ghana's industrial base through the processing of timber, cocoa and minerals, to add value to them before they were exported. It was hoped that, through this strategy, more foreign exchange could be earned. Emphasis was also placed on the development of basic industries such as iron steel, oil refinery, and petrochemicals. With these, industrialization thus gathered speed with the implementation of the 7YDP. By 1965, about £41 million was spent on industrial development, two large cocoa processing firms were constructed in Tema, an interlocking Food Complex was built by the German firm Drevicchi and a steel plant by a British firm was rolling 30,000 tonnes of steel from scrap iron annually.

Ghana also gravitated more towards the ideological camp of the Soviet Union and Eastern Europe. This was evident in the contraction of economic, scientific and cultural links with the former Soviet Union, the countries of Eastern Europe and China. These developments were not, however, for purely ideological reasons, but also due to the regime's desire to broaden Ghana's trade links with the outside world. Experts from the socialist countries were invited to help Ghana undertake various development projects. Projects completed under bilateral agreement with Eastern European countries included, fruit and vegetable plants, shoe factories, a leather tannery, cement factories, a rubber and tire factory, a pharmaceutical complex, a meat packing factory, two sugar factories and a gold refinery.

In 1964, the government set up the State Enterprises Secretariat to assume full responsibility for control and evaluation of these industrial enterprises. By 1965, state-owned enterprises amounted to over 50 in different sectors producing at different levels of complexity. Some 90,000 workers were employed in manufacturing alone. To the Nkrumah regime, industrialization is the key through which Ghana could reduce its dependence on the external world and eschew material poverty. As he noted "the momentum of [Ghana's] development will increase so that in a relatively short period Ghana will become a modern industrial nation, providing opportunities for all and a standard of living comparable with any in the world" (Nkrumah, 1961:242). Nkrumah maintained that state enterprises were the main economic pillars upon which to build socialism (cf. Asante, 1993:264). In order to cope with the extensive role the state was to play in socioeconomic development, the bureaucracy was also expanded. Such, then, was the state of government's efforts at development by the mid-1960s. Yet, in spite of substantial material progress, the Nkrumah government was overthrown in a military coup in 1966. Why was this the case?

Many answers have been given to this question. Indeed, much has been written about the CPP regime's achievements and failures than any other in Ghana, except perhaps Rawlings' regime in contemporary times. There are those who think that Nkrumah's development strategy was correct and it was external factors, such as low commodity prices for Ghana's cocoa that brought the strategy into crisis (see Green, 1971). Others see a conspiracy of the West and the CIA (the Central Intelligence Agency of the US) as responsible for the military coup mainly because of the regime's anti-imperialist and West posture (see Hutchful, 1987). Still others blame Nkrumah and his regime's policies (see Killick, 1978; Berg, 1971). The differences in interpretation are often linked to different assumptions about what was possible and desirable at the time, sometimes with ideological overtones. Beckman (1976:14-28) has also provided a summary of other views expressed by analysts about Nkrumah's downfall (see also the debates in Foster and Zolberg, 1971).

Surely, the CPP, more than any regime to date, had a very ambitious development programme for Ghana, but never lived to see it through. To argue that, but for the coup, the CPP government would have achieved its goals with its so-called socialist development strategy with or without modification, would be a counter-factual claim and a matter of mere speculation. Why did some section of Ghanaian population supposedly benefiting from the socioeconomic development rapidly unfolding in the country, come to overthrow a government responsible for such a programme, when it encountered a temporary economic crisis? This is an interesting question that needs to be addressed empirically.

As it will be shown in the next chapter, personal motives (agency) and perhaps as some analysts argue international conspiracy did play a role in the genesis of the 1966 military coup, but there was a structural factor or rather conflict operating in society that also seemed to have played some role in the regime's overthrow. This structural conflict was created by or rather associated with CPP's development strategy which exhibited some intriguing paradoxes and contradictions. The next section tries to explicate how this structural conflict contributed to Nkrumah's fall. Although some of the arguments developed here overlap with others already made by other writers, there is no claim to an exhaustive treatment of all the factors that possibly brought Nkrumah down. This issue cannot be settled in this study.

The Acceleration and Crisis of Development

The period from 1951 to 1960 before the 7YDP was imposed can be described as one of rising expectations. GDP grew over 5 percent between 1951 and 1960 (Esseks, 1975; Killick, 1978:66). This economic growth was reinforced by favourable world market prices for primary products in the late 1950s, although, by the early 1960s, world market prices for cocoa had gradually begun to fall. This enabled the state to assume a leading role in socioeconomic transformation. The government was able to meet many of the social expectations. The importation of consumer goods was easy. The running down of foreign reserves and the appropriation of cocoa surpluses from farmers through the CMB (see below) also permitted the funding of infrastructural projects and exceptionally high investment in industry. In the early 1960s, industrial output grew at 13 percent, employment at 8 percent per annum, and industrial production at one stage attained a level equivalent to over 14 percent of the GDP (cf. Roe and Schneider, 1992). Though the country continued to depend on imports, there was a significant shift in the composition of imports from consumer to capital goods and raw materials. An industrial base was thus being laid for Ghana.

Yet, by the early 1960s, the Ghanaian economy began to show signs of weakness, and this continued so that, by 1966, when the military overthrew the CPP regime, the economy was in crisis. There are internal and external causes of this crisis. But for now, the concern is with some of the problems and weaknesses associated with the agricultural and industrial programmes, leaving the paradoxes, tensions and conflicts generated by the development strategy as well as other internal and external causes of the crisis for later sections. No systematic discussion is made around these factors, but it is hoped that they become clearer in the course of discussion.

One salient feature of Nkrumah's development programme was its search for a quick economic "take-off" and rapid structural transformation. This led to the quicker use of resources than if development had proceeded at a slower pace. By the early 1960s, development of economic infrastructure and productive capacity was consuming 80 percent of planned investment (Ray, 1986:120). Also, although the development of an extensive "egalitarian" social sector—free education, health and other services—could be seen as indicator of social development, it also meant that the government was undertaking what, for lack of a better term, can be described as "re-distributive social production system", entailing high recurrent expenditure. Much of this was financed through deficit budgeting leading to high inflation (Kraus, 1971:54). Consequently, from 1951 to 1954, government expenditure annually exceeded revenue, and this continued through to the early 1960s. Within a relatively short time of the nation's history, much of the financial resource had been expended, and the nation was beginning to experience "fiscal crisis".

This problem could have been eased, had the socialist production system yielded reasonable returns at this time. The socialist agricultural programme's performance was, however, mediocre and unsatisfactory. The State Farms that were established with massive doses of financial resources and with liberally supplied capital equipment and technology (bull dozers, tractors, etc.) were not profitable. For example, at the end of 1964, State Farms alone were reported to have lost £150 million; a figure estimated to have doubled in 1965 (Seidman, 1978:178). In spite of the heavy investment made in state farms, they showed little food output for their activities; about 98 percent of food production continued to be in the hands of peasants.

The main reasons for the poor performance of state farms were underemployment, political interference in their activities, lack of adequate management and technical skills, over-staffing, corruption and nepotism. Also, much of their Soviet or Eastern European agricultural technology was not suitable for Ghana's tropical conditions, and thus frequently broke down. Ghana, however, did not have qualified indigenous technicians to repair them. The lack of foreign exchange affected the importation of critical components for agricultural and industrial machines. Also, most farms were set up for political, rather than economic reasons, without adequate feasibility studies and consideration for agronomic requirements of the crops to be cultivated. No proper accounts were kept of the operations of the farms. There was also lack of discipline and commitment. It was reported that some leaders of the YFL were political agents for the CPP and were cultivating their own private farms while receiving salaries and allowances from the government. The co-operatives established by the UGFC were equally afflicted with the above-mentioned problems and did not operate as a union of co-operatives. These factors increased the overhead costs of the State Farms and Co-operatives established to transform agriculture (cf. Frimpong-Ansah, 1991:83; Dadson, 1993).

Seidman provides a summary of the problems that afflicted State Farms in Ghana. The programme, she notes,

> can hardly be said to provide an adequate test of the ultimate effectiveness of large-scale state-owned farms in Ghana or elsewhere in Africa. The State Farms Corporation expanded far too rapidly for the country's limited technical personnel. Management was clearly inadequate on the farm level...

> A major hinderance... was the policy of hiring large numbers of redundant labourers in an effort to reduce unemployment. Clearly no organization could be expected to function efficiently or economically with such a surplus labor force of as much as twice the required size... Economic criteria must be established for all large scale state agricultural projects (Seidman, 1978:183-4).

The SFC, WBs, the farmers' co-operatives, and the other units engaged in agriculture also became instruments through which the CPP could dispense patronage and, in return, garner political support and control society. For example, the activities of the agents and co-operatives set up by the CPP to buy cocoa for the CMB were politicized and this in turn undermined their viability as an effective instrument for managing the cocoa affairs. Following several complaints about this state of affairs, an Investigation Commission was established by the government to look into the situation. The Commission, headed by Justice Jibowu, found that the Cocoa Purchasing Company, which was established in 1952 by the CMB, was used to build support for the CPP to the extent that, when finally the company was abolished, the CMB lost £800,000 out of the £1,2 million it had invested in the company (cf. Seidman, 1978). The commission also found that the UGFC that was to serve the interest of farmers was actually headed by CPP activists and many of them became corrupt and received loans that they never paid back (cf. Austin, 1964:65).

Another shortcoming of the 7YDP was that it ignored the activities of peasant farmers who, constituted about 60 percent of the population. This is, however, astonishing given the fact that peasants have historically demonstrated through the cocoa industry that, given the proper incentives, they could increase production. Apparently, they were not considered important in the 7YDP that was to provide abundant and cheap food for the nation. The peasant mode of production, which has provided food for the nation since time immemorial, was now considered archaic and was to be transcended by mechanized agriculture.

While the state spent colossal amounts of money to bring in sophisticated and foreign agricultural machinery, peasants continued to cultivate the land with traditional tools such as cutlasses, hoes and mattocks and got little or no access to improved seeds and fertilizer. In the 7YDP, socialized agriculture was to account for at least 10 percent of the total area cultivated, but it accounted for only about 1 percent. Yet peasants continued to control 98 percent of the total area cultivated and produced the bulk of the nation's food. By the time the regime was overthrown, 60 percent of machinery supplied to the state farms was beyond repair and the level of under-utilization was estimated to be 80 percent (Dadson, 1993).

The industrialization programme was also fraught with problems. To begin with, the break-neck speed at which Ghana was to be industrialized also led to the establishment of industries without adequate feasibility studies and consideration to cost-effectiveness. Also the industries did not develop pragmatically, therefore there were little forward and backward linkages among them. Their establishment was often crowded into narrow time spans, and this led to mistakes. The Asutuare Sugar Factory, for example, could not start production because there was not yet sugar cane plantation and water ready. Ghana also had not yet developed either the competent local technological base, adequate technological know-how, or the managerial skills that could support this rapid industrialization programme. For these reasons and others, most state enterprises operated at loss and had to be subsidized. This poor performance was acknowledged by Nkrumah when he noted that:

> we have already established many industrial projects and enterprises, as a means of securing our economic independence and assisting in the national control of the economy. I must make it clear that these state enterprises were not set up to lose money at the expense of the tax payer. Like all business undertakings they are expected to maintain themselves efficiently, and to show profits. Such profits should be sufficient to build up capital for further investment as well as to finance a large proportion of the public services which it is the responsibility of the state to provide (Nkrumah quoted in Asante, 1993:263-4).

In addition, the impatience with which general development was pursued overstretched bureaucratic capacity. Although the bureaucracy was expanding rapidly as a result of the government's accelerated educational programme, development was actually running ahead of it. The enthusiasm of a young nation to become industrialized within an incredibly short time and within Nkrumah or his regime's life-span generated its own problems: the resources (both human and material) were overestimated by the CPP Government and the tempo of events and actions were too swift to enable the bureaucracy to learn to implement and evaluate programmes and policies, to correct mistakes and be efficient. Thus, although the bureaucracy was expanded, it did not mean it was always efficient in its administration. This problem was also compounded by the CPP's interference in the appointment, promotion and dismissal of civil servants (Rimmer, 1992), a strategy through which the CPP sought to make the bureaucracy responsive to its programme.

Surely, industrial development is a critical stage in the transformation of societies, from the stages characterized by low-productivity and low income to those characterized by high productivity, high income and economic diversification. Indeed, all developed nations are more or less industrialized. Industrialization is relevant in Ghana given its dependence on a single primary export and the volatility of its price on the world market. Industrialization has the potential to diversify economic productive structure, generate employment and raise living standards and welfare for the people. As such, the CPP industrial programme was not misdirected.

But successful industrialization does not only mean the acquiring of industries or adding to physical industrial capacity.

Technology and industries also have soft components: skills, such as organizational abilities, the ability to utilize equipment and facilities fully and efficiently, to innovate, to improve productivity over time, and to diversify as conditions warrant. Setting up factories is one component in the industrialization process, but factories need to be manned by trained manpower. Factories, per se, achieve neither growth nor structural benefit if they are inefficiently utilized and if they are incapable of establishing linkages with other sectors of the economy. Any country with enough wealth can build or acquire industries (the hardware of technology), but this does not mean that the industries in question will automatically be efficient and productive. The quality of industrialization counts more than the quantity.

As the historical review of colonialism suggests, colonial rule did not leave Ghana with a developed indigenous scientific and technological capacity to adapt, imitate and even copy foreign technology. One major problem that afflicted the industrialization programme was that, at this early stage of development, Ghana did not have all the necessary skilled manpower to support an accelerated industrialization programme of the kind Nkrumah pursued. The so-called human resources alluded to at the outset of the study were only in relation to other African countries and not enough for a rapid industrial development programme (Killick, 1978).

The CPP government was, however, not oblivious to the problem. Indeed, the establishment of the University of Science and Technology at Kumasi and the School of Administration shortly after independence was meant to deal with the lack of technical and administrative skills, but whatever manpower was being made available by these training centres was simply not commensurate with the speed of the rapid industrial and structural transformation that Ghana was undergoing. Anecdotal evidence from the older generation of graduates of the Universities of Ghana suggests that, before they completed their final examinations, jobs were already awaiting them or companies were already at the university campuses recruiting personnel. This view, if indeed true, only lends credence to the argument that there were inadequate managerial and technical skills during this period. As many earlier studies have indicated, one major reason for the poor performance of State Owned Enterprises (SOEs) in the Nkrumah period was mismanagement resulting from the lack of adequate managerial skills (Seidman, 1978; Killick, 1978).

Another problem that indirectly affected the industrial programme was that the educational system, which was considered the pride of the nation and which was consuming much of the national resource, remained essentially elitist; couched in British traditions with emphasis on law, medicine and the liberal arts. Elitism in the educational system meant that much human resource was not devoted to the pursuit of disciplines such as agriculture, sciences and engineering at the intermediate levels. The University of Science and Technology was producing "office scientists" and not "practical scientists" who could do things with their hands. There was little or no linkage between industry and the university, neither was there systematic planning to target education to the various specific needs of the country. University education was expanded for its own sake. This situation continued throughout the post-Nkrumah years, so that, when the economy collapsed, most university graduates could not find decent employment. This resulted in a great number of university graduates emigrating to Europe, North America and other more prosperous countries. There was also little, if any, attempt to harness the country's scientific talents or local technology.

Furthermore, many industries, like the state farms, were over-staffed as a way of solving unemployment problems. They also operated below capacity because of the lack of critical input (Seidman, 1978:210-214). The major foreign investment, VALCO, was a successful venture but was an industrial enclave, and, as such, did not bring substantial benefits to Ghana. As already noted, it was capital intensive and required very little labour power and its terms of agreement disadvantaged Ghana. But, whatever problems afflicted the industrial programme, it is important to note that the regime's life-span was curtailed, perhaps prematurely, and most of the industries it set up were in their infancy, needing time to learn to produce effectively. It is thus unfair to argue definitively that they were unproductive. Their immediate profitability might not be an adequate measure of success at this time. Thus, what future governments made of Nkrumah's achievements and legacy must be of interest.

The economic, administrative and technical problems that afflicted the agricultural and industrial projects contributed to the poor performance of the economy in the early to the mid-1960s. This poor performance and many other factors translated into economic crisis that increased resentment against the CPP regime. But beside this crisis in the socialist economy, the CPP socialist development programme also created tensions and conflicts with the peasants and the emerging bourgeoisie.

The Tensions Between the State, Peasants and the Bourgeoisie

According to the principles of scientific socialism, in a truly socialist society, all the means of production are nationalized by the state on behalf of the proletariat. The CPP regime's attempt to construct a socialist state did not go far enough in this direction. This was evident in the tension between the CPP state and peasants and the paradox that defined its relation to private economic actors. The implementation of the 7YDP was replete with paradoxes and tensions, and consequently conflict, which was, among other factors, precipitant in the downfall of the regime.

As noted earlier, cocoa had become a major source of wealth for the state and the farmer. It is, therefore, understandable that both state and farmer would like to protect their interests. The CMB, established after World War II to market cocoa, was mainly to guarantee prices to farmers by lowering the domestic price that farmers received and retaining the difference between this price and the one that prevailed on the world market. This difference was to be accumulated as a fund to offset local prices in times of difficulties, that is, when world market prices collapsed to depress the domestic ones. In this way, farmers were to be shielded from the vagaries of the international market prices and thus safeguarding the cocoa industry.

The CMB was also mandated to use cocoa money for research, control of diseases, credit and amenities that could benefit cocoa producers and the cocoa industry, in general. The CMB was an extractive instrument, but was not supposed to engage in a comprehensive development of the colony or be a profit-making institution. By 1951, when Nkrumah and the CPP were allowed to participate in government, the CMB had accumulated as much as £71 million in the fund. The CMB fund was part of the approximately £200 million that Ghana inherited at independence. This fund was supposed to be used only for the interest of farmers and the cocoa industry, but in practice, it was used for the development of Britain.

For Nkrumah, however, the cocoa industry was a national asset and should be socialized, i.e. taxed, to support broader development goals for the whole nation. In this regard, in 1954, the CPP regime modified, through an Act of Parliament, the purpose of the CMB so that the state could borrow money from the CMB fund for its development programmes. This led to the imposition of a producer price ceiling of 72 shillings for a 60lb load of cocoa in 1954 (see Frimpong-Ansah, 1991:84-5). Appropriating rural resources for the purpose of industrial or modern development by developmental elites as in Ghana was not a new phenomenon. Indeed, as Moore (1968:506) in his sociological-historical comparative study of the making of the modern world, sombrely declares, "at bottom of all forms of industrialization so far have been revolutions from above, the work of a ruthless minority", and in most cases it is the peasants who bear most of the cost of such rapid social transformation.

The capturing of cocoa surplus by the government is thus one common historical approach to bringing about modernity. But, it is also important that the cocoa industry which was now the backbone of the Ghanaian economy, was protected from collapse so that, in spite of government taxation, farmers continue to produce. In the Ghanaian context, this calls for at least three things: (i) that local prices are not so depressed to become a disincentive to investment in and/or maintenance of cocoa farms; (ii) that local prices do not fall vis-à-vis the producer prices in neighbouring countries to promote the smuggling of Ghana's cocoa; and (iii) that the cocoa resources so extracted are actually used productively in the modern sector, which, in turn, could support the viability and modernization of the rural-peasant economy. Most of these conditions were not met in much of the postcolonial era. Indeed, by the early 1950s, Bauer (1954) was already warning about the impending adverse effects the operations of the CMB could have on the cocoa industry and the general the economy.

The nationalistic desire of the CPP to use cocoa resources to build the whole nation, however, met with local resistance. The decision to lower cocoa prices infuriated cocoa farmers throughout Ghana, but especially those in the Ashanti region, which had become a major cocoa producing area. Following their defeat by the British and the subsequent lost of control over their gold mines, the Ashantis turned to cocoa production as a major economic activity, which had by now become a major source for wealth for many people. Indeed, Ashanti chiefs themselves became cocoa farmers. The Ashantis have, thus, come to see their well-being as closely connected to the cocoa industry. Consequently, cocoa interests deeply penetrated local politics.

The Ashantis saw the CPP's policy to peg cocoa prices as an attempt to weaken them since they had already begun resisting what they saw as the CPP's "dictatorial rule". The chief spokesperson for the Asantehene at this time, Baffour Osei Akoto, put across Ashanti suspicion when he spoke of the government's highhanded manner of pegging cocoa prices, which affected the Ashantis more than any other group, and of its intention to dismember the Ashanti nation-state (cf. Frimpong-Ansah, 1991:85). This became one reason for the Ashanti state's demand for federalism and desire to break away from independent Ghana (see below).

In order to counteract Ashanti threat, the CPP government initiated a political solution: it adopted a "divide and rule" tactic that could contain Ashanti nationalism. It sought support from kings and chiefs in the Brong and Ahafo regions where the Ashantis had invested much in the cocoa industry. It did this by promising to give the Brongs their independence, which was lost to the Ashanti State in the nineteenth century. The Brongs and also the Ahafos, in turn, openly accepted the CPP's cocoa price and accused the Ashantis of trying to break up the aspiration of the country by demanding a federal system of government. On 13 August 1954, the CPP regime passed a Bill in Parliament that created a separate region for the Brongs and the Ahafos, thus making them autonomous of the Ashantis. The defeat of the Ashantis, however, intensified Ashanti state and farmers' hatred for the regime.

The victory over the Ashantis gave the CPP full mandate to lower the price of cocoa and use cocoa funds for the entire development of the country. But this development nipped in the bud an historical opportunity for state and cocoa farmers to find a mutually beneficial formula to protect their respective interests, then and in the future. The CPP success atomized cocoa farmers and gave it and other successive regimes (see later chapters) undue advantage over cocoa matters vis-à-vis cocoa farmers. But the CPP formed the UGFC to be the sole buyer of cocoa and representative organ of cocoa farmers. This organ was to protect the interest of cocoa farmers but it also became politicized and ended up as a patronage organization for the CPP. Its operation was riddled with financial losses resulting mainly from massive corruption and the granting of loans to mainly CPP activists, to the detriment of the interest of the cocoa farmers.

Overall, the amendment of the CMB mandate meant that producer prices for cocoa were not to be determined solely by market forces, but also by administrative policy determined by governments through the CMB. With this, the CPP ensured a reliable fiscal base for the state and was able to fund its rapid infrastructure and industrial development. In 1959, the price of cocoa was reduced from 72 to 60 shillings for 60 lbs cocoa (headload) by a so-called voluntary contribution towards the financing of development programme volunteered by the UGFC on behalf of cocoa farmers. In 1961, the price was further reduced to 54 shillings by a compulsory savings scheme; in 1965, the price was further lowered to 50 shillings in order to provide funds for the Agricultural Credit Bank, and for the 1966 crop year, it was fixed at 40 shillings. These reductions were responses to falling export prices, but they also reflect the struggle to hold up public receipts from cocoa so as to underwrite development programmes in the wake of declining export prices and rising marketing costs (cf. Rimmer, 1992:78-80).

This practice of administratively setting prices enabled successive governments to tax cocoa farmers as they wished without necessarily caring about the viability of the cocoa industry and the interest of the cocoa farmer. Consequently, they ended up destroying, indirectly, the very lifeblood of the nation: an issue to which we shall return in later chapters.

There is no doubt that cocoa resources enabled the creation of physical infrastructure, social services and most of the industries established by the CPP regime which many people enjoyed. But cocoa funds were also consumed in the form of increased imports and salaries for the bureaucracies, whose size was increased to plan and execute state-directed projects—many of which were poorly executed and unproductive (e.g. Workers Brigades and State Farms). They were also consumed by expanding free social services such as education, health, potable water, whose contribution to development lay more in the future than in the present. This means that the state, at this time, assumed much social responsibility and was forced to take more from the farmers than was perhaps desirable or necessary. The beneficiaries of these increased social programmes were undoubtedly Ghanaians but ironically when the regime experienced economic crisis and was ousted by the military most people expressed jubilation.

In the 1950s, as a result of relative economic viability and boom in imported consumer commodities facilitated by its cocoa wealth, Ghana was able to attract through smuggling, cocoa from Cote d'Ivoire and Togo. But, by early the 1960s, the difference between the price of cocoa in Ghana, on the one hand, and Cote d'Ivoire and Togo, on the other, was becoming significant enough to encourage the smuggling of Ghana's cocoa to these neighbouring countries (Nugent, 1991; see also Bequele, 1985). This trend did not only continue but intensified throughout the post-Nkrumah years, until perhaps 1983, when the IMF/World Bank, through their SAP and the Ghana government, decided to increase the producer price as a way to end the smuggling of cocoa (see later chapters). The CPP and other successor regimes have thus alienated cocoa farmers whose resources they critically needed for their own reproduction and their development programmes.

On the other hand, although the CPP regime gave the impression of being more nationalistic than its rival, the opposition party of the UGCC descent, its development strategy that emerged from 1964 onwards was not nationalistic in one sense. This is evident in the way it handled the interest of the local bourgeoisie. Colonialism, as already noted, threw on the historical stage social groups—incipient indigenous merchants, an independent civil service with British traditions and values, and independent actors such as lawyers, doctors and rich cocoa farmers—with interest in continuing the liberal colonial economic order. All these groups had expected the independent state to protect and advance their interests. These interests could have been better protected had the postcolonial state continued past economic processes and had it encouraged Ghanaians to gain a strategic role in the national economy. This approach would also have been nationalistic, so long as independence politics was used to "seek the economic kingdom for Ghanaians". At least this seemed to be the vision of the UGCC members who opposed the CPP (see the thrust of Busia's programme in Chapter 6).

Economic nationalism, in this sense was, however, not part of the CPP's development strategy. The regime did not wrestle the state from the colonialists to give independent Ghanaian economic actors the chance to capture the "commanding heights" of the economy. Nkrumah feared that, since indigenous capitalists were small in number and weak and could not industrialize society, if the state did not assume a leading role in the economy, foreign investors would come to dominate Ghana's economy (Nkrumah, 1963; Killick, 1978:40). The fear of neo-colonialist domination seems to be one reason for the state's declaration of socialism to enable it control the major means of national production. This fear, as we shall see later in the study, continued to excessively plague national perception of the causes of Ghana's economic decline in the post-Nkrumah era, so that domestic politics came to be conducted in a manner (albeit perhaps inadvertently) that was destructive of many economic activities of citizens and was, consequently, injurious to Ghana's productive infrastructure (see Chapter 7).

The CPP's development strategy, however, exhibited two intriguing paradoxes, which created conflict in society and antagonism towards the regime. The first was that industrialization and large-scale commercialization of Ghanaian society was to be carried out solely by the state and not by Ghanaians acting as private economic actors. But interestingly, the Nkrumah regime "tolerated" foreign capital operating in the country. Threats of nationalization were often made but the regime did not nationalize any important foreign company (Frimpong-Ansah, 1991:74). This seems rather patently inconsistent with socialist ideology, which advocates complete state ownership of the means of production on behalf of the workers.

Given the dominance of British and western economic interests in the economy as a result of colonialism, the extent to which the CPP regime had room to manoeuvre in socializing all productive means but failed to do so is not clear. What was, however, clear was that Ghana's socialism was conceived to accommodate foreign capital because it was premised on the belief that expanding and effective state sector activities would squeeze out the foreign companies. The establishment of the GNTC to participate in the commercial sector could be seen in these terms. By 1965, the GNTC had as many as 210 retail stores with warehouses in the country handling about two-thirds of the import trade in Ghana (Seidman, 1978). The items it dealt with included consumer goods, raw materials, spare parts, motor cars and capital equipment, competing with foreign capital, such as British UAC and UTC, and French SCOA, that dominated Ghana's commercial and trade sectors. But the spread of so many stores all over Ghana indirectly blocked the possible economic space that the indigenous business class could occupy in their own country. The establishment of financial institutions and state-owned industries could be also seen in these terms but at this moment they were not profitable enough to challenge the threat of foreign capital (Killick, 1978).

The second paradox is closely related to the first, for although foreign capital could operate in Ghana, indigenous capital and capitalism were not to be encouraged. The activities of the incipient, indigenous productive bourgeoisie in the postcolonial economy were considered insignificant and even antagonistic national interest. Nkrumah accorded a "compradorial" status to the Ghanaian business and industrial class, noting that their participation in foreign enterprises had proved to be simply a front for the manipulation of Ghanaian commerce by colonial business ventures, and such participation would be prohibited in the future (cf. Markotvitz, 1977:257). They were to be limited to small-scale activities "provided they [were] not nominees of sleeping partners for foreign interest" (cf. Killick, 1978:37).

In March 1964, when Nkrumah launched the 7YDP, he indicated his uncompromising stand towards Ghanaian business class in this way:

> Mr Speaker, we would be hampering our advance to socialism if we were to encourage the growth of Ghanaian private capitalism in our midst. This

> would, of course, be in antipathy to our economic and social objectives (see *National Assembly Debates*, 11 March, 1964).

Nkrumah went on to acknowledge quite correctly that Ghanaians have vast entrepreneurial energies but should direct them towards state-sector activities. He also noted that the government intended to place far greater emphasis on the development of co-operatives rather than encourage Ghanaians to start private businesses. In the CPP's Draft programme *For Work and Happiness*, the government recognized five sectors, namely: (i) state enterprises; (ii) enterprises owned by foreign private interests; (iii) enterprises jointly owned by the state and foreign private interests; (iv) co-operatives; and (v) small-scale Ghanaian private enterprises. Further in the programme, the government stated clearly that it accepted the operation in the country of large-scale enterprises by foreign interests. But when it came to indigenous Ghanaians, the government's position appeared that they were to be allowed to engage in small- and not in large-scale enterprises. This contradiction in the CPP's development programme provoked a policy statement: *What the People Want*, from J. B. Danquah, in which it was stated that "...while the Government's policy [maintain] that the Nigerian, Togolese, Algerian, French, Indian and other foreign entrepreneurs would be allowed to come in and do business on large scale, the same facility is to be denied the native-born Ghanaian..." (Danquah, 1997:174).

Consequently, a hardening of attitude towards indigenous entrepreneurs began to emerge from 1961 onwards. Earlier, in October, 1960, Nkrumah had announced that his Government would give no further assistance to local private business and maintained that it "must stand on its feet and not rely on the Government for its development" (cf. Esseks, 1971). While this, as Esseks notes, implies that the government would not hinder the efforts of the local businessmen to develop through their own initiatives, in a country like Ghana where one needed foreign exchange allocation and import licence from the government to do business, and where even state-owned banks and expatriate banks such Barclays Bank and Standard Chartered Bank, preferred to give credit to foreign companies, the position of Ghanaian businessmen was not an enviable one. This paradox was to lead to conflict between the independent state and the emerging bourgeoisie; a conflict that became manifested in Nkrumah's overthrow and apparent in the new military rulers' attempts to reverse the CPP's development strategy in favour of the interest of the indigenous bourgeoisie (see Chapter 5).

From 1957 to the end of 1960, Nkrumah stressed the importance and desirability of the private sector in Ghana's economic development but gave it only modest actual assistance. From 1961 to 1966 when Nkrumah was overthrown, the CPP government neglected and even took actions that some analysts see as discrimination against indigenous businessmen. In

March 1961, for example, the government compelled Ghanaian private buyers as well as foreign companies to withdraw from the internal marketing of cocoa. In their stead, the government established a co-operative affiliate to the CPP to buy cocoa. The state's loan programmes, established for Ghanaian businessmen before independence, received only moderate financial support from the CPP regime. The input of the Nkrumah government into this scheme plus the original £100,000 invested in a short-term credit facility for timber producers amounted, to only £825,064; while A. G. Leventis, a Greek Cypriot businessman said to be close to Nkrumah who had considerable commercial interest in Ghana, obtained a government loan of £2 million in 1959 (cf. Esseks, 1971:14). The government also established the Capital Investment Board and the Ghana National Investment Bank in 1963 to provide long-term financing for approved development projects run by foreign capital (cf. Markovitz, 1977:257).

Indigenous capital was systematically discriminated against through the use of a variety of administrative and fiscal mechanisms—import licensing, foreign exchange controls, etc.—that were imposed in Ghana from 1961 to regulate the economy. The purchase by the government of the Commonwealth Trust Limited, which owned forty-five stores, in late 1961, was a calculated move to inhibit the rise of the bourgeoisie. Consequently, the Ghanaian business class was disenchanted with the independent state's anti-business policies.

Businessmen and opposition leaders became increasingly vocal in their criticism that government policies had created confusion in the commercial life of the country. Some of them were among those imprisoned under the Preventive Detention Act (PDA) of 1958, which was meant to detain political dissidents. The 7YDP programme also barred Ghanaians from buying shares in large enterprises (cf. Markovitz, 1977:256; Garlick, 1971; Killick, 1978). Thus if the colonial state had benignly neglected the interest of the emerging business class, by now the independent state can be seen as actively working against it.

Nkrumah's pronouncements and actions, however, did not deal a death blow to private indigenous enterprise. Private business activities no doubt might have suffered, but some managed to survive, mainly because the regime did not frown upon petty trade, which was the basis of the informal economy. Though Ghanaians were formally prohibited from developing large-scale commercial ventures, the government tolerated petty traders (sometimes without permanent stores and fixed capital investment) engaged in breaking down commodities and re-selling them in quantities ranging from small packages of cigarettes to dozens of tomatoes or a few pounds of powdered red pepper. Generally, the government did not see the informal sector dependent on small capital as menace to state socialism, but, rather, the activities of the emergent big businesspeople.

The emerging middle class with modern skills—doctors, lawyers, architects, and other private actors whose private interests obviously could not advance under socialist economic and political order—also detested the monopolizing state. In *Consciencism,* Nkrumah wrote that "capitalism at home is domestic colonialism" (Nkrumah, 1964:74). And according to CPP's development programme, Ghanaians were not to be encouraged to become large-scale capitalists. This meant that foreigners who were "allowed" to be capitalists were also by implication "allowed" to exploit Ghanaians. This created a friction between the local bourgeoisie and the CPP Government.

This resentment became obvious when the regime was overthrown and a pro-indigenous business economic order imposed. Since the CPP regime never nationalized any significant foreign investment and also "allowed" Lebanese, Indians and Syrians to do business in Ghana, it can be argued that foreign capital indirectly became government's allies working against the emergence of indigenous business. In other words, the nascent state was inadvertently "colluding" with foreign capital to stifle the development of the indigenous Ghanaian businesspeople. For the Ghanaian business community, then, the struggle for independence was not yet over.

On the other hand, however, Nkrumah explained his regime's "uncaring" attitude towards the indigenous capitalists by saying that he was disappointed in their reluctance to utilize previous government assistance (cf. Esseks, 1971). But Mr Ayeh Kumi, Special Consultant on Economic Affairs to Nkrumah, explained after the fall of the CPP regime that Nkrumah told him that, "if he permitted African business to grow, it will grow to the extent of becoming a rival power to his and the party's prestige" (cf. Iliffe, 1983:77; Garlick, 1971:119-131).

The threat an economically powerful bourgeoisie could pose to Nkrumah and the CPP rule was thus recognized. The concentration of power in Nkrumah, so as to make CPP rule dominant, also made the political system neopatrimonial. With these anti-local capitalist statements, it can be contended that Nkrumah perceived the rise of the national bourgeoisie as a danger to his power, confirming Weber's classic argument advanced in Chapter 2 that patrimonial rulers are often hostile to autonomous economic actions of their citizens. Apparently, the "new men" in politics and in business were not necessarily one and the same, and not always in sympathy with each other. Nkrumah and the other like-minded state-actors failed to see how the nation's interest and prosperity as well as their own careers would be also well-served by accepting demands for greater assistance to Ghanaian business groups.

Nation-Building, Political Centralization and Patrimonialization of the State

Ghana, like many African countries, is made up of many heterogeneous ethnic groups who speak no fewer than 75 languages and dialects, each coinciding more or less with a distinct ethnic group. The largest of these groups are the Akans, Mole-Dagbani, Ewe and the Ga-Adangbe, which constitute respectively 44.1, 15.9, 13.0 and 8.3 percent of the population. Non-Ghanaian minorities include British and other Europeans, Lebanese, Syrians, Indians and other Africans (Boateng, 1994).

Before the advent of colonialism, these ethnic groups lived under different traditional states with different degrees of socioeconomic and political development. Though these groups share some similar cultural and cosmological values (Nukunya, 1992:11), these were not sufficient to provide them a national identity. They were, rather, sub-nationalities that sometimes harboured historically-rooted antagonisms against each other. Colonial rule brought an end to much of this antagonism, but did not totally melt their "separateness". Indirect Rule, which "preserved" traditional political institutions, also indirectly preserved the relative coherence and distinctiveness of these groups.

At the same time, the limited socioeconomic development that occurred under colonialism was skewed in favour of those regions that first came into European contact or had resources that interested the colonialist. Consequently, some regions had a head start over others in development and modernization. For example, in general, southern Ghana was more "developed" than the north. This geographical unevenness in initial socioeconomic development often translates into vertical inequality whereby southern people assume feelings of superiority over their northern counterparts.

Yet, in spite of these actual or perceived differences, the peoples of the Gold Coast united to fight a common enemy: the colonialist. But, once the colonialist was gone, the anti-colonial coalition began to disintegrate. Old animosities and new conflicts re-emerged. Various ethnic groups became sharply aware of their relative distinctiveness and disadvantageous positions vis-à-vis others and sought political space to manoeuvre. Structural conflicts and regional differences were exploited and translated into political action leading to uneasy alliances between groups seeking to challenge the hegemony of the CPP. This was easy because the advance of capitalism under colonialism did not induce a class society that could anchor the representative government that emerged in the immediate independent era. Given this, ethnic, religious and regional cleavages easily became the pivot around which much of the postcolonial politics revolved.

These cleavages underlined the new political groups that emerged to contest the CPP. Among them were the Togoland Congress (TC), formed in 1951 by the Ewes of British Togoland, who felt that they were economically ignored and, for that matter, sought re-integration with the Ewes of French Togoland; the Muslim Association Party (MAP) formed in 1953, to advance the interest of the Muslims; the Northern People's Party (NPP), formed in 1954, though ethnically diverse, was nonetheless, a regionally based party seeking not only the socioeconomic advancement of the North but also to forestall the dominance of "socialist" and "anti-chief" southern politicians of the CPP.

In the south, too, the cocoa farmers of the Ashanti region did not take kindly to the CPP's lowering of producer prices for cocoa. Farmers' organizations in the 1940s gave much support to the nationalist movement, but later broke up with the CPP over the marketing issue. Besides the cocoa industry, which the Ashantis specialized in, and which was perceived as being overtaxed by the CPP, the Ashanti state also possessed rich mineral resources in the form of gold, diamonds, bauxite, and other forest products such as timber that the central authority was also controlling. Conscious of their wealth and autonomous past, the Ashantis sought to re-emerge as a nation independent of Accra and the new politicians.

As noted, the independence struggle had already created a rift between the new elites: the CPP (made up mainly urban youth, the unemployed and the relatively less educated elements of society) and the UGCC (the small merchant-professional bourgeoisie, lawyers, architects, doctors, businessmen, educators, rich cocoa farmers). The UGCC had worked for a non-revolutionary path to independence and had wanted to continue the liberal economic and political order begun by the British. The CPP's preference for militant nationalism, state socialism, and its distaste for capitalist development, infuriated many elements of the UGCC and exacerbated the inter-elite conflict.

Having lost political power to the CPP, the merchant-professional class re-grouped to form a new party called the Ghana Congress Party (GCP) and denounced the CPP leaders and members as "verandah boys" (people with low educational attainment who slept under the verandas of the rich people in the cities because they could not afford their own accommodation) who have stolen the UGCC's "political birthright": to lead the Gold Coast into independence. As already noted, this conservative group never forgave Nkrumah and the CPP elements. Thus, inter-elite and ethnic conflicts became intertwined into a new political movement.

In 1954, the National Liberation Movement (NLM) emerged in the Ashanti region bent on securing complete autonomy from the national centre and sharing the national assets. Later, the NLM explicitly demanded a revision of the 1954 constitution to establish a federal political system for Ghana (see Boahen, 1975:183-186). This was fiercely resisted by the CPP, which was articulating mobilizing, nationalistic and ethnically neutral politics. The NLM did not only rally behind the Ashanti and Akan identity in defence of neo-traditionalism, but also brought together opposition parties and the business class throughout Ghana to pose a challenge to the emerging state. Thus an "unholy" alliance between the secular "modern-bourgeoisie" of mainly the UGCC origin and sacred traditional-aristocratic elites was brought against the CPP regime (see Crook, 1990; Kraus, 1971).

The CPP felt greatly threatened, and perhaps justifiably so, since its formal appearance as a mobilizing party was not actually based on a substantially large number of dedicated followers. At least Nkrumah was not unaware that he had dissociated himself from a significant proportion of the newly educated elites who were canvassing for independence before his arrival in the country. In the 1956 elections, the CPP managed to pull only 57 percent of the votes cast and the opposition parties got 43 percent. Furthermore, the CPP was out-voted in the Ashanti and northern regions, making it clear that Ghana was not a single party (CPP) dominated state, as suggested by 1951 elections (Austin, 1964:30). This was suggestive of a formidable opposition emerging to challenge the CPP, and certainly would have worried Nkrumah and his party. The CPP, for its part, was also determined to undermine effective opposition to enable it to rule.

Between 1954 and 1957, politics was characterized by a high level of intimidation and violence to an extent that there were simply no acceptable norms for regulating political groups. The independence constitution that emerged out of the early political crisis made room for regional assemblies and social diversity and had constitutional safeguards for these arrangements within a unitary political system. As such, it was a compromise document to give some space to various competing interests. But this did not solve the crisis. The opposition parties never considered the Nkrumah regime as legitimate and hardened their attitude towards any constructive agreement that could secure an elite consensus on the goals and strategies of economic and political development. They boycotted many parliamentary sessions and made governance difficult for the CPP (Kraus, 1971).

The need to maintain political order and strengthen territorial control became accentuated. Nkrumah now found the 1957 conciliatory constitution "unworkable" and took a plethora of actions to maintain order. The constitution was revised, giving Nkrumah sweeping powers to decide the framework of political activity. In 1958, the Preventive Detention Act (PDA) was promulgated, which allowed the government to detain political dissidents without trial. The Act, however, also acted as a potential threat to parliamentarians who would have deviated from mainstream CPP's political thinking and opposed the regime. Between 1958 and 1966, some 2,000 people were imprisoned under the PDA (Rimmer, 1992:71). In the heat of this political acrimony, some oppositional members joined the CPP. Others, such as J. B. Danquah of the United Party (the opposition party in 1960 presidential elections), who held steadfastly to their views, were imprisoned.

Furthermore, traditional rulers who opposed the CPP regime or supported opposition elements were discriminated against in the allocation of state resources. Some were destooled and replaced by CPP supporters most of whom did not hold traditional qualifications for chiefly positions. The CPP increasingly employed unconstitutional means to stabilize the ruling process. It exploited its political incumbency through a variety of means to keep the opposition out of active politics. Opposition leaders who were not jailed went into exile. In 1960, a referendum was held that "accepted" the CPP's proposal for a one-party state. By 1961, the CPP was the only party of any significance. Effective opposition had declined in Ghanaian politics, not so much because opposition parties lost popularity, but because of the CPP's determination to see them blighted. These actions further diffused opposition and *hegemonized* CPP's rule. In 1964, the Nkrumah regime proclaimed Ghana a one-party state. Nkrumah justified this as African, deriving from precolonial political practice where there was no expressed opposition, (Austin, 1964:35-37) even though this was not usually the case. He also argued that the strains experienced by an emergent country call for measures of "totalitarian kind" (Nkrumah, 1957:x).

The CPP regime also attempted to *hegemonize* its rule over society. It sought the essential unity of the emerging nation by demanding that Ghanaians belong to only one party: the CPP. It co-opted civil associations and organizations, such as the United Ghana Farmers' Associations, Ghana Trade Unions Congress (TUC), and Ghana National Women's League. It also created other groups—Young Pioneers, the National Association of Socialist Students and Workers Brigades for the unemployed—and turned them into CPP networks (Apter, 1972:372-4). Civil society was thus politicized and made dependent on the CPP.

The CPP also interfered in the appointments, promotion and dismissal of bureaucrats. The failure or inability of Nkrumah's regime to ensure the "independence" of the civil service as a technocratic machine and the subsequent personal domination of the state brought about low morale, internal indiscipline, excessive bureaucratism, rent-seeking and self-serving behaviour into the bureaucracy. This was one reason stated by the military for the overthrow of the CPP regime. Predictably, the bureaucrats resented this encroachment. Like the opposition politicians, the civil service was also at loggerheads with the CPP regime, though this was not always openly expressed. Feeling that the civil service was not receptive to its "socialist" policies, or that it was frustrating its ambitions, the CPP regime attempted to develop a parallel civil service, or what Amonoo (1981) describes as "institutional dualism", just before the 1966 military coup (Crook, 1990). The government also attempted to indoctrinate the military and police with the necessity of socialism. The overall strategy of the CPP regime was to fuse state and government in order to hold society together.

On 12 June, 1959, during the celebration of the tenth anniversary of the formation of the CPP, Nkrumah noted:

> Comrades, it is no idle boast when I say that without the CPP there would be no Ghana... The CPP is Ghana. Our Party not only provides the Government but is also the custodian which stands guard over the welfare of the people (Nkrumah quoted in Amonno, 1981:13).

Later in 1961, he added that,

> The CPP is a powerful force, more powerful, indeed, than anything that has yet appeared in the history of Ghana. It is the uniting force that guides and pilots the nation. Its supremacy cannot be challenged. The CPP is Ghana and Ghana is the CPP (Nkrumah, 1961:209).

The identification of the CPP with state and the larger society made legitimate opposition difficult to sustain, while the paucity of opportunities outside the state's apparatus made the loss of office for politicians and bureaucrats as well as other persons dependent on the state a matter of extreme seriousness. Political centralization and personalization of state power had now emerged in Ghana, and the paradox in development strategy is suggestive of a political system with neopatrimonial dimensions. Much power was invested in Nkrumah, who, by now, had become an executive president.

Like typical patrimonial leaders, Nkrumah could now shape policies according to his wishes and control his staff. He had power to appoint, promote and dismiss civil servants, judges, the military and police. Executive, Legislative and Judicial powers were fused in him. His office at the Flagstaff House acquired responsibility for a wide range of matters affecting security, the press and the radio, economic planning, African affairs, the civil service, regional and local government, state enterprises, education, research and the mass organizations absorbed within the CPP. In 1961, an enactment of laws made it a criminal offence to engage in any activity that brought hatred, ridicule and contempt to Nkrumah (cf. Rimmer, 1992:70-71; Asante, 1993). The mass media came under state control and the freedom of the press also curtailed.

The lack of press freedom and constitutional "checks and balances", as well as effective opposition in parliament meant that the CPP was unaccountable to institutions that restrain the use and misuse of political power. In the absence of effective opposition, press freedom, etc. many national policies were not subjected to rigorous debates and government increasingly did not benefit from alternative viewpoints to its policies. Consequently, the government was unable to respond swiftly and effectively to the emerging economic crisis. It hesitated in its decision to match development projects to existing resources. The CPP remained a very visible entity, the strong arm of the government in each locality, the ears and eyes of the central committee; it seemed strong and enduring but it was said to have little substance and the reports its regional committees sent back to the capital, on which policy was often based, were frequently wishful thinking tailored to the desires of senior figures in the administration (cf. Hodder-Williams, 1984:123).

Having parted company with a significant section of the well-educated people, Nkrumah was forced to rely on people with little capabilities and competence who out of their own limitations became sycophants to CPP rule and willing to cultivate a personality cult for Nkrumah. In return, these people exploited their political positions for personal gains, rather than use them to serve the public. For example, Mr Adamafio, in 1961, suggested that "we must look to the Osagyefo [Nkrumah] as our Saviour and Messiah. Let us exploit him for the benefit of the nation. Let us make an institution out of him" (*Ghanaian Times,* 10 April 1961:6). The National Assembly also tabled a motion that Nkrumah should become President for the rest of his life, although Nkrumah rejected it. But this serves to illustrate the degree of sycophancy that had emerged among politicians in this period. Given this, CPP members turned to be above the law and criticisms and this indirectly contributed to Nkrumah's inability to demand from his own ministers the accountability and bureaucratic discipline that are required in the performance of public duties.

Corruption, nepotism and discrimination became rampant in the upper echelon of society as evident in Nkrumah's broadcast in April, 1961, and in numerous inquiries (cf. Apter, 1972:376). These inquiries often condemned political corruption, but it was also generally tolerated by officials, a phenomenon that Ayi Kwei Armah (1968) parodies in his novel, *The Beautyful Ones Are Not Yet Born* [sic]. Corruption was one means through which the CPP ensured political followers. The patrimonialization of the state was also manifested in patron-client relationships that made the survival of civil organizations dependent on the state and the CPP.

No serious allegations had ever been levelled against Nkrumah that he personally amassed wealth, but commissions of inquiries noted that he made no firm distinction between public money and party funds. Numerous inquiries into the alleged malpractice of the regime indicated that corruption had become extensive and obligatory in the business of state administration. It involved ministers accepting the proverbial 10 percent of the value of contracts awarded, the buying of golden beds, and the demanding of monies for import licences so that businessmen were compelled to pay bribes to survive (cf. Rimmer, 1992). Le Vine (1975), who has done work on corruption in Ghana, has argued that, by 1966 a culture of political corruption had been institutionalized in Ghana. Owusu (1970) has also regarded Ghanaian politics at this time as highly instrumental rather than programmatic geared mainly towards socioeconomic development.

The government's main concern at this time was to keep the CPP in power. The regime refused to seek re-election, and the 1960 constitution did not specify the term of office for president. This meant that Nkrumah could rule for life. Although its social base had diminished partly because of its monopolizing politics and the emerging socioeconomic crisis, the regime survived mainly through patron-client relationships that permeated most civil societies and state organizations.

The fusion of state and government and the personalization of the state also impacted negatively on the implementation of government plans and policies. The 7YDP, in particular, was not followed as planned. Indeed, each minister sought to implement what he deemed fit. For example, according to Kraus (1971:63), the deputy minister for Works and Communications was said to have thought it an excellent idea for the Eastern Region to have its own Seven-Year Development Plan. This idea was accepted by the CPP Regional Steering Committee, extended to other regions, and later fitted into the national 7YDP. It was also reported that there was a lack of coordination and orderly decision-making among the numerous administrative units: 31 ministries and 50 state enterprises (cf. Killick, 1978:140). Their functions and responsibilities often overlapped. They were often reshuffled, budgets were wrongly estimated by departments, new projects that had never been previously envisaged appeared in the development plan; projects were wrongly appraised and located or over-capitalized, and contractors paid commissions to push for projects (cf. Rimmer, 1992:86-91).

Furthermore, the centralization of power in the CPP enabled the regime to steer the state's priorities to attain Pan-African ambitions. The 1960 constitution was framed to surrender Ghana's sovereignty in the interest of Pan Africanism. A number of government departments also came under the direct control of the president's office. An example was the Budget Department of the Ministry of Finance. Nkrumah also established the Consolidated Fund of £2 million for supporting Pan-Africanism[2] and decolonization movements in Africa (Kraus, 1971:59), but no further records were ever kept of this Fund, so it is difficult to know how much was actually spent on this foreign policy objective (Frimpong-Ansah, 1991:76).

In addition, an African Affairs Secretariat, separate from the Ministry of Foreign Affairs, was established. The CPP regime also established the Kwame Nkrumah Ideological Institute to propagate socialism based on the African experience. In furthering solidarity among socialist-inclined African countries, the CPP regime gave as loan £10 million to Guinea and Mali, but Guinea mainly used the money to cut neocolonial ties with France. Ghana, Guinea and Mali formed a union (Rimmer, 1992). In 1958, there was a conference in Accra of eight independent African states and an

"All African Peoples Conference" attended by nationalist leaders from the colonial territories. By 1960, it was evident that Nkrumah was devoting more attention to international rather than domestic affairs (Austin, 1964:41).

These noble objectives of Pan-Africanism undertaken by Ghana on behalf Africa gave pride and the dignity to Ghanaians and Africans and further awakened the continent to the evils of colonialism. But, they diverted resources away from national development and indirectly skewed domestic development and economic policies into areas not immediately needed. In 1965, for example, the only item on the development agenda was the completion of the expensive State House to the host the October, 1965, OAU conference. As recalled by one of his economic advisors, in 1965, Nkrumah was concerned that the government filled the shops with imported items so that "every African who visited Ghana should be made to see the fruits of independence" (Frimpong-Ansah 1991:76). This inflated the balance of payment problem in 1965 (Rimmer, 1992). African Unity engaged much of Nkrumah's attention and Ghana's resources but was not seriously on the agenda of many African leaders who jealously guarded their newly acquired independence. Nkrumah was thus ahead of his contemporaries.

The centralization of power, however, did not end the inter-elite conflict in society. On the contrary, it inflamed it. Oppositional activities went underground and sought to destabilize the political system. A number of attempts were made to assassinate Nkrumah, the most notable being the Kulungugu and Accra bombing attempts in August, 1962, and January, 1964, respectively. Opponents to Nkrumah were blamed for these incidents. The senior military and police officers implicated in these attempts were made to proceed on early retirement. Nkrumah now distrusted the nation's military and police institutions and created his own army: the President's Own Guard Regiment of 1,142 men (Kraus, 1971:60), which the army and police saw as a parallel institution and a blow to their influence. In 1965, the Ghana Army was being prepared for war against the white minority government in Rhodesia (now Zimbabwe) which had proclaimed unilateral independence from Britain. To aggravate an already strained relationship between the government and the military, Nkrumah was persuading the army to accept equipment and training from the Soviet Union and Eastern Europe, something that they never liked.

Economic Crisis, the External Environment and Domestic Adjustment

The emerging crisis of the socialist economy pointed out earlier was aggravated by an external factor: the decline of world prices for cocoa in 1965. World prices for Ghana's cocoa had been on the decline for some time. Between 1959 and 1966, annual cocoa export earnings losses amounted to £110 million. This was mainly due to increased production on the world market, to which Ghana had also contributed. In 1957/58, Ghana produced 206,000 tonnes of cocoa, and by 1965/66 production had risen to 572,000 tonnes. This increase, which was gradual from 1958, might have depressed world market prices and translated into a balance of payment crisis. In an effort to manage the crisis, the government introduced administrative controls in 1961 (up to 1961 there were unrestricted imports), for example, quotas on imports and import licensing to regulate the use of scarce foreign exchange. This led to a sharp increase in prices. Between 1954 and 1961, imports of dairy products increased by 145 percent, sugar by 174 percent, wheat flour by 99 percent and motor cars by 100 percent (Esseks, 1975:37). These shortages could not be alleviated domestically because Ghana's import substitution industries were just beginning to produce on a limited scale and some had not even started because of a shortage of raw materials and technical problems. The administration of these measures, as is usually the case in many developing countries, was tainted with corruption and inefficiency.

The collapse of world market prices for cocoa in 1965 worsened the balance of payment problem further. Ghana could not meet its creditors' demands for payment for imports. Ghana had depended on short-term credit, rather than bilateral (government to government) loans, which normally carry lower interest rates and are repayable in the long-term, mainly because western governments did not take kindly to the CPP regime's anti-imperialist rhetoric and Ghana could not lessen its dependence on western goods, particularly consumer items. Control measures were thus tightened, and this affected luxuries and basic items such as sugar, flour, soaps, milk, drugs, tires and vehicle parts. Goods became scarce, hoarding became a practice of most commercial houses and traders, and consequently prices soared, undermining real incomes. Consumer prices in 1966, the time the regime was overthrown, were estimated to be 75 percent higher than they were in 1960 (Berg: 1971:188). For a nation long accustomed to the consumption of foreign goods and the easy-life made possible by the economic boom of the 1950s, the impact of the restriction of imports meant a sharp decrease in living standards, and this induced resentment against the Nkrumah Government.

By the beginning of 1966, the economy was insolvent and Ghana became less credit-worthy. The nation's foreign reserve had dried up, mainly as a result of the search for a quick economic take-off. External debt was estimated at £120 million (Esseks, 1975:37), and the major loans contracted for earlier period were due for repayment. Considerable arrears had built up on imports because Ghana was unable to pay its import bills, major suppliers were reluctant to offer new credit facilities and ship goods (capital and consumer). The "economic crisis" also had a more serious impact on the general performance of the economy. Ghana's young import substitution industries, which were set up as the main motor of structural transformation, were starved of foreign exchange to import critical inputs.

This crisis was, however, not totally unexpected. By mid-1964, the Finance Minister, J. H. Mensah (the earlier finance minister, Mr Gbedemah, had left Nkrumah mainly because of ideological reasons), was counselling that private investment could, after all, be useful in development, and that Eastern European countries might not be appropriate models for Ghana. This view was condemned in the government's newspaper, *The Spark*, J. H. Mensah was labelled a capitalist ideologue and later resigned (Kraus, 1971:64).

The government, however, approached the western governments for new loans and a temporary moratorium on debt repayments in 1965, but none was willing to make new aid/grant commitments on substantial scale that would "rescue" Ghana. The refusal of the West was mainly ideological, since the Nkrumah Government had at least formally moved Ghana out of the West's ideological influence. Furthermore, Ghana's flirtation with socialist ideology, the support for independence struggles in Africa, the anti-imperialist rhetoric in both state-controlled mass media and in Nkrumah's own writings, were, in all probability, taken by western powers as signs of risk for western investment in Ghana, or at least as a nuisance to the West's quest for global dominance in the Cold-War context. Ghana had become the spokesperson for Africa, seeking a continental unity and railing against imperialism, neo-colonialism, the West.

Nkrumah also believed that Ghana's independence was continuously haunted by imperialist exploitation and neo-colonialism and argued that aid was the latest strategy of holding back the development of new countries—the paraphernalia of neo-colonialism—and had referred to the US as the world's leading imperialist power (see Nkrumah, 1965:56 and Chap. 18). This helps to explain the CPP's ambivalent attitude towards foreign investment: at one time, the regime claimed that it was necessary for Ghana's economic progress; at another, it was denounced as neocolonialist trap that was to be avoided. This ambiguity might have also contributed to the relatively low foreign investment in Ghana at this time, for it possibly sent a signal to potential investors that Ghana was not interested in them. In denouncing western nations as imperialistic and in flexing its muscles as Africa's leader and spokesperson, Ghana was caught in super-power rivalry. The Communist countries that had become the regime's allies could not offer the hard-currencies Ghana needed to pay the West, nor could they provide facilities for more imports, as Ghana had already over-used such facilities with them.

In the wake of the failure to persuade Western governments to offer debt relief, Ghana approached the IMF as last resort for a stand-by credit. The IMF and the World Bank team that assessed Ghana's "economic crisis" criticized the "socialist" development strategy, in general, and, in particular, the operation of state farms, industries, and macroeconomic management as detrimental to economic growth. The IMF/World Bank argued that there was a need for economic stabilization measures and de-acceleration as well as a re-orientation of Ghana's development strategy to conform to "acceptable" approaches to economic development. Also mentioned was the need to attract foreign investment into Ghana (cf. Hutchful, 1987:15).

The IMF/World Bank proposals were critiques of the structural transformation Ghana was supposed to be undergoing. They sought a diminished role for the state in socioeconomic processes. This was unacceptable to the Nkrumah government, which still believed that structural transformation could be undertaken without sacrificing equity or re-distributive goals. The IMF/World Bank stuck to its guns, and so the Nkrumah Government and the IMF/World Bank parted company.

A change in the regime's attitude, however, became discernible by mid-1965. The government was on the way to adopting its own solution to the problem and even to implement some of the IMF and World Bank's recommendations. Statements were made to the effect that Ghana was passing through economic difficulties but not stagnation, and that Ghana had depended too much on short-term suppliers' credit and that this needed to be reversed. The government accepted that the 7YDP's expenditure had to be trimmed to depend on Ghana's own resources, and the following year's budget was also to be limited. In particular, the Minister of Finance indicated that whether or not the IMF/World Bank team made these recommendations, the President and the Government had agreed that there was need for financial prudence. According to the Minister, they were putting their house in order; an order that was, nevertheless, coming too late (cf. Hutchful, 1978:45-53).

These statements were an official acknowledgement that the social production system set in motion by the CPP regime had encountered some difficulties and needed to be reformed to make it more viable. On 1 February, 1966, Nkrumah administered a prophetic warning that "it is not the duty of the army to rule or govern because it has no political mandate, and its duty is not to seek political mandate" (*Ghanaian Times,* 2 February, 1966). On 24 February, when Nkrumah embarked on a peace mission to Hanoi, Vietnam, the military and police exploited the economic crisis and discontent in society and staged a coup that ousted Nkrumah and his CPP regime.

Conclusion

The period from 1951 to 1966 was epochal for the breathless attempts made by Nkrumah and his CPP regime to develop Ghana and restructure socioeconomic processes through a rapid and comprehensive industrialization programme in order to create a socialist society. Nkrumah had wanted his government to be judged by the welfare of Ghanaians. Ghana was to be industrialized and its agricultural production system modernized along socialist lines within the shortest time possible. Towards this goal, Nkrumah worked assiduously. In terms of welfare, the regime had much to its credit: physical infrastructure and free education and health services reached many people. This social advance undoubtedly improved the quality of life and also the human resources of Ghana, which were necessary for long-term development.

But Nkrumah's development strategy was not sustainable mainly because it was too ambitious and aggressive; and not always in tune with the nation's limited human and material resources: the so-called fragile economy. Development was pursued at a high speed with impatience, which put an enormous burden on the bureaucracy which was just developing and expanding. This partly accounts for the massive inefficiency and mismanagement that attended many state projects and investments. Too much was being attempted at once without proper consideration to project feasibility, cost, efficiency, scale and viability. Consequently, serious mistakes were made. The CPP development programme also called for financial resources that were not always available. In the absence of resources, the Nkrumah state had to depend on what cocoa farmers saw as unreasonable taxation of their produce, and this created antagonism between them and the state.

The quantitative expansion in social services or the re-distributive social production system is laudable for its egalitarian impact on society, albeit the main beneficiaries were in the urban areas. But, sadly, it saddled the state with a large, recurrent expenditure yearly. In early stages of development, there is always a tension between how to balance social goals and economic growth. While development means more than economic growth, the latter is still a *sine qua non* of the former. Without consistent increases in economic quantities (economic growth) over long periods of time with consequent modifications in economic structures, the attainment of broader developmental goals is impossible. But a common dilemma many governments in the development world face, given the enormous material poverty of their populations, is whether the pursuit of economic growth should come before social re-distribution or whether both should be pursued simultaneously. Nkrumah sought to balance the two, but social re-distribution and employment generation seem to have outweighed consideration to growth and efficiency; thus, his regime came to experience economic crisis, which was seized to topple his regime.

In addition to this crisis, the regime was antagonistic towards the rise of indigenous bourgeoisie. Indeed, in a sense, the Nkrumah state was "unrepresentative" of vested interests: cocoa farmers, businesspeople traditional rulers and military elites, hence its claim to monopolize power and represent national interest as well as its development ideology were contested by these groups. In the end, in spite of the enormous socioeconomic progress made by the CPP regime, it was overthrown.

The efforts to lay the material foundations for modern development were commendable and justified, but the manner in which the development programme was implemented created the impression that Ghana had enough resources and skills to develop an industrialized and prosperous economy within a short time, after independence. The programme ignored the fact that a young nation needed time to learn to modernize, correct mistakes and develop gradually on a sustainable basis. On the other hand, the personalization of the state and the fusion of development strategy with Nkrumah's own philosophy meant that development strategy was not always guided by "objective" rationality or criteria. This undermined the state's capacity and effectiveness in the development process. Although the centralization of power in Nkrumah and the CPP led to political de-participation of many Ghanaians, this feature had one advantage: it enabled the CPP to consolidate the nascent nation and prevent it from being torn asunder by forces based on regions and ethnicity.

All in all, however, the CPP regime took the development of Ghana very seriously. Its achievements are unparalleled so far in Ghana's postcolonial development history, and contrast sharply with the underdevelopment of the immediate colonial past, showing the extent to which the colonial authorities had deprived Ghana of socioeconomic development. Nkrumah had maintained that the success of his regime should be measured by the welfare of Ghanaians and the happiness with which they managed their affairs. Certainly, Ghanaians achieved more welfare, but it is not clear if all of them were happy about the manner in which they were supposed to manage their affairs under Nkrumah's rule. Despite its shortcomings, the CPP regime did lay a base for industrialization and modernization. Ghana succeeded in establishing national control over the economy, although the instruments of control were sometimes weak and incoherently applied. The CPP's energization of Ghanaians to action has given pride to them that they are capable of taking care of themselves, even though, as a young nation, much had to be learnt by way of devising effective development strategy. As such, future development strategies and how successive governments built upon Nkrumah's achievement must be of interest.

Notes

1 The project cost £70 million. The Ghana Government financed half of the cost of the project (£35 million). The other half was financed by the following lending agencies as loans as follows (i) The World Bank, £16,785,000 at 5.7% in 25 years; (ii) United States Agency for International Development (USAID), £9,643,000 at 3.5% in 25 years; (iii) US Export and Import Bank, £3,572,000 at 3.5% in 25 years; (4) United Kingdom Exports Credit Guarantee Department £5,000,000 at 6% in 25 years (see Seidman, 1978).

2 Pan-Africanism began as a movement for redressing the grievances of Africans and peoples of African descent throughout the world. Its emphasis shifted towards a concern for the unity of Africa and the liberation of those parts of the continent that were under colonial rule. The most significant instrument of Pan-Africanism today is the Organization of African Unity (OAU), which was established on 25 May 1963. Among other activities, the OAU seeks peaceful settlement of inter-state disputes, a continental approach to economic and social development and solidarity among member states.

5 The Search for a New Socioeconomic and Political Order (1966-1969)

Introduction

The military and police coup of 1966 marked a turning point in Ghana's postcolonial development history. As argued in the last chapter, the coup can be conceived as the culminating point of the antagonism between one elite group seeking to radically evolve a new set of institutions and model of development or socioeconomic production system for society and another seeking to hold on to the inherited colonial socioeconomic and political order. This chapter focuses on to the activities of the new military leaders whose rule lasted from 1966 to 1969, when they handed over political power to a new group of elites: mainly Nkrumah's previous political opponents. The chapter examines the new military rulers' motives and justifications for the coup, the reasons for policy reversal and/or the dismantlement of Nkrumah's economic and political order, and the impact thereof on society. The chapter's main argument is that the systematic attempts made to reverse or dismantle Nkrumah's socialist production system reflect the fundamental conflict within the state and between two main elite groups over ideology as to what direction Ghana's political and socioeconomic development should follow. Furthermore, it argues that the policies of the new military regime not only brought about an abrupt discontinuity in the development process unfolding under Nkrumah and the CPP regime, but also led to the outright abandonment and sale of many important projects and national assets critical for Ghana's long-term development.

The National Liberation Council (NLC) and Its Justifications for the 1966 Coup

The National Liberation Council (NLC) that resulted from the coup consisted of eight members: Lieutenant General Joseph A. Ankrah, as the head of NLC, Colonel Emmanuel K. Kotoka, Brigadier Akwasi A. Afrifa, General Albert K. Ocran, as army officers. The police officers were Inspector General of Police John W. K. Harlley, Deputy Inspector General of Police, Bawa A. Yakubu, John E. O. Nunoo and Anthony K. Deku. The main objective of these coup makers was to remove Nkrumah from office and purge Ghanaian society of socialist ideology.

The broadcast that announced the coup, or what the NLC itself called a "revolution", gave reasons such as economic chaos, abuse of human rights and liberties, dictatorship, capricious use of power by Nkrumah, etc. as justifications for the coup (cf. Dowse, 1975). Allegations and accusations about corruption were made, especially after the commissions of inquiry established by the NLC to look into the "chicanery" of Nkrumah's state started making adverse findings against past political leaders (Pinkey, 1972). But the utterances and publications of the memoirs of some of the architects of the coup indicate that the causes of the coup were murkier and more complex than what was expressed in the broadcast.

In his book, *The Ghanaian Coup: 24th February 1966*, which came out soon after the coup, General Afrifa accused Nkrumah of having sold Ghana to the Russians, and lamented Nkrumah's "destruction" and "corruption" of the "decent" political system of government that Britain bequeathed to Ghana. Afrifa acknowledged that there were development projects here and there, but maintained that Nkrumah and his government got their priorities wrong. He also accused Nkrumah of detaining most of the opposition leaders and forcing others into exile. Other accusations were: that Nkrumah made a steady assault on the Army with a determined programme to indoctrinate it with the ideology of "Nkrumahism"; that the Presidential Guard Regiment received kingly treatment while the Regular Army was left in tatters; that the cost of living for the ordinary soldier was high; that Nkrumah interfered in the internal affairs of the army by dismissing army Generals (see below); that Nkrumah was a dictator and extravagant; that he paid lip-service to Ghana's membership in the Commonwealth; that Nkrumah had eroded from the army the wonderful traditions it inherited from Britain; and that the participation of the Ghanaian Army in the Congo Crisis in the 1960s and the plans to send them to solve the Rhodesian "crisis" of 1965[1] were just designs to further Nkrumah's ambition for African Unity (Afrifa, 1966:12-15, 93-144).

Similarly, but less ideologically, General Ocran, in his memoir, *A Myth is Broken* (1968), also contrasted the plight of the regular army with the privileges of the President's Own Guard Regiment. In particular, he complained that the regular army's clothes were torn, their canvas shoes were worn-out, and they had no shoe polish. As he further noted, "one day they [the Army] were to pay for electricity; the next day they were to lose their travelling allowances; the following day, they were to lose their travelling facilities... We all wondered what was happening to us... When the British were here our interests were better protected" (Ocran, 1968:41-44). Later, Ankrah told reporters that Nkrumah was trying to turn Ghana into a vast public service machine, and the NLC's "revolution" was meant to cure this ideology (cf. Hutchful, 1987:15), which accounts for the military's hostility towards Nkrumah and the CPP regime. This hatred enabled the NLC to achieve a sense of purpose in what they wanted to do with power: to "de-Nkrumahnize" Ghana.

Afrifa and Ocran's statements leave no shadow of doubt that the army generals did not take kindly to the civilian politicians' anti-West political posture and "unBritish" attitude of tampering with military appointments, promotions, cadet training and the restrictions placed on military budgets. Nkrumah had sought, under the one-party state, to bring the national army under civilian control; a CPP branch was inaugurated in the military academy and cadets were obliged to apply and join. Nkrumah also dismissed Generals Ankrah and Ocran because their loyalty to his regime was doubtful as a result of the assassination attempts on his life. These army officers thus seemed to have cause to believe that a similar fate might attend them. It has been noted that Colonel. E. K. Kotoka, one of the minds behind the 1966 coup, was known not to get along with General C. M. Bawah, who upon the directives of Nkrumah, had replaced Generals S. J. Otu and J. A. Ankrah who was dismissed from the army. Kotoka was also believed to have been doubtful about his own professional future under his new Commander (cf. Decalo, 1973). Thus, the army generals seem to have acted not solely out of concern to improve socioeconomic conditions, but also out of resentment over the decline in the material standards of the military and fear of losing their jobs (Dowse, 1975:16). They also felt marginalized as a result of the establishment of the Presidential Guard Regiment. Thus, by these accusations, they were indirectly equating their well-being with that of the larger society.

Accordingly, Generals Ankrah and Otu were reinstated in the army and made members of the new regime. Soon after seizing power, the NLC wrote to the Ministry of Finance demanding that the conditions of service and pay of members of the Armed Services should be improved. This, it noted, was to make good its promises to the Armed Forces on the eve of the NLC's so-called revolution. Their demands included total exemption from (i) Income Tax; (ii) quartering charges; (iii) payment of electricity, water and conservancy charges; and (iv) the restoration of pensions rights of the officers and men of the Armed Forces, which were removed as a result of the coming into effect of the Armed Forces Act of 1962. On the issue of tax exemptions, the National Economic Committee, which the regime had set up (see below), noted that the army's request would create a dangerous precedent in the nation's history that could make it difficult to administer tax laws in the country. But, it went ahead to approve the request, noting that, "a loss of 4.8 cedis million [yearly] was undoubtedly not too high a price to pay for the true freedom that has been brought to the nation by the Armed Forces and Police" (cf. Hutchful 1987:229-231).

The extension of these privileges meant that the state was expanding the portion of its budget devoted to the maintenance of the army. As we shall see in the next chapter, the attempt to reduce such privileges by the successor civilian regime of the Progress Party in the wake of further economic difficulties was to be one factor that precipitated the 1972 coup. This assertion of army interests questions the NLC's claim that it was acting unconstitutionally to "liberate" society. Clearly, the army was acting for army reasons and this confirms Decalo's general argument about many African armies that:

> once in power, an internal settling of accounts within the army, increases in military budgets, and promotions for the major participants in the coup have been nearly universal. The country's 'most powerful trade union' rarely pauses for a decent interval before appropriating the financial and material needs commensurate to its professional self-image (Decalo, 1973:119).

Given the size of the army and the economic crisis of 1965-66, some of these so-called hardships should not come as unexpected. At the time of the coup, the Ghanaian Army was the largest in West Africa, numbering about 600 officers and some 14,000 in other ranks and the cost of its maintenance was the fifth largest on the continent (Dowse, 1975). The army expanded very rapidly since 1960, mainly for nationalistic and ideological reasons. Just as the Ghanaian civilian politicians and civil servants welcomed independence because it enabled them to gain access to political power and jobs, which hitherto were exercised and occupied by Europeans, so did Ghanaian army officers welcome the Africanization programme of the army.

In 1957, Ghanaian army officers constituted only 10 percent of the entire army officer corps and there was pressure on the civilian regime to accelerate the Africanization of the army, just as it did for the civil service. But, expatriate domination of the army made it difficult for the army to meddle in politics, a situation that worked to the advantage of CPP politicians. But the politicians also wanted a fully Africanized military institution to reflect the sovereignty of their new nation and their

"Africanness" by taking the politically inexpedient action of removing the expatriates. By the early 1960s, the entire officer corps of the army was Africanized and Ghana had established its own Military Academy. These actions, dictated mainly by nationalist aspirations were, however, a Trojan horse in the demise of the civilian regime, for it enabled Ghanaian soldiers to have some control over the state's instrument of coercion.

On the other hand, the establishment of an academy did not in any way obliterate the army's so-called colonial mentality. Nor did it bring an end to further training in Western military institutions. Ghana Army was a direct descendant of the colonial army created by the British (the Royal West African Frontier Force: the RWAFF). Many of the army officers of 1966 were soldiers under the British colonial administration and had been exposed to many British military traditions. Indeed, most of the military and police officers (i.e. NLC members) were trained in British institutions, the most notable being Sandhurst. The army's structure of organization and operation was British; its equipment, uniforms, and other accoutrements came from Britain and other western countries (Etsiah, 1989/90:38). Most of its trappings were, thus, simply carried over from the colonial period. This helps to explain their admiration for British traditions and values, which Nkrumah and his fellow CPP members officially appeared to dislike. The significance of this connection was acknowledged later by Nkrumah when he observed that the fact that John Harlley and others found it fairly easy to get support in the army may also be explained by what has been called the "Sandhurst mentality" of certain officers. These men trained in various English military establishments, he further noted, prided themselves on being more British than Ghanaians, and tended to frown on everything in Ghanaian way of life which did not conform with English customs and traditions (Nkrumah, 1968:45).

Given this military socialization, it is not surprising that the NLC members (the so-called Sandhurst graduates) considered themselves more western and did not approve of the CPP politicians' official distaste for western liberal thoughts. For instance, they believed that Ghana should practise western democracy and should not have a one-party state or be ruled by an authoritarian leader. They also believed that their institutions, that is, the military and police, should not come directly under the control of politicians, particularly as regards the appointment, promotion and dismissal of army officers.

The military establishment may be separated geographically from mainstream society, but it still remains an integral part of the larger society that it is supposed to serve. In this respect, any economic difficulties that the country had to endure were bound to affect not only civilian society but also the military establishment. The economic hardships in the country, affecting virtually all Ghanaians, apparently drove the military to act against the Nkrumah Government. But internal grievances in the army, personal ambitions, fears, insecurity and ideological differences also seem to have played a role. Thus, the economic crisis affecting mainstream society (peasant farmers, civil servants, workers, teachers, etc.) and political authoritarianism that had emerged in the CPP's effort to build a unified nation-state were used to justify the 1966 coup.

Many Ghanaians, like electorates everywhere who suffer temporary economic recession, turned away from the CPP politicians. The jubilation that greeted the 1966 coup, and the fact that the coup makers met with little resistance in their unconstitutional removal of Nkrumah and the CPP, can be interpreted as a loss of popular support for the Nkrumah state. The coup makers saw themselves as "liberators" from Nkrumah's "authoritarian rule", and this was reflected in their regime's name, National Liberation Council (NLC). They also saw themselves as providing a transitional "corrective" government that would restore "economic order" and also create the conditions that would re-vitalize Ghana's development and democratization process (cf. Afrifa, 1966). They maintained that, once these objectives were attained, they would withdraw to the military barracks.

Bureaucratic Authoritarianism and Reversal of Nkrumah's Economic Order

As noted earlier, Ghana's economy has been afflicted by shortages of imported consumer items and capital goods, raw materials, spare parts and other basic items originating from Western countries. This has undermined the productive capacity of the economy. Also, essential foodstuffs were in short supply and domestic prices had risen very rapidly. By February 1966, debt-service alone was consuming 28 percent of annual export earnings. The international confidence in Ghana's economic prospect reached a low ebb (cf. Hutchful, 1987:61-62). Thus, the major economic task that confronted the NLC was how to revamp confidence in Ghana's traditional commercial suppliers, mainly in the West, so that they could supply consumer and capital goods to Ghana. The easiest way was to demonstrate to the western world that Ghana was leaving the socialist camp in this Cold-War era and returning to the western orbit where it perhaps "traditionally" belonged.

Consequently, the NLC formally broke Ghana's links with the Communist countries. Officials, diplomats, and technical personnel from the former Soviet Union, Eastern Europe and China, most of whom were advising on diverse development projects were expelled. This measure brought some immediate economic dividends. Western governments extended credit and grants that enabled the NLC to import essential commodities such as sugar, wheat, milk, and drugs to flood the Ghanaian market. The importation of commodities brought some relief to an economy that had, in recent memory, been riddled with scarcity. This increased social support for the regime. The more fundamental economic problems relating to debt re-payments, low productivity and inflation, were addressed within the framework of the IMF stabilization programme.

Shortly after the coup, the NLC set up a seven-member National Economic Committee (NEC) to give advice on economic matters. This committee made up of civil servants was headed by E. N. Omaboe, who had earlier been head of the government's Central Bureau of Statistics, (now Statistical Service of Ghana), Albert Adomako, Governor of Bank of Ghana, R. S. Amegashie, the Director of the Business School at Achimota, two principal secretaries of the Ministry of Finance, and two officials from the Planning Commission, all of whom had held posts under the Nkrumah regime (Hutchful, 1987:43).

The appointment to this committee of civil servants and the retention of the principal secretaries and heads of departments who had served under the Nkrumah regime was quite a surprise, since the civil service, as argued by the members of the NLC regime, was subjected to political interference from the Nkrumah Government. Indeed, most of these civil servants were appointed by the Nkrumah Government; some were the "experts" who drew up and implemented the 7YDP. Given the politicization and corruption of the civil service and its size, which, as a number of critics noted, was large, one would have expected the NLC regime to call on civil servants to account for their "rule" and also reduce the size of bureaucracy to lower cost and improve its efficiency. But no measure was taken to this effect. General Ankrah's argument that, at independence, the civil service was excellent and free from politics, but became demoralized by Nkrumah (cf. Dowse, 1975:20), can be interpreted as claiming that political interference from the Nkrumah government was the main reason for its deficiencies and corruption.

The reliance on civil servants by the NLC, however, manifest a common problem military regimes everywhere must deal with: the problem of how to administer a nation after political power has been successfully captured, since soldiers are not trained in the skills of public administration. The NLC was astute enough to recognize its administrative limitations and sought civilian professional expertise. The NLC members were officers with long army service, little education and limited political sophistication and outlook; they were inadequate to the task of administering a modern state. Until perhaps recently, when a noticeable number of university graduates have undertaken military careers, the military had attracted the relatively less educated elements of society whose skills were obviously not adequate for state administration. Furthermore, soldiers are not trained to run governments and administer national economies, as their chief *raison d'etre* is the defence of their nations against external aggression.

The NLC thus allowed the civil service considerable latitude in the formulation of policy and management of the public realm. It was the era of "bureaucratic authoritarianism": the civil servants provided the bureaucratic competence and the military supplied the political authoritarianism for economic reform. This coalition created the impression that the military came to find technical solutions to the country's economic problems and was not as ideologically engaged as Nkrumah and the CPP regime. Theirs was not to create a "political" but an "economic" society based on a pragmatic actions aimed mainly at economic efficiency. This outlook gained the NLC regime a measure of respectability from the western world, particularly when such pragmatism was more oriented towards capitalist than socialist development.

At home, the NLC also enjoyed legitimacy at least from the civil servants whose "independence" from politics seemed to have been thwarted by neopatrimonialism and socialist policies of the immediate past. In addition, although the NLC judged that the public sector, in general, was bloated, and was to dissolve the WB, YFL and other organizations, it refused to downsize the bureaucracy, which would have meant the loss of jobs and increased antagonism towards the regime. This decision was left to the incoming civilian regime of Busia. But as the next chapter will show, when the Busia regime implemented this decision, it lost its social base and was overthrown by the military.

By May 1966, the National Economic Committee and the IMF reached agreement on an economic "stabilization" programme for Ghana, a programme that the Nkrumah regime was hesitant to accept. This programme entailed a drastic reduction of the budget deficit. It also demanded that budgets should be limited to receipts of revenue and borrowing should come from non-inflationary sources. Throughout its rule from 1966 to 1969, the NLC maintained "austerity budgets" and the economy was subjected to monetarist solutions. The Central Bank lending to the government and loans advanced to the private sector and commercial banks were restricted. The Bank's lending rate, for example, was raised from 4.5 percent to 7 percent, and commercial bank lending was reduced by between 25 percent and 33 percent, depending on the priority of areas (Hutchful, 1987:18-19). Though these figures were not always strictly adhered to, the NLC made efforts to keep government expenditure within orthodox limits.

There were also cutbacks in the state's expenditure on infrastructure development. Between the fiscal years of 1966 and 1969, capital expenditure, which was already being reduced by the Nkrumah Government, was further reduced by as much as 54 percent. The state's role in productive activities was also reduced. Commercial relations with the former Soviet Union, Eastern Europe and China were reduced, and, in some cases, terminated. Central planning of the economy was replaced by the market. State control of foreign exchange was relaxed, import licensing for foreign trade and payments was de-regulated, and wage increases were limited to 5 percent (Hutchful, 1987). Through these programmes, the NLC sought to rectify Ghana's imbalance in foreign payments, rescue the economy from inflationary pressure, and create the conditions that could make the economy more productive and efficient. By implication, the NLC was re-defining the role of the state in socioeconomic development as "minimalist" and was reversing, if not dismantling, the domestic macroeconomic management system and institutions the Nkrumah regime had set up.

In return to these seemingly bold liberal economic measures, Ghana got a standby credit of £13 million from the IMF. Following the agreement on stabilization, a deal was struck whereby Western creditor nations accepted a temporary moratorium on debt payment until after the IMF sponsored meeting in December 1966. These creditors also agreed that Ghana could postpone 80 percent of the debts originally due in July 1966 to December 1968. This deferment of debt repayments was a relief and important breakthrough, for, Ghana was asked to pay only £11.6 million instead of £58 million, the equivalent of five months' import. And another meeting in 1968 further extended the payment of the 80 percent already re-scheduled to 1972 (cf. Esseks, 1975:45). Similar agreements were negotiated with the communist countries to which Ghana was indebted. In July 1967, the currency was devalued 30 percent against the US dollar, with the hope that the cost of imported goods would rise relative to exports to enable Ghana to solve its escalating balance of payment problem. Furthermore, between 1966 and 1968, CMB increased the price for standard sixty pound load of cocoa almost by 100 percent. The CMB was also given import licenses that enabled it bring more insecticide and equipment to improve the cocoa industry. In essence, by implementing these liberal economic measures, and others of similar kind (see below), market economy or capitalism, which had been viewed with suspicion as "too complex" for Ghana by the Nkrumah regime, was now wholeheartedly embraced.

The Abandonment and Sale of the Eastern European Projects

As noted above, one reason offered by the military for overthrowing the CPP regime was that Nkrumah got his priorities wrong, which means that the existing socioeconomic order was to be changed for other priorities. The 7YDP was abandoned. Between 1966 to 1969, when the NLC accepted an IMF "stabilization programme" and pursued liberal economic policies, it also took measures to de-emphasize the state's role in the economy and also "de-socialize" the economy. It confined the state's role to areas such as public utilities, industries of strategic importance, i.e. to pioneering industries which private capital or initiative would avoid because of profit motives. The measures the regime took to reduce state sector activities included: abandonment of projects which it (including its policy advisors) considered as "pork-barrel", outright sales, mergers, or partnership with foreign investors. The contraction of the state sector was based on the assumption that the private sector (both domestic and foreign) would expand, particularly in areas such as industry, agriculture, commerce and housing to compensate for the state's withdrawal from these sectors (cf. Hutchful, 1987).

The reduction of the state's role in the economic development process affected several projects that were being constructed with the help of technicians from the Soviet Union and Eastern Europe and China. Hutchful's (1987) recent documentation of the projects that were either abandoned or sold illustrates the extent to which the NLC sought to undo Nkrumah's order. For example, a concrete panel factory that was being rehabilitated by the Soviet Union, costing at the time, 2.3 million cedis in machinery and equipment and 1.5 million cedis in civil works, which was completed and required only a working capital to commence production was abandoned (cf. also Esseks, 1975). Another project being built with Soviet assistance was the Tarkwa Gold Refinery. This project cost at that time a little over £500,000. About 90 percent of the project's work was completed and was scheduled to be finished by June 1966. This project was also abandoned. In Nkrumah's plan, however, the Tarkwa project was meant to refine gold ore and give added value to Ghana's gold.

It was also part of Nkrumah's long-term development objective to bring the nation's second major export product—gold—under national control. This was to be achieved by 1986, when the lease for the Obuasi mines of Ashanti Goldfields would have expired. But, as Hutchful argues, the management of Ashanti Goldfields Corporation and other foreign interests had been opposing this attempt to make Ghana self-sufficient in gold refining. The coup then provided British Lonrho (a transnational gold-mining company) an opportune chance to extend control over Ghana's mineral industry, and, by 1968, it took control over the Ashanti Goldfields Corporation and the Ghana government became a junior partner.

Also, a ferro-manganese project intended to eliminate under-invoicing by the African Manganese Company (the Ghanaian subsidiary of Union Carbide), and to create conditions for the forward integration of the industry was shelved. Instead, the NLC entered a joint venture with Union Carbide through the National Investment Bank (NIB) for the manufacture of batteries. The NLC also reached a deal with the multinational company United African Company (UAC) and Ghana Textile Manufacturing Company (GTMC) to take over the Chinese integrated textile mill at Juapong. This measure also enabled the UAC to control a substantial share of domestic textile manufacture. In addition, the silos being built by the Nkrumah regime at Tema to enable Ghana to store about 200,000 tonnes of cocoa so as to influence world market prices were abandoned. To this day, the silos at Tema stand semi-completed and decaying in the tropical weather.

Furthermore, the State's Fishing Complex at Tema, being built at the cost of £2.3 million (cost of machinery, equipment and civil works), was halted. This project was nearing completion; almost 20 percent of the machinery and equipment was on the site and about 45 percent of the civil work was completed. A Soviet crew of 350 personnel working with the State Fishing Corporation were abruptly expelled in March 1966. In April 1966, following protests from the Secretariat for the Committee of Economic Cooperation with Eastern Countries (CECEC), the National Economic Committee tried to salvage the fishing industry by writing to the Chairman of the merchant company, UAC requesting it to provide personnel to operate the Russian vessels of the State Fishing Corporation. This request was rejected by UAC, because according to them, they had no experience in operating merchant vessels (cf. Hutchful, 1987).

Thus, the activities of the State Fishing Corporation were grounded, and some 28 fishing vessels belonging to it and other private Ghanaian fishing companies laid idle standing in water when their 350 Soviet crew and technicians were expelled in March 1966. In 1968, two years after the coup, the National Economic Committee attempted unsuccessfully to sell 10 Soviet vessels on the pretext that they were uneconomical to operate. In 1970, the Busia Government also tried to sell the 10 rusting vessels to a private foreign buyer for $120,000, which would have meant a loss of some $3.6 million to the Ghanaian state since their original cost was $3.72 million. This was also unsuccessful, for the High Court discovered a fraudulent transaction involved in the deal. The vessels remained unsold and continued to depreciate because of lack of maintenance. During this period, one of the vessels sank and was re-floated. By the time of the 1972 military coup, eight vessels were still standing idle and rusting in the waters of Tema Harbour, occupying docking space and obstructing shipping. Finally, Mankoadze Company and Ocean Fisheries (private fishing companies) had to take over the vessels without even the provision of a formal contract (cf. Hutchful, 1987:20-22).

Other important projects were either abandoned or stopped. The NLC returned 18 Illyshin planes to the USSR; a stop was put to two tractor assembly plants that were to be constructed by Yugoslavian firms in Kumasi; the former German Democratic Republic which had built a Printing Press for Ghana at Tema and had won a contract to build a Pulp and Paper Mill which should use raw material Bagasse from the Sugar Factories at Asutuare and Komenda and Rice straw from the State Farms was also stopped. Other projects were defined as "prestigious" and abandoned. For example, the Axim-Half-Assini Road, Tamale Airfield, and the Gold Mines at Bibiani (Western Region) under construction were stopped. Work on the Axim-Half-Assini Road, however, resumed after considerable pressure from chiefs and other groups. Other state-owned rural industries such as including bakery, furniture and joinery factories were also abandoned.

The Privatization and Dismantlement of State-Owned Industrial and Agricultural Enterprises

State-Owned Enterprises (SOEs) were held by the CPP regime as the motor of Ghana's industrial development. But, by the time of 1966 coup, it was estimated that state investment in industry was around £39.8 million. There were mixed opinions as to the economic performance and viability of the state enterprises. Although the SOEs were "young", and, at an early stage of development and for that matter their mediocre contribution to the national economy cannot be used as a test of their ultimate viability, there was general feeling that they were operating at a loss and some would not have existed without government subsidies. By the end of 1965, profit for 32 of the 35 state enterprises stood at £1.2 million while, accumulated losses were £15.1 million. Out of this figure, three companies combined: Ghana Airways, the State Mining Corporation and State Farms lost £13.6 million, while the remaining 29 lost £1.5 million (Hutchful, 1987:27-39). Ghana Airways, for instance, was flying uneconomical routes to improve communication between African countries because political, rather than economic consideration was one factor that influenced its operation. The State Mining Corporation's loss was due to the high cost of production vis-à-vis fixed world price for gold and investment in mine development. The poor performance of the state enterprises has been attributed to the general problems that afflicted State Farms, such as the lack of adequate managerial and supervision skills, under-utilization of capacity and lack of technical skills (see Chapter 4).

As a result of these problems, the NLC through its National Economic Committees, negotiated with a number of foreign companies for the participation in and virtual takeover of many of these state-owned enterprises. It also mandated the National Investment Bank to look for foreign investors who could supply investment capital and managerial as well as technical skills for the operation of these state ventures. This enabled some foreign companies to have a wide-ranging management and policy control over a number of the state companies. For example, in addition to the acquisition of Lonrho of the Ashanti Goldfields mentioned earlier, the Norwegian Cement Export A/S (Norcem), which had 25 percent in Ghana Cement Works, came to have equal shares like the Ghana Government with wider and more favourable policy control. The Rubber Estates and Tire Factory at Bonsasa, built with Czechoslovakian help, were transferred to joint ownership with Firestone Corporation of USA, and Switzerland took over the match Factory at Kade. The state's Pharmaceutical Industry had to be idle because the American company, Abbott Laboratories, which had expressed interest in it, pulled out at the last minute over problems of negotiations.

About 28 industries that could not be sold or run on a partnership basis because they did not attract the interest of foreign investors, or had other problems or both (e.g. the Cannery Factories at Nsawam for pineapple, Wenchi for mango, and Pwalugu for tomatoes; the Electronic Industry at Tema; the Cocoa Products Industry of Takoradi and Tema; Vegetable Oil Mills at Essiama and at Denu for producing copra oil, Atebubu, Tamale, and Bawku for groundnut oil; the Paper Conversion Corporation), were retained under the state sector through the newly created Ghana Industrial Holding Corporation (GIHOC).

The modification or change of ownership rights of the nation's infant industries meant that the Nkrumah regime's long-term objective of integrating and transforming the productive structure of Ghana's economy was not to occur. For example, as Hutchful argues, it was planned by the CPP regime that the two cement factories at Takoradi and Tema would be developed to utilize Ghana's proven limestone deposits at Nauli (400 million tonnes reserves), Buipe (100-200 million tonnes), and Bong-Da (15 million tonnes). This would be done by using local charcoal for the production of clinker which would then be mixed with *gypsum* in the final crushing and milling stage. The company, Norcem, which took over these cement factories in partnership with the Ghana government, was not interested in these long-term integration. One consequence of the failure to develop integrated industries is that many factories ended up importing semi-finished industrial products with high import components. The absence of foreign exchange and balance of payment difficulties Ghana experienced often made the operation of such factories haphazard. Linking Ghana's industries to local resource base and relieving pressure on balance of payment was one serious challenge that continued to face Ghana's industrialization programme (cf. Hutchful, 1987:29-36).

The rapid disposal of some of the nation's fixed assets, however, led to a serious conflict between the two strategic organs shaping policy at the time: the National Economic Committee (NEC) and the National Political Committee (NPC). The conflict became manifest over the Agreement that NEC signed to lease the state's prestigious hotels (namely Ambassador Hotel, Star Hotel, Meridian Hotel, City Hotel, Continental Hotel and others) to the Inter-Continental Hotels Corporation of America. The deal was to be a joint venture between this company and the Ghanaian Government, but according to the Agreement, the government would have no say in the management of the hotels. The Political Committee was against the deal because, as its own investigation showed, the main problems facing the hotels were over-staffing (mainly because of political appointees forced on the hotels by the previous regime) and lack of working capital of about 500,000 cedis (the estimate of the manager of Ambassador Hotel as against 700,000 cedis estimated by NEC). These problems in the estimation of the Political Committee could be solved without foreign expertise, since workers could be retrenched and the capital provided.

The Political Committee was also incensed that the Economic Committees did not scrutinize the legal implications of the deal or bring legal experts into the negotiation process of such an important state sector enterprise. According to the Political Committee, the Agreement signed was misleading. It pointed out, for example, that there was a clause in the Agreement, which at first glance read that the lease was to be for twenty years, but the true position was that it was effective for 40 years, unless the Inter-Continental Hotels chose not to renew it after 20 years. This option was not available to Ghana. Again, a clause gave the impression that the Government could terminate the Agreement in the first year if it was not satisfied, but the true position was that only the Inter-Continental Hotels Corporation could do that. According to the Political Committee, this situation meant that, although Ghana would advance much fixed capital in the deal, it would exercise little control over the venture.

As a result, the Political Committee recommended to the NLC that the National Investment Bank be given the mandate to run the hotels. But it also recommended that, should the NLC go ahead with the deal, the clauses should be modified to give Ghana the option to cancel and renew the Agreement in the first year. The lease for the first twenty years' period should be reduced to fifteen and the provision for renewal should be struck out altogether. The Political Committee and Economic Committee were thus in conflict over the fate of national assets. The conflict illustrates the division that was emerging over the NLC's de-nationalization programme. But, more importantly, it is also an indication that a nationalistic ideology in terms of greater state control of the economy still found favour with some Ghanaians. This somehow underlies the policies of the two military regimes that were to rule Ghana from 1972 to 1979.

The rapid sale and abandonment of national projects also upset the Committee for Economic Cooperation with Eastern Countries (CECEC). Consequently, its General-Secretary sent a memorandum to the NEC urging that, as a matter of national interest, a handful of Soviet and Eastern experts should be allowed to stay in the country to complete projects they had started. The secretary also drew attention to various long-term Economic and Technical Aid agreements that Ghana was benefiting from with these countries and how Ghana would lose should these contracts be abrogated. The CECEC also advised the NLC that, since much of the machinery of ongoing projects was from Eastern Europe, some of the technicians from these countries should remain in the country to install the equipment and supervise the projects. The Secretary to the CECEC further noted that the State Farms and the Training Centres could go on without the Russians; that Ghana could get experts from other countries to continue the Geological Survey; that since the Gold Refinery was a specialized venture with Russian technology, at least 3 Russian technicians should be allowed to remain and help in the installation of equipment; that the Reinforced Concrete Panel was near completion and must be continued, that the State Farms could go on without the Russians; and that he had nothing on record about the Atomic Reactor and the Tamale Air field (cf. Hutchful, 1987). But such persuasions were not heeded as there was much ideology at stake at the moment.

The agricultural sector was also restructured. Soon after coming to power, the NLC established the National Agricultural Committee (NAC) to examine the agricultural problems facing the nation and advise the National Economic Committee on agricultural policy. The NAC's terms of reference included feasibility studies of all the agricultural projects that were being undertaken and recommendations for their maintenance or abandonment. In its Report presented in November, the NAC noted many technical and administrative problems that afflicted the agricultural sector, problems already specified in the previous chapter.

Consequently, it recommended that the State Farms Corporation be abolished while some of its farm projects were retained and assigned by the Ministry of Agriculture to the following organizations: (i) the Oil Palm Corporation was to take over 4 major oil palm projects; (ii) Rubber Corporation was to be given 9 rubber projects; (iii) Fibre Corporation 10 fibre projects; (iv) Extension Services Division was to allocate 3 rice schemes to farmers; (v) Animal Husbandry Division was to operate 7 livestock projects for supplying improved breeds to farmers; and (vi) Cocoa Division was to recover its 12 cocoa stations for use as seed gardens and for conducting experiments. Some 20 projects were to be given back to the landowners or farmers whose lands were confiscated in the agricultural mechanization process and some 12 should be offered for sale (cf. Hutchful, 1987).

The rapid abandonment or transfer of state farms that occurred between 1966 and 1968 lost Ghana 6.6 million cedis in net value (cf. Hutchful, 1987). But the actual cost of the projects (in machinery and manpower) can never be adequately assessed, since no systematic evaluation of their cost and potential value was made. The NLC later closed down some 105 state farms, and the farming machinery of the State Farms, the Co-operatives, the WB and the YFL were abandoned so that the countryside became littered with rusting Soviet and Eastern European machinery. No significant new projects were initiated in this era.

Although the NAC noted that State Farms had failed to provide sufficient food to justify the state's huge investment in them and that the peasant farmer continued to provide 90 percent of the domestic food, the NLC, like the CPP regime, made no serious attempts to modernize the technology of peasant production, viz: the development of mechanized agriculture, irrigation and storage facilities. Ghana's food production system continued to be essentially rain-fed and characterized by fluctuations from bumper harvest to lean seasons, with significant post harvest losses. General agricultural production was often so low that there were usually shortfalls in domestic food production. The failure of the NLC to promote domestic agriculture was also evidenced in the inability of the sector to provide raw materials for the country's agro-based industries. The World Bank's report of 24 March 1968 on the Economic Stabilization in Ghana, summed up the state of affairs of agriculture:

> Effective management for the two sugar refineries and plantations has yet to be obtained and it will take considerable time and additional investment to grow sufficient input for the factories. Processing plants for pineapple, tomato and mango are still idle for lack of raw materials and no concrete proposals to grow the necessary inputs are yet in sights... Other programs such as the

> encouragement of production and marketing cooperatives, the improvement of extension services, the distribution of farm supplies and programs to improve water management, to mention only a few, are equally important and programs in these areas have been most uneven. Program priorities and operational plans are generally not yet established much less implemented. Instead of this type of work, it appears that too much time and resources are devoted to unproductive [sic] on-going programs like farm mechanisation and land settlement or the discussion of policy in very general terms without due regard to requirements for their implementation (World Bank in Hutchful, 1987:27).

The State and the Private Sector in Economic Development

A central pillar of the NLC's economic "reconstruction" programme was the private sector, particularly foreign investment, whose participation it sought to encourage in the industrial development of Ghana. This was consistent with the World Bank's advice, which, among other things, suggested a 60 percent reduction in diamond marketing fees and mining royalties, elimination of market fees and duties on timber, reduction in other mining royalties, elimination of the surcharge on company taxes earned in the future and remitted abroad. The NLC for its part, also gave at least four reasons why it favoured foreign investment: (i) foreign capital was necessary to close the trade gap related mainly to import demand resulting from economic development; (ii) because of mounting pressure of recurrent consumption needs, combined with other reasons, domestic savings could not be mobilized to a sufficient level to permit high rate of investment; (iii) foreign capital could ensure a steam of modern industrial technology into Ghana; and (iv) that it constituted a medium through which the practices of labour and management that make for high productivity in the advanced countries could be imported into Ghana and learned by the indigenous people.

Consequently, the government declared that it would welcome foreign investment, joint ventures and partnerships. Towards this end, it established the Capital Investment Board to administer the Capital Investment Act, which stipulated the following incentives: a ten-year tax holiday, exemptions from import and excise duties, deferment of registration fees and stamp duties, reduction of taxes on profits from 50 to 45 percent and on dividends from 15 to 7.5 percent.

In addition, there were assurances that there would be no nationalization of companies and no political interference in their operation. There would also be no restrictions on remittances of capital or repatriation of capital to the origin of an investment in the event of the sale or liquidation of projects. Furthermore, the administration of government agencies that handle project proposals was to be streamlined to process information expeditiously so that investors would no longer be bothered by vexatious controls in these agencies. The NLC's radical divestment of the state sector was highly criticized particularly by Leftist intellectuals, students and other radical elements of the Ghanaian society.

The NLC regime also sought to encourage domestic private sector activities. It established a Committee to recommend the Promotion of Ghanaian Enterprises, noting, among other things, that the industrialization of the country was first and foremost the responsibility of Ghanaians (cf. Hutchful, 1987:100). Consequently, in December 1968, it passed the Ghana Enterprises Decree to protect local businessmen against foreign competition in some selected sectors of the economy, in which it judged that Ghanaians had the resources and competence to participate. Under the decree, retail trade was to be reserved solely for Ghanaians. Credit facilities were also to be set up to help Ghanaian businessmen. Greater economic space was also sought for Ghanaians through the reservation of areas such as food cannery, animal husbandry, garment manufacturing, printing, timber and ammunition production. But, with regard to the timber industry which depended so much on foreign machinery, some exemptions were made to allow foreigners to participate. Other areas to be exclusively reserved for Ghanaians were distillery, laundry and dry cleaning, bakery, polythene, film extrusion and plastic moulding. Foreigners already operating in these areas were to remain, provided they showed evidence that the capital for establishing their industries was brought from outside Ghana, that they employed at least 100 Ghanaians, that they had been in the industry for at least 5 years, that they had paid an average tax of £6000 or C12,000.00 per year for six years and had employed Ghanaian accountants and managers and did not operate retail shops.

Another area to be reserved for only Ghanaians was taxi service, which was dominated by Lebanese and Syrians. The Public Works Department was also restructured to give preferential treatment to Ghanaian contractors, and, by 1967, they were awarded about 97 percent of the total contracts. Import licences, were also made more readily available to Ghanaian merchants. In the process, foreign merchant houses and the GNTC, which had dominated the commercial life of Ghana, saw their import licences decrease from about 30 percent to 21 percent in 1967 (Esseks, 1975:47).

The NLC's Committee on the Promotion of Ghanaian Enterprises had also recommended that the large international companies withdraw from retail trade, but the Political Committee drew attention to the possibility of such an action creating an economic vacuum that would be filled by Lebanese, Syrian and Indian traders. The Committee thus counselled the government to embark on a definite, bold and effective plan or legislative action to enable the Ghanaian business class to gain greater control over the ownership of the national economy. This was to provide the basis for the Aliens Compliance Order that the civilian government of Busia implemented in 1969. In comparison to those of the Nkrumah Government, these measures mark a significant shift in the attitude of the state towards the local business community. These Ghanaianization measures were, however, to be accomplished within a time framework of 4 to 5 years. This meant that most of these policies were to be implemented by the successor civilian regime of Busia.

Since the 1966 coup was anti-CPP, the NLC's Ghanaianization policy of the economy moved in another direction. The regime's philosophy was that Ghana's economic strength depends on how vibrant indigenous entrepreneurs are and how effectively their economic activities are mobilized by the state to contribute to the development process. According to the NLC, the test of the economic maturity of an industrial system was the extent to which its own nationals could run local industries. The NLC lamented the indigenous business class's lack of capital, entrepreneurship and technical and managerial expertise, as well as how over the years it had been "crowded out" by the larger and better financed international trading companies (for example, UAC, SCOA, UTC, G. B. Ollivant, etc.), as well as by the Lebanese, Syrians and Indians, and was forced to operate in farming, small- to medium-scale retailing of manufactures and primary commodities and other services, notably transportation.

Unlike the Nkrumah government, the NLC sought to give special and practical encouragement to Ghanaians to organize themselves more efficiently and to assume an increasing role in the industrial and commercial development of the country. It stressed the importance of the domestic private sector to national economic viability, noting that it would remain in the foreseeable future the largest sector, in terms of number engaged and gross output. On this new initiative, Mr R. S. Amegashie, a member of the National Economic Committee, noted that Ghanaians wanted economic independence and not something abstract, like more power to the state. He further noted that the operation of small-scale enterprises protected from expatriate competition by Ghanaians would serve as fruitful apprenticeship for managing larger enterprises (whether private or state) (cf. Hutchful, 1987).

There were, thus, deliberate efforts to advance the interest of the indigenous business class. In Nkrumah's view, the state, through the GNTC, was to compete with foreign concerns, outgrow them and ultimately drive them out. But they proved resilient, with Lebanese and Indians moving into retail trade, hotels and taxi services. The NLC, however, was counting on the indigenous private sector to play a critical role in national development, a task that the CPP laid rather more on the state.

Ghanaian merchants gained access to more foreign exchange. More construction contracts were given to Ghanaians, and Ghanaian businesspeople moved into the retail sector and the transport industry. The Ghanaianization policy was meant to weed out foreigners, particularly the Lebanese and Indians, who had dominated these sectors and were said to be doing things that Ghanaians could easily do, like buying and selling goods. They were also said to have constituted a serious drain on the economy and balance of payment accounts because of their personal remittances abroad, illegal transfers and currency smuggling.

Consequently, a number of business associations were formed or revived to advance the interest of the Ghanaian business community. Notable among them were the Association of Ghanaian Businessmen, Indigenous Ghanaian Manufacturers' Association, the Crusade for the Protection of Ghanaian Enterprises, and the Ghana National Contractors Associations. The progress made by these occupational groups was cogently summed up by the President of the Indigenous Ghanaians Manufacturers' Association, Mrs Esther Ocloo, who said that her organization had metamorphosed from a "dinner club" into a formal interest group, largely in response to the Ghanaian Enterprise Decree and its promise of decisive protection for "indigenous entrepreneurs" (Mrs Ocloo, quoted in Esseks, 1975:49).

These associations also sought political power. For example, during the 1969 election, Mrs Esther Ocloo, Mr E. Acquah-Harrison, President of the Ghana Manufacturers' Association, Mr K. Amuah-Sekyi, Treasurer of the Association of Ghanaian Businessmen and J. C. Yeboah, Secretary of the Ghana National Contractors' Association, all stood for election as parliamentary candidates, either of the National Alliance of Liberals or of the Progress Party, both of which, given their liberal ideology, endorsed the Ghanaian Enterprise Decree and promised substantial aid to the business class (Esseks, 1975). With these political actions, it can be argued that the bourgeoisie was thus attempting to capture political power, to advance its occupational or class interests.

The Impact of the Economic Policies of the NLC Regime

The main rationale underlying the NLC's radical reversal of Nkrumah's economic order is found in liberal theory of political economy, which holds that free market policies lead to better use of resources, efficiency and consequently economic growth. To the NLC and policy-makers of this era, the socialist economy had been saddled with low productivity, inflation, waste, scarcity of goods and inefficiency. Thus, the resort to the market and liberal policies were meant to cure some of these economic problems. What, then, did the rule of the NLC achieve?

The NLC re-kindled some degree of confidence in Ghana's commercial suppliers, which improved the supply of essential commodities to Ghana. It renegotiated the repayment dates for outstanding national debts and got new credit to solve some short-term economic problems. However, the deferment of debt payment and contraction of new loans by the regime means that the NLC was passing on to the incoming civilian administration a debt burden (about £394 million) far heavier than the one it had inherited from the Nkrumah regime. This would pose more economic difficulties for the civilian regime. The budget deficit was also reduced and balance of payment was improved, but the price was that public investment fell by 17 percent in 1966, 20 percent in 1967, and, by 3.5 percent in 1968. GNP grew at the rate of 0.6 percent in 1966; 1.6 in 1967; and 0.8 in 1968. Government expenditure was also brought within manageable limits. For example, throughout its rule, the NLC limited capital expenditure. No new wholly state-owned enterprises were launched (cf. Hutchful, 1987:38-39).

The currency devaluation led to temporary output recovery in some sectors of the economy. Between 1966 and 1969, the volume of timber export rose by 46 percent. This greater output increased foreign exchange earnings by 54 percent, in spite of the devaluation that lowered the price of Ghana's timber exports. The cocoa industry also responded favourably to the CMB's price incentives and the downward trend in production evident in the last years of Nkrumah's rule was minimized. Though there was a temporary drop in cocoa production in 1967-68 mainly due to bad weather, good harvest was recorded in the country throughout the following three years.

But food production continued to suffer, mainly for reasons already suggested. The interest of the small-scale farmer in food agriculture was again neglected. The NLC retrenched the state's agriculture sector, but had no agricultural development programme to stimulate production in the private and co-operative sectors. The result was a shortfall in food production that was met by resorting to food imports, aggravating the balance of payment problems of the country. From 1966-67 to 1968-69, in spite of "austerity budgets", fiscal allocation for agriculture fell by 27.2 percent, while that of defence rose from 22 percent to 41 percent (Gyimah-Boadi, 1989; Decalo, 1973). According to Hutchful (1987), the total budget for military salaries rose by 230 percent between 1965 and 1969.

Throughout the 1966-69 liberalization period, private investment, particularly the foreign investment which was to fill the vacuum created by the dramatic withdrawal of the state from the economy, did not materialize. The NLC encouraged foreign capital, already operating in the country, to exercise greater control over some strategic national assets. The only significant achievement was the new contract it signed with multinational oil companies to prospect on Ghana's off-shore areas and the arrangement to sell electricity from the Volta River Project to the neighbouring countries of Togo and Benin, despite the fact that about 60 percent of Ghanaians—mostly in the rural areas and even some regional capitals—were without hydro-electric power.

The NLC's contraction of state-sector activities led to unemployment problems. The closure of State Farms, Workers Brigades, and other enterprises led to some 10,000 people unemployed. All in all, it was estimated that about 10 percent of the labour force, i.e. some 66,000 persons lost their jobs. Among those dismissed, only 7,400 were relocated. Government's retrenchment policies also led to a decline in real incomes, particularly for the urban working class. Real per capita GNP fell from 142 cedis in 1965 to 135 cedis in 1969. Trade unions' dissatisfaction with these measures and the apparent insensitivity of government culminated in a series of confrontations and disruptive strikes between 1966 and 1969, meant to protect and improve living standards and conditions of the working class. These strikes were met with police force and often resulted in lock-outs, dismissals and casualties. Worried and nervous about the widespread industrial unrest engulfing the country, the NLC in 1967 decreed even the mere "incitement to a general strike" as a capital offence. The regime also sought to modify the Ghanaian Industrial Act that created workers' Trade Union Congress (TUC) movement as an umbrella organization (cf. Hutchful, 1987).

The main beneficiaries of the 1966 coup were the military and police, who saw an expansion in their privileges and bettering of service conditions. The coup makers were promoted to new ranks, their incomes and allowances increased and they gained new privileges. The army and the police institutions were also provided for in terms of arms, uniforms and better barracks. At this time, the Border Guard Division of the army was transferred to the police. It is thus not surprising that the Central Bank later reported a major drain on foreign reserves, owing largely to imports whose purported purpose was to strengthen the internal security and defence of the country (cf. Opoku-Agyeman, 1988:411). Since the bureaucracy was not retrenched and was given much power over national policies, it is reasonable to conclude that NLC rule was beneficial to them, too. In this respect, Hutchful's and also Ninsin's arguments, noted at the outset of this study, that the bureaucracy benefited from military coups is valid for this period. On the other hand, the NLC rule brought a new chapter in the management of socioeconomic and development processes. The opening up of economic space for the domestic bourgeoisie to occupy and reproduce themselves indicates that the state was now committed to promoting indigenous capitalism in Ghana.

Political Restructuring: The Depatrimonialization of the State

The NLC's revolution was not only concerned with economic restructuring, but also changing the patrimonial political order that had emerged or was embryonic during CPP's rule. The main motive behind this political change was to prevent Nkrumah and his CPP from ever taking power again and also Ghana coming under authoritarian rule. Thus upon seizing power, the NLC proscribed the CPP and dissolved parliament, the 1960 Republican Constitution was suspended, and all political activities in the country were banned. Ghana was administered through a series of military decrees and proclamations. Political prisoners (some 500) held under the PDA of 1958 were released, but J. B. Danquah, one of the founding fathers of the UGCC, had died in detention. The NLC was, however, worried about the possibility of Nkrumah's return or attempts by CPP supporters to re-capture power.

Though Nkrumah's handling of the economy in the last days of the CPP regime left many people disenchanted, and, for that matter, the NLC gained popular support, it, however, did not naively assume that its social base was secure. It was aware that "Nkrumahism" could not go away overnight. It thus took steps to "de-Nkrumahnize" Ghana's political life. The army-police alliance proved advantageous, since, unlike the army, which was stationed in only some regional capitals, the police had branches all over the country and could monitor and control those who could oppose the NLC regime. The alliance was consecrated through the expansion of privileges to the military and police. This was important, since, unlike civilian politicians who are elected by constituencies upon which they could draw for support, illegitimate military rulers must justify their rule and create their own legitimacy. This fruitful alliance was to be re-enacted, though on a low scale, during the National Redemption/Supreme Council (NRC-*cum*-SMC) rule from 1972-1979.

Following the allegations levelled against the CPP, the NLC set up a series of inquiries and commissions to probe the old politicians and many aspects of Ghanaian life. The findings suggested that past politicians were corrupt. For example, the Azu Crabbe Commission of Inquiry reported that many previous CPP ministers and officials were demanding 10 percent of the value of contracts they awarded to their favourites. The NLC did not hesitate to use these "negative" findings to vilify and defame Nkrumah and his regime. The state-controlled mass media was the main instrument used in this campaign, and, in many instances, books written to "glorify" the Osagyefo were burnt. In Accra, soon after the coup, Nkrumah's statues were pulled from their pedestals and destroyed. The bank accounts of former ministers, leading CPP members, and MPs, who were found "guilty" of corruption, were frozen and investigations into their assets were made. Through these measures, the regime sought to introduce a sense of accountability into public life, but with one exception: the principal secretaries and heads of departments who had served in the civil service in the Nkrumah and the CPP regime, enjoyed immunity from such actions (Dowse, 1975:20).

The NLC also sought support from those social groups that had suffered political dispossession under the CPP regime. It allied itself with opposition elements from the NLM and NPP and chiefs, lawyers and local businesspeople who felt eclipsed by the policies of the Nkrumah state. It would be recalled that the Nkrumah Government did not take kindly to chiefs who supported its opponents. This was especially the case when the chiefly estate of the Ashanti became the bulwark of the NLM, and, later the United Party. Surely, these traditional elites, most of whom were also members of the NLM and NPP, had axes to grind with the Nkrumah Government. The NLC exploited this situation to its advantage by elevating the chieftaincy institution into prominence, as it was in the colonial period, and sought to give chiefs an important role in the development of the nation. Chiefs who were "dubiously" enthroned or recognized by the CPP were destooled and the rightful ones installed. In December 1966 alone, as many as 200 chiefs recognized by the Nkrumah government were demoted. Some 134 had their chiefly status withdrawn, and, in February 1967, a further 92 chiefs lost their recognition. The NLC members stressed that they would respect the institution and role of chiefs as an important aspect of national life and development (cf. Dowse, 1975:21). This newly acquired respectability of the chiefly estate was matched by reforms in local government institutions that allowed chiefs to deal with traditional issues. Chiefs were also allowed to serve on the various Commissions set up to advise on the political direction of the nation.

In addition, the NLC established a Commission to draw up a new constitution, and an electoral ordinance was enacted to effect a re-democratization process. A centre for Civic Education was also established to educate citizens about civic responsibilities and obligations and to raise the political awareness of an electorate who, in the NLC's judgement, had been deprived of opportunity to learn civic values and norms of liberal political culture as a result of the patrimonialization of state power by

Nkrumah and the CPP members. The Civic Education Centre was also to instil bureaucratic discipline and work ethics in the workforce. Lawyers were appointed to Commissions of inquiries, to the Constitutional Commission, and the Constituent Assembly. The legal profession felt that its expertise was recognized and respected. The NLC re-constituted the Judiciary; appointed a new Chief Justice, Mr E. Akufo-Ado, to the Supreme Court, and terminated the appointment of 17 judges (cf. Dowse, 1975). Many of those who served on the Political Committee and the other Commissions were mainly from groups that had opposed Nkrumah and his CPP regime. For example, Dr. K. A. Busia, one-time opposition leader, became Chairman of the Political Committee from 1966 to 1968, when he resigned to campaign for election.

As already noted, the NLC was determined to obliterate the Nkrumah political heritage. But Nkrumahism still found favour with many, particularly workers who had "benefited" from the extensive employment opportunities implicit in the Nkrumah economic order that had been severely curtailed by the NLC. In opening politics for contestation in 1969, the NLC sought to exclude major former CPP activists from participating in politics and holding public office for some time in the future. The NLC, perhaps like the Colonial administrators, preferred to see the political Right capture power. But, whereas the colonial administrators left their preference to chance and thus failed to realize their ambition, the NLC took actions. In January 1968, Decree No. 223 was passed to this effect, banning certain categories of persons from holding public office for ten years. Furthermore, through a series of legislation, decrees and, perhaps, deliberate omissions, the NLC disqualified many CPP politicians. In other words, the political Left was to be prohibited from making any further inroads into the political arena.[2]

However, the NLC Decree No. 223, which banned people connected to Nkrumah and the CPP, contained something of a Trojan Horse. For it established an Exemption Commission that could free from the ban those who could prove that they were forced by circumstances beyond their control to work for the CPP or had opposed CPP from within. It was here that Mr K. A. Gbedemah, one-time protégé of Nkrumah, got the chance to become an aspirant for power in the upcoming election. The Exemption Commission, presided over by Mr. Justice Apaloo, after hearing favourable testimonies from people of diverse ethnic background, such as Nana Ofori Atta II, William Ofori Atta, the former United Party leader, Mr. A. L. Adu former Secretary to the CPP cabinet and P. K. K. Quiadoo, a CPP rebel, decided to exclude Mr Gbedemah.

The exclusion of Gbedemah from party politics originated from the fear that he could resuscitate the now defunct CPP. But Gbedemah had parted company with Nkrumah before the CPP regime fell. General Afrifa (an Akan), now the NLC Chairman (Ankrah had resigned from the NLC following allegations that he was forming a political party to contest the election in 1969), probably worried about the possible resurgence of the CPP via Gbedemah (an Ewe) or about the so-called Ewe hegemony in Ghanaian public life, became infuriated at the exclusion of Gbedemah and condemned the whole exclusion exercise as futile and defeating the political objective of the NLC's revolution (Dowse, 1975). This divided the NLC members along ethnic lines.

In February 1969, the NLC issued Decree No. 332, which superseded Decree 223. This new decree listed the names of the people now exempted from the ban. Gbedemah's name did not appear, and here again, ethnic motives were read into the action. It was thought that there was some kind of Akan conspiracy bent on stifling Ewe political aspiration. Gbedemah was believed to be a friend to other Ewe members on the NLC, and Busia (Akan) was thought to be favoured by Afrifa (Akan). The political contest between Busia and Gbedemah, the leaders of Progress Party (PP) and National Alliance of Liberals (NAL), the two main parties to emerge, was re-enacted within the NLC itself, which was already experiencing ethnic factionalism. But the NLC generally agreed that political power should not go to the radical elites any more. Indeed, although by no means drawn from the older elite families, the NLC members (as their reasons for the coup make explicitly clear), like the emergent civilian politicians, were also anxious to prevent the resurgence of the Nkrumah-type populist politics in Ghana. The possibility of the CPP re-incarnating itself loomed large, hence the determined efforts by the new establishment to take no chances.

In April 1969, two new Political Party Decrees were passed within two days of each other. The first reinstated some of those disqualified and still banned Gbedemah, but the second exempted him. Given the large support the PP had from mainly Akan communities, it is doubtful that National Alliance of Liberals (NAL) could have won the 1969 election. Yet, there was no doubt that the NLC's political tactics and "gerrymandering" weakened the NAL election campaign. Because long before the election campaign, Dr. Kofi Busia was allowed to use his position as Chairman of the Political Committee and the Centre for Civil Education, and as a member of the Constituent Assembly, to travel all over the country to propagate his ideas. The leaders of other political parties did not have such advantageous positions to exploit. Also, Ewe military officers were transferred from strategic command posts in the army and replaced by Akan officers, so as not to endanger PP's rule (see Baynham, 1984/85; Brown, 1982; Philips, 1979; Chazan, 1982).

On 1 May 1969, the ban on political activities was formally lifted. Five major political parties emerged out of an initial twenty to participate in the coming election. They were: the Progress Party (PP); the National Alliance of Liberals (NAL); the United Nationalist Party (UNP); the People's Action Party (PAP); and the All People's Republican Party (APRP); among these only the PP and NAL were of major political significance. The PP was led by Dr. K. A. Busia, a professor of sociology, and NAL by Gbedemah, a businessman. The election was competitive, but mainly along ethnic lines. It was held in August, 1969 in a relatively fair atmosphere. Out of the total vote cast, the PP obtained 58.68 percent, the NAL, 30.44 percent, and the other parties 10.88 percent. Out the 140 seats in the parliament, the PP won 105, or 75 percent; the NAL, 29 or 20.71 percent; and the other minor parties, 6 or 4.29 percent (cf. Twumasi, 1975).

Ideologically, all parties were basically the same; they cautiously avoided populist manifestos characteristic of the CPP party. They were, in varying degree, elitist although they enjoyed more support from the ethnic communities from which their leaders hailed; they endorsed the NLC's vision of Ghana being a democratic country and following a capitalist path of development. They all emphasized the need for a rural renaissance and promoting Ghanaian business enterprises, the importance of such commonplace goals as building more roads, hospitals, improving food production and marketing systems, expanding educational opportunities, and creating an enabling or favourable environment for foreign investment. As far as these parties were concerned, ideology seemed dead.

Given this apparent consensus on ideology, it can be conjectured that had the Left been allowed to contest the election, most of these parties would probably have been forced into coalition. The Second Republic, headed by Dr. K. A. Busia, was inaugurated on 1 October, 1969, within the framework of Westminster-type parliamentary democracy. The NLC members did not return to the barracks, but retired from the army to their comfortable homes in Accra to play tennis, golf and to ride horses; only Yakubu returned to the Police Force and became the Inspector General of Police (Pinkey, 1972);

Conclusion

The NLC era witnessed the reversal of the previous development strategy and efforts to impose a new economic and political order on Ghana. The coup that led to the initiation of this process itself seemed to have been mainly driven by dissatisfaction within the army, the decline of living standards and personal insecurity, but was linked to the larger socioeconomic problems of the society. The military-*cum*-police regime believed that the country's economic problems arose from socialist ideology and what can be regarded as patrimonialization of the state. They reduced the state's role in the development process and sought to rely on the private sector to activate the development process. They introduced IMF supported stabilization programme and sought monetary reform and fiscal probity. They devalued the domestic currency and liberalized trade and the economy. They sold or abandoned many state-owned enterprises they judged to be uneconomic and unprofitable. They brought Ghana back into the western orbit and re-introduced multi-party democracy. The 1969 Constitution was carefully designed to make it difficult to return to dictatorial rule. They drew up a favourable investment code to attract foreign investors into the country, but also sought to create and strengthen a national productive bourgeoisie. In essence, state-society relationship was re-constituted, and a new macroeconomic management system, as well as institutions, were established.

But the rapidity with which they disposed of projects without proper assessment of their contribution to long-term economic development gave the impression that in their quest to change the Nkrumah economic legacy they might have thrown out the baby with the bathwater. Nothing could be more damaging for Ghana's progress at this moment than the outright neglect and abandonment of such expensive, ongoing development projects and machinery.

The NLC's rule also re-constituted the pre-independence Rightist opposition groups made up of chiefs, lawyers, professionals and the bourgeoisie and gave them power. The actions to reverse the Nkrumah economic and political order only serve to show the degree of elite conflict over development strategy. It suggests that the rules of the game and institutions were not yet established, let alone agreed upon. Ghanaian elites, so far, have been unable to establish a "settled politics" or an acceptable political framework, let alone, a single development strategy. By deliberately marginalizing the Left and reversing its programme, the NLC did not bridge the elite divide that emerged during the independence struggle. Rather, it inflamed it, and this was to lead again to a reversal of development strategy in the future.

Notes

1 Nkrumah regarded Ghana's independence as a prelude to the total liberation of Africa from colonial domination. In 1966, Nkrumah was talking about sending Ghana Army to Northern Rhodesia, now Zimbabwe, to fight the white minority regime of Ian Smith which had declared a Unilateral Declaration of Independence (UDI) from Britain in November 1965, indicating that the whites were there to stay forever.

2 For example, in an interview to the *Legon Observer* of 30 August 1968, one of the traditional chiefs, Nene Azu Mate Kole, maintained that, just before independence, it was generally believed that the well-educated people and the chiefs of substance, could operate a system of government and rule Ghana like gentlemen, but in the course of events, the lower strata captured political power. He also warned about the possibility of this history repeating itself.

6 The Failure of the Bourgeois Democratic Revolution and the Resurgence of Nkrumahism (1969-1979)

...if gold ruste, what shal iren do? For
if a priest be foul, on whom we truste,
No wonder is a lewed man to ruste—
Geoffery Chaucer, *The General Prologue*

Introduction

In October 1969, the National Liberation Council (NLC) disengaged itself from politics. It seemed satisfied that the socioeconomic and political order that Nkrumah established was at least restructured, if not dismantled. On handing over power, the NLC noted that Ghana had achieved a sufficient measure of economic stability and the conditions for new growth had been restored, but much had to be done by the incoming civilian regime (cf. Killick, 1978:56). As far as future development was concerned, the NLC was confident that the new regime would continue its neoliberal economic policies and never subject Ghana to what it saw as a socialist pattern of development and dictatorial rule. Yet, in 1972, barely three years after handling over power, the army staged another coup and ousted the civilian regime it had installed.

This chapter discusses the rule of the civilian regime, the Progress Government (1969-72), and the factors that led to its overthrow. It also focuses on the rule of the military regime (the National Redemption Council and the Supreme Military Council (NRC-SMC 1972-1979)) that replaced it. The chapter argues that the main reason behind the 1972 coup, was the civilian regime's introduction of an austerity budget that sought to curtail the privileges and allowances of civil servants and the military as a way of addressing the crisis in the economy. Somehow indirectly, the impact of the budget amounted to a further attack on Nkrumah's socioeconomic order, which provided employment and free education and medical services for many Ghanaians. Consequently, the regime became unpopular and was overthrown. The coup thus brought into power military elites more interested in the maintenance of a re-distributive social production system. These new military leaders, however, ended up patrimonializing the state, thus undermining its the bureaucratic integrity and weakening its developmental capacity. This set in motion a self-reinforcing socioeconomic decline from the mid-1970s to 1979, when the regime was violently overthrown by junior military officers.

The Second Republic (1969-1972): A Bourgeois Democratic Revolution?

The Second Republic was different from the first in at least two respects: personnel and economic orientation. Its leadership was made up of mainly liberal elites, led by Prime Minister K. A. Busia, a professor of sociology. Prominent among Busia's cabinet were ministers such as J. H. Mensah (Minister of Finance), J. Lamptey (Minister of Defence), William Ofori-Attah (Minister of Education, Culture and Sport), Victor Owusu (Minister of Foreign Affairs), and K. Safo-Adu. These ministers were all drawn from the middle class and belonged to the group of elites that lost in the independence struggle. Of the 17-member cabinet, eight were lawyers, four were businessmen, and four were teachers. Like the former NLC members, nearly all had some education in Britain and the United States, which helps to explain their liberal outlook as far as economic orientation was concerned. The Busia regime was a reincarnation of the UGCC (from which Nkrumah broke), the National Liberation Movement (NLM) and the United Party, (UP)—the main opposition parties during Nkrumah's CPP era (cf. Austin and Luckman, 1975).

By the time the Busia Government assumed office, the Ghanaian economy had entered what can be described as an economic recession. The radical "dismantlement" of Nkrumah's economic order and the "reversal" of policies brought a sudden rupture with the past state-led socioeconomic development. This affected the performance of nearly all facets of the national economy, the main symptoms being massive unemployment and slow economic growth. But the Busia Government was convinced that a return to the so-called socialist production system was not in the interest of the nation. To the regime, market forces represented a better mechanism for managing the national economy.

Thus, the state's role, unlike in the Nkrumah era, was defined as mainly regulatory and not to be radically transformative of society. Again, unlike the CPP regime, Busia's Progress Government accepted social pluralism and sought compromise with all types of groups. It also sought the enhancement of civil rights and obligations of citizens as well as individual liberty. It did not regard the expansion of the state sector as political and social good in itself, but simply as a necessary expedient. The government was minimalist, concerned with the provision of what can be termed an enabling political framework: rational rules, predictability in the administration, protection of property and means of production, as well as incentives for promoting capitalist development. It accepted that the private sector and not the state-sector should be the engine of economic development. It did interfere in the economy, but only in areas where spontaneous foreign and private initiatives were lacking, for example construction of roads and the rural electrification programme.

The Busia Government, however, did not content itself with providing infrastructure and an institutional framework for capitalist development, but it also initiated policies that could enable Ghanaians and specifically indigenous capitalists to gain a strategic hold in the national economy. The government also sought to diversify agriculture to expand export earnings, and to attract foreign investors into the country. In its first phase of rule, the Busia Government showed some determination to make liberal democracy survive, but later became undemocratic in its relation to other branches of government, particularly the judiciary and workers' organization: the Trade Union Congress.

Soon after coming to power, the Progress Government introduced the "Open General Licence": a de-regulated system of importation. This brought a large import boom to society. In order to accelerate the Ghanaianization of the economy, the government in June 1970 passed the Ghanaian Business Promotion Act, which superseded the NLC's Ghanaian Enterprise Decree. Its main objective was to further protect Ghanaian businessmen from foreign competition and to develop an authentic productive bourgeoisie in Ghana (cf. Killick, 1978:57). In 1970, the government deliberately reduced taxes on businesses and further prohibited foreigners from engaging in retail trade. It also established terms for the withdrawal of foreign partnership in trading companies. Further economic space was created for Ghanaian businessmen through the prohibition of foreigners from areas such as the production of cement, road construction, produce brokerage, banking, printing (other than textile) beauty culture, advertising and publicity.

Ghana's economy during the colonial period had attracted many migrants from the neighbouring countries. The CPP government also maintained a liberal immigration policy given its pan-Africanist ideology and concern to make Ghana the leader of African Unity. Many migrants had thus come to settle in Ghana and some had not bothered documenting their stay in the country. Most of these immigrants were a source of labour for the cocoa industry, but some were also successful businessmen, mainly in retail trade, hotels, printing and transportation. In addition, most of their children were also beneficiaries of Ghana's egalitarian social system, which provided almost "across the board" free education and health services. There were also many successful Lebanese, Syrians and Indians in retail trade. As it is usually the case in many countries that have a significant number of foreigners, these immigrants and their situation did not go unnoticed by the indigenous population. These so-called illegal immigrants were said to be dominating retail trade, repatriating monies home and marginalizing Ghanaian businessmen. Consequently, the government enacted an Aliens Compliance Order, which stipulated that, by April 1970, all immigrants without valid residence and work permits should leave Ghana. The order was enforced within two weeks of 19 November 1969, and this led to the summary expulsion of some 155,000 businessmen, craftsmen, and unskilled workers, mainly from neighbouring African countries (Esseks, 1975:50).

The expulsion of illegal immigrants however, earned the Busia Government the displeasure and anger of the governments of countries such as Burkina Faso, Cote d'Ivoire, Togo, Mali, Niger, Nigeria and Benin, whose citizens were affected by the order. The expulsion measure was, however, popular at home, since it was carried out by the government on behalf of the Ghanaian business community. Conceivably, the state had now emerged under the Progress Government as an instrument in the hands of the indigenous bourgeoisie.

But it was not only on the West African regional level that Ghana's image suffered. Ghana was also isolated on the continent because the Busia Government violated the Organization of Africa Unity (OAU) and United Nations resolutions by seeking "dialogue" with the Apartheid regime of South Africa. Ghana's prestige as leader of Africa nationalism and a founding member of the Non-Aligned Movement also suffered. Ghana's relations with the former Soviet Union and Eastern Europe remained cold as the "Eastern projects" abandoned in the 1966 coup were not continued.

Ghanaians made inroads into the sectors exclusively reserved for them. They became transport owners, small-scale industrialists, bakers, owners of printing houses, etc. Unlike Nkrumah, Busia shared the belief that the indigenous business class could contribute to the development of society. The Progress Government's policies were nationalist, but in another direction: not state-ownership, but indigenous private ownership of the means of production. Again, unlike Nkrumah, who might have distrusted Ghanaian businessmen and perhaps regarded them as imperialist stooges and incapable of genuine national development, for Busia, they were a capable category that needed to be encouraged by the state so as to gain strategic control over the ownership and management of the national means of production. By trying to be liberal democratic and deliberately aiding the ascendancy of the productive indigenous bourgeoisie, the Progress Government was attempting what can be called, for lack of a better word, a "bourgeoisie democratic revolution". Like the mercantilist leaders of premodern Europe, Busia recognized that the economic kingdom of Ghana depended on the economic advancement of the bourgeoisie and the citizens, in general. Thus, the Progress Government's development programme not only envisaged the maintenance of the NLC economic heritage but also sought to deepen it.

Although the Busia regime favoured a free market, it did intervene in the economy. In 1970, it established the Ghana Agricultural Development Company with four Italian firms to develop poultry and livestock industry in Ghana. This venture was akin to the state farms initiated by the Nkrumah regime, except that, at this time, it was a joint venture. The government also set up a rice farming project in northern Ghana and also sought scientific and technical advice from the World Bank to rehabilitate the cocoa industry. It also encouraged educated private individuals to enter agriculture, and it provided monetary incentives such as special retirement benefits to public servants who sought to take up agriculture.

One of the achievements of the CPP regime, as noted in Chapter 4, was in education. The expansion of education in the late 1950s and early 1960s resulted in the production of a large number of school leavers (mainly middle and secondary) looking forward to employment in the towns and cities. Rural-urban migration had now become a problem in Ghana, in that it not only deprived the rural societies of human resources and undermined their economies but also put pressure on social services in the urban areas and created unemployment. The Progress Government sought to redress this development imbalance by redirecting the focus of development on the rural areas. This was evident in the manifesto of the Progress Party in August 1969, and in Mr J. H. Mensah's pronouncement that the principal task of the Progress Party was to search for social justice in Ghana, in particular to raise the levels of rural income as a whole vis-à-vis that of the urban centres (cf. Killick, 1978).

Thus on assuming power, the Progress Government launched a rural development programme and instituted a Development Fund supported by a 10 percent development levy on incomes to underwrite it. In 1970 alone, the government constructed 450 miles of feeder roads. Rural health centres were also expanded and medical personnel and teachers were posted to the rural areas. Furthermore, advances were made in rural electrification, potable water supply and an improvement in rural housing through loans for building materials (cf. Gyimah-Boadi, 1989). In the urban areas too, and particularly in Accra, the Busia government initiated a programme that was meant to modernize the country's sanitation system. For example, the disposal of toilet waste was to be centralized and brought underground to avoid the situation whereby every household and office had to privately dispose of its waste as it chose. This long-term programme was, however, to be abandoned when the government was overthrown in 1972. And up to today, Ghana does not have a centralized underground sewage and public waste disposal system. Consequently, many cities are unable to provide a clean and decent environment for the inhabitants. Waste now is a major polluter of the cities' environment.

Yet some of Busia Government's policies brought about some positive changes in the general economy, as well. Between 1969 and 1970, there was recovery in output in manufacturing, fishing and food agriculture sectors. The value of manufactured goods grew in constant prices by an annual average of about 25 percent compared with an average yearly increase of 13 percent between 1962 and 1966. Industrial efficiency also was improved and 19 of the 20 industries under GIHOC were making profits. For the same period, fish output increased by about 45 percent. Maize output also increased from 206,000 tonnes in 1969 to 900,000 tons at the end of 1970. The GNP grew at a rate of between 3.0 and 4.0 percent throughout the rule of the Progress Government (Esseks, 1975:51). This modest economic performance could have befitted a liberal democratic regime that did not pretend to radically engage Ghana's underdevelopment, as the CPP had done. But such a performance left many social problems unresolved: unemployment continued to soar, the cost of living remained high, etc.

While the free-for-all "Open General License" system of importation brought a large amount of imports to the society, it drew the Ghanaian economy into new balance of payment problems. In 1971 alone, the value of imports exceeded that of exports by 16 percent (Petchenkine, 1993:52). Furthermore, much of Ghana's short-term foreign debts amounting to about half the value of annual exports was due for payment. To aggravate the situation, in 1971, the world price for Ghana cocoa, the lifeblood of Ghana's economy collapsed, leading to a reduction in national earnings.

The regime decided to make some hard policy choices to accommodate the crisis. Thus in 1971, it introduced an austerity budget which, from the benefit of hindsight, anticipated some of the elements of the structural adjustment programme going on in Ghana today. It also accepted an IMF stabilization programme and devalued the currency by 90 percent in local terms. Interest rates were raised, petrol and kerosine prices were increased, the free medical care system introduced by Nkrumah was abolished, a tax of 10 to 25 percent of the value of imported goods was imposed, and some imports—particularly luxuries—were curtailed. Prices of basic commodities were also increased. Civil servants were to lose some of their amenities (car and rent allowances), likewise, the army officers, whose privileges and amenities were expanded by the NLC. A student-loan scheme (interest free and payable over 12 years) was to be introduced to replace Nkrumah's free college and university education. These austerity measures would raise the general cost of living throughout the country.

Furthermore, the government decreased public expenditure by contracting state-sector employment, particularly the civil service, which, in its opinion, was over-staffed. Its retrenchment exercise affected some 568 civil servants (Petchenkine, 1993), the bulk of whom were Ewes and Gas, and not even one was from the Akan ethnic group to which Busia and most of his ministers belonged (Pellow and Chazan, 1986:55). This ethnic bias did not go unnoticed by the general public.

The Progress Party won the 1969 election not so much of because of its liberal ideology (since the opposition was also equally liberal) but because of Akan (the greatest ethnic group from which Busia came from) support. The National Alliance of Liberals (NAL), the main PP opponent, also had its greatest support from the Volta Region because its leader K. A. Gbedemah was an Ewe. The PP won no seats in the non-Ewe areas of Ghana, and, likewise, the NAL won no seats in the Akan areas (Frimpong-Ansah, 1991:100). Because of this ethnic bias, the Progress Government did not enjoy enough nationwide support. When the government retrenched mostly Ewes and Gas, it was accused of favouritism on the basis of ethnicity, friendship and kinship. This was very damaging because the regime also had no single Ewe in its cabinet. This "ethnic motif" made the regime unpopular, at least among the Ewes who saw the retrenchment exercise as a deliberate attempt to curb what was often referred to as the Ewe dominance of the public sector. In spite of all these unsettling accusations, which were widely publicized in the media, the Progress Government appeared committed to democratic governance. The mass media remained largely free from government control and government tolerated some of the criticisms, even though they were directed against it.

The test of the government's commitment to democracy came in the famous court case of Mr Sallah versus the Busia regime. Mr Sallah was the Executive Manager of the state's monolithic trading company, the GNTC, during Nkrumah's rule. He retained this position during the NLC rule, but was now dismissed by the government. Mr Sallah took his case to the Court of Appeal, which ruled in his favour and requested the government to reinstate and compensate him. The outcome of the case infuriated the government and was further taken to the Supreme Court, which also returned the same verdict on constitutional grounds. In liberal democracies, the Executive branch of government should have accepted the verdict of the Judiciary, but the Busia government did not, and, rather, accused the Judiciary of setting up opposition to its rule. In the eyes of the public, this confrontation confirmed the Progress Government's anti-Ewe stand, since Mr Sallah was Ewe.

As a result of the court's decisions, the Progress Government instigated a series of noisy demonstrations under the banner of slogans such as "Down with the lawyers!" and "Let's purge the courts!" and Busia himself, in a radio and television broadcast, threatened to persecute the court officials who decided "to play political games" (Petchenkine, 1993:49-50). These threats brought some political compromise. The judiciary became more cautious in subsequent cases. The government also refrained from further retrenchment it had planned. But this event can be interpreted as civil servants' resistance to the government's retrenchment exercise, since there were others beside Mr Sallah seeking to bring their cases to the courts.

Domestic Economic Adjustment and the Deterioration of State-Civil Society Relationship

As already noted, by 1971, Ghana was once again in an economic crisis, and the Busia Government reacted by "re-ordering" its economic priorities. But the actions it took to deal with the crisis infuriated many urban social groups and civil associations that had earlier rendered support to the Progress Government. Their reactions were swift and dramatic. University students were indignant over the government's proposal to impose a loan scheme and demonstrated; civil servants hated the government's retrenchment exercise; the judiciary was in disarray over the famous Sallah case; workers went on strike to protest against unemployment and the rise in the cost of living induced by the currency devaluation. At Samreboi, for example, in the African Timber and Plywood Factory in the Western Region, there was a ten-day stoppage of work accompanied by violence. The state's security forces' reaction to this strike left three people dead.

As industrial unrest spread throughout the major cities, the government froze the assets of the Trade Union Congress (TUC), an umbrella organization of workers' unions, and, under a Certificate of Urgency, enacted a new Industrial Relation Bill in an effort to dismantle the TUC, although not its constituent autonomous unions. The government also threatened the leadership of students. As social discontent grew, so did pro-Nkrumah sentiments. In the cities, people began to sell the portraits of Nkrumah, who had, since the 1966 coup, been living in exile in the West African country of Guinea. In August, 1971, the government rushed a bill through Parliament that made the advocacy of Nkrumah's return a criminal offence.

The austerity budget also affected the police and the military. When the Minister of Finance presented the budget to the parliament, he argued the need to adjust the domestic economy to realities of the era. The budget abolished a number of allowances for army officers and also reduced expenditures for the Armed Forces. Free water and electricity supplies were stopped. Telephone allowances were reduced; rent payment was increased from 7.5 percent to 15 percent of officers' salaries. It was also rumoured that further cuts were underway. During the budget presentation, the Minister of Finance told the parliament that the government had been faced with the problem of what to do about a level of defence expenditure which was clearly burdensome for a small country such as Ghana (cf. *Daily Graphic,* 29 July, 1971). The cost of maintaining the military was by now consuming about 10 percent of total annual budget (cf. Etsiah, 1998/90:41), but there was no enemy in sight to justify the "huge" expenditure on the military institution. Thus in the new budget, the government reduced the expenditure on the Ministry of Defence from the 45 million cedis of the 1970 Budget to 40.4 million cedis, of which 31.4 million cedis was for recurrent expenditure and 9 million for capital expenditure (cf. Bennett, 1975:300).

National armies exist primarily to serve as instruments of foreign policy objectives (i.e. to defend a country against foreign aggression or to initiate a war) and their existence is easily justified by the existence of competing states. In the absence of state-competition, the army's role as an instrument of defence was still unclear in the postcolonial era. Apart from a brief campaign in the Congo crisis of the 1960s (a role that was rather mainly to keep peace), Ghanaian soldiers had never mounted or halted an invasion and had never fired a shot at a soldier of another nationality. Ghana's army thus appeared to be an "idle" institution. It is no wonder that in the wake of economic crisis, civilian politicians began to question the utility of the army to the nation.

Meanwhile, since coming to power, the Progress Government had been using the army to play a more positive role in national development. Beginning somewhere in April, 1970, the military was supporting the Border Guard Unit of the police in anti-smuggling campaigns along Ghana's national borders and Army-Police operations against crime were also initiated. The Army Medical Corps was also active in anti-cholera campaigns and the Air Force provided relief aid in the northern regions of Ghana. In spite of these functions, some sections of Ghanaians felt that the army could still be used in a more productive way to justify the large expenditure the state was incurring on it. For example, in mid-1970, the National Union of Ghanaian Students (NUGS) suggested that the technical branch of the army be used in agricultural and construction work. But this was rejected by the then Minister of Defence, Kwesi Lamptey, with the argument that the army would only be used for defence purposes. A year later, when B. K. Adama became the new Minister of Defence and when the country was in financial straits, the Minister of Finance announced the government's new realism that it should be possible to combine training with substantial contributions to national development. He also noted that the government had planned that whenever soldiers went out on training exercises they would seek to leave behind some positive development projects in the form of roads, sanitary buildings, etc. (cf. Bennett, 1975:301).

In most modern societies, military institutions are always subordinate to civilian power. This situation, however, did not exist in Ghana. The retreating military regime of the NLC did not reorganize the military in ways that would make their intervention in politics difficult in the future. Kraus' (1970:242) observation that "...a fundamental problem of future civil-military relations is how a civilian government can handle and maintain, in a subordinate status, a military and police which have been ruling directly and retired willingly", was to be prophetic. The NLC members did not go back into the army. The command structure of the army had thus passed on to new officers, some of whom had not been promoted by NLC and were disgruntled. Thus, any economic weakness to be exhibited by the civilian regime was likely to be exploited by the army, as one common feature of the military is that, once it intervenes in politics, it begins to redefine its role in society. It becomes politicized and sees itself as the guarantor of political stability and national integrity, and consequently acquires an irresistible taste for power. Thus, as Diamond (1988:24) notes, with each coup, the threshold of military intervention is lowered and new coups become more likely.

The cuts in financial allocations were not well-received by the army. The economic crisis and demonstrations continued throughout 1971. This accentuated the spectre of a military coup. The viability of Ghana' emerging democracy did not hinge only on bread and butter issues, but also on the degree to which the military was willing to subordinate itself to civilian power. Events in the country made the need for such subordination more urgent. The Prime Minister was said to have received a letter which threatened a further coup if conditions within the army did not improve (Bennett, 1975). At this juncture, B. A. Yakubu, the only member of the NLC to return to his former post, retired as head of the police. But, instead of appointing a new Inspector-General from within the police institution, the Busia Government appointed a lawyer civil servant, R. D. Ampaw, to succeed Yakubu. This angered the police also.

The government was thus becoming increasingly unpopular as a result of the way it was handling the economic crisis. Yet, at the same time, however, government officials were perceived as displaying conspicuous consumption patterns in a context of economic difficulties. One of the weaknesses of the Progress Government was that, perhaps as a result of his academic background and liberal outlook, Prime Minister Busia allowed his cohorts considerable latitude in the administration of the country. There were cases that some cabinet members who could argue persuasively carried the day and some were allowed to implement policy freely. This laxity in the decision-making process permitted government officials to give preference to their supporters in awarding contracts and government projects. Soon social evils such as corruption, favouritism and a "matter-of whom-you know" tendencies became discernible among the civilian politicians. All of these undermined the government's credibility.

It is often said that when the army enters politics, so does politics comes into the army. Once army officers have controlled the state and directly made political decisions, the military's "insulation" from politics, and also the supremacy of civilian power, become compromised. The 1966 coup and subsequent actions taken by the NLC in many ways politicized the army. Some soldiers were promoted while others were not. The coup also disrupted the army's command structure. The members of the NLC did not go back into the army when they handed over power to the civilian regime of Busia. This means that they were not available to shield the civilian regime from possible military coups. The army was thus politicized before the Progress Government assumed power. The loyalty of many new officers was doubtful (Bennett, 1975:302-303) and this certainly worried the civilian regime.

In the mist of all these "noises", the Busia Government became less confident about its social base and worried about the possibility of another military coup. The economic crisis and the unrest in the urban areas became occasions for calls for a return to the "good old days" of Nkrumah's "re-distributive" social production system. Posters and pictures of Nkrumah were now openly sold in the cities, defying the government ban.

In order to pre-empt a possible coup, the government made further changes in the command structure of the army. Trusted officers (mostly Akan) were brought to command positions, which was seen as dominated by the Ewes. By early 1972, in a military that had a large number of Ewe officers, only one Ewe—Col. Tevie—was in any position of senior rank (cf. Bennet, 1975:305). In this reshuffle of command structure, also came the fateful appointment of Lieutenant Colonel Ignatus Kutu Acheampong as leader of the First Infantry Brigade, which provided an opportunity for a military coup against the regime. This interference in the internal affairs of the military was an attempt by the civilian politicians to subordinate the military to their rule and to forestall future coups. Ironically, however, it precipitated the coup.

On 13 January, 1972, after barely three years of civilian rule, the military, led by Lt. Col. Acheampong, once again seized key installations, including the radio station, the airport, the cable office and the seat of government, and announced that they had taken over power. The Prime Minister was in London undergoing medical treatment at the time of the coup, reminiscent of the time of the 1966 coup when Nkrumah was on his way to Hanoi and Peking. The rest of the army supported the coup. Nobody defended the 1969 Constitution, the basic document for political behaviour of the land, or for that matter, the constitutional government. The coup shattered the "bourgeois democratic revolution" and any attempt to provide minimal conditions or institutions for capitalist development. Once again, the state had come under military rule.

The National Redemption Council/The Supreme Military Council (1972-1978): NeoNkrumahism

The 1972 military coup resulted in the regime of National Redemption Council (NRC). The Constitution of 1969 was abrogated; parliament was once again dissolved and all political activities were banned. Ghana again came to be administered by military decrees. In a speech announcing the coup, Lt. Col. I. K. Acheampong noted that the army had intervened in politics to "redeem" Ghana (hence the name National Redemption Council) from its "problems": arbitrary dismissals, arrests and detention without trial, economic mismanagement and incompetence. Lt. Col. Acheampong also claimed that the problems that the Armed Forces had tried to eradicate in 1966 had re-surfaced under the civilian regime (cf. *Speeches and Interviews by Colonel Acheampong*, Vol. 1, n.d.). He further maintained that Busia's economic mismanagement had created suffering throughout the country. In this speech reminiscent of statements made by the leaders of the 1966 coup, he accused Prime Minister Busia of interfering in the affairs of the military and policies. As he noted:

> the first people Busia put his eyes on were the Armed Forces and the Police. Some Army and Police officers were dismissed under the pretext of retirement. Some Officers were put in certain positions to suit the whims of Busia and his colleagues. Then he started taking from us the few amenities and facilities which we in the Armed Forces and the Police enjoyed even under the Nkrumah

> Regime. Having lowered morale in the Armed Forces and the Police to the extent that officers could not exert any meaningful influence over their men so that by this strategy coming together to overthrow his Government was to him impossible, he turned his eyes on the civilians (*Speeches and Interviews by Colonel I. K. Acheampong*, Vol. 1, p. 2, n.d., 1973?).

That the NRC was to be subjected to reshuffles, manoeuvres and would be eventually personalized was somehow discernible at this early stage of its rule, even though few people perceived the regime that way. This was evident in the constitution of NRC membership, which did not go smoothly. In his broadcast to the nation, Lt. Col. Acheampong said his cabinet would have a broad social base including political leaders, representatives of religious bodies, the Council of Chiefs, trade unionists, representatives of farmers' associations and local governments. This statement was, however, meant for the gallery, because soon after it, Col. Acheampong announced that the NRC would consist only of the military. After some hesitation, the first eight members of the NRC emerged as: Colonel Acheampong (as Chairman); Major Kwame Baah (Second in Command of the Fifth Infantry Battalion in Accra); Commander Boham; Major Agbo (Second in Command of the First Infantry Battalion in Tema); Major A. Selormey (Second in Command of the Calvary Regiment); Lt. Col. Barnor Lt Col. Benni; and Mr E. N. Moore, (the Attorney General), as the only civilian.

The vacillation in nominating the regime's members indicates that Col. Acheampong had not developed a network of like-minded conspirators. Indeed, Bennett (1975), who has studied Ghanaian military regimes, has suggested that the NRC appointees had nothing in common except that they were new officers who had now found themselves in positions vacated by older officers and did not belong to the generation of officers that made up the NLC. These NRC appointees were simply willing to accept political office because of the opportunity created by the coup. Most of them, including Col. Acheampong, had attained their positions through military career, rather than education. Some of them suffered in terms of promotion under the NLC rule because they were not affiliated to any member of the group that brought off the 1966 coup against Nkrumah, and, as such, were by-passed for the NLC's promotion. Col. Acheampong was said to have complained earlier that Afrifa of the defunct NLC had been a Company Commander under him, but had retired as Lt. General. Indeed, contemporaries of Acheampong, such as Twum-Barima, had been Lt. Colonels by 1967 and had attained Brigadier status, while he remained only a

Colonel. This resentment was alluded to in a speech to students shortly after the coup, when Col. Bernasko, one of the NRC Regional Commissioners emphasized that the "escalating promotions" that occurred under the NLC were unlikely under the NRC (cf. Bennett, 1975:305). The coup then had become a means through which the military was resolving its internal contradictions and grievances.

In 1966, when the soldiers came to power, they made it clear that they would be returning to the barracks. They also humbly acknowledged their inexperience in managing the affairs of the nation and chose to collaborate with the bureaucrats. The strategy allowed the NLC to bring bureaucratic competence to bear on the ruling process. Furthermore, because soldiers did not dominate the public sector, they were not unduly exposed to corruption and other vices that political positions can easily generate. In the case of the NRC, however, none of these conditions held. Col. Acheampong and his colleagues were evasive about the date for returning power to civilians. Soon after the coup, the NRC passed Decree No. 2 establishing a consultative committee on economic matters. But, four days after the committee met, the NRC made it clear that it intended not to share political power and decision-making with civilians (Petchenkine, 1993:59). These consultative committees lost effective influence on the regime. The regime saw the bureaucrats as necessary for implementing policies, not for policy-making.

The NRC's decision to govern and make policy alone thus brought a new phenomenon of military rule into Ghanaian politics. Whatever shortcomings were to arise from this strategy were to be compensated for by the principle of collective decision-making within the NRC. This principle was, however, violated as the regime came under personal rule. By refusing to share decision-making with civilians, the NRC seems to have believed that societal problems were clear and their solution simple, and all that was needed were "the right orders". This view led the regime to have a very simplistic and idealized perception of society as an object that could be easily moulded into a "good society", which, in its view, was to return to the "good old days" of Nkrumah, even though a new economic context had emerged in Ghana that seemed incapable of supporting such a generous, Nkrumah-type, re-distributive socioeconomic order.

The NRC also involved the military in public life. Many army officers were appointed or seconded to high positions in the ministries, government departments and other state institutions. Among these were, for example, the appointment of Commander Addo to the Cocoa Marketing Board, Commander Boham to the State Fishing Corporation, Colonel Akwagyram to the Ministry of Agriculture and Colonel Tetteh to the Meat Marketing Board. All previous civilian ministers of Ghana's regions were replaced by military personnel now known as commissioners. The main criteria for appointing soldiers to such strategic posts were determined largely by Col. Acheampong, and it is doubtful if merit was the main consideration. The NRC regime also allied itself with the police, though not as closely as the previous military regime, the NLC did. The domination of public institutions by the military was, however, to expose many of the officers to wealth, privileges and power, which they did not hesitate to exploit as public accountability was eroded in the political system as a result of the re-emergence of patrimonial rule.

A common problem that faces many military regimes is how to govern after power has been successfully captured. Those who usurp power illegitimately need to legitimatize their rule and establish recognition for their rule at home and abroad. One way of dealing with this problem, if the last chapter on the NLC rule is anything to go by, is to denounce economic policies and development plans of the previous regimes as the cause of existing crisis and to seek allies abroad by revising foreign policy or to assure the outside world that things remain the same.

Foreign Policy: The Return of Eastern Europe and Ideals of Pan-Africanism

In seeking a favourable image for itself, the NRC re-defined Ghana's external relations as based mainly on "non-alignment" and "positive neutralism". The notion of dialogue with South Africa, championed by the previous government, was abandoned and Ghana was to resume its prominent role in African affairs. The NRC re-established links with the socialist countries, which Ghana broke as a result of the 1966 military coup. A ten-member Committee for Economic Co-operation with Socialist Countries (CECSC) was formed, charged with the responsibility for revamping trade, scientific and technical co-operation between Ghana and Yugoslavia, Poland, the former Czechoslovakia and Rumania. Relations were resumed with the People's Republic of China, and the Embassy of the former Soviet Union in Accra regained the influence it lost since the 1966 coup. The government hoped that an association with Eastern Europe would give it foreign support against the West and also lead to the rehabilitation of Eastern projects abandoned by the NLC and the Progress Government. By these actions, the NRC returned Ghana to the socialist camp that it had left since the 1966 military coup. Eastern European countries welcomed this gesture, but were less enthusiastic than before about what they saw as inconsistency in Ghana's foreign policy.

Ghana's commitment to the ideals of Pan-Africanism, which had diminished after the overthrow of Nkrumah, was also re-kindled. As Acheampong noted:

> Our foreign policy will, first of all, be based on a vigorous and dynamic African policy. We intend to foster the closest and most cordial relations with all African brothers still not free and who suffer deprivation, degradation and indignity on the continent of their birth. Accordingly, we shall vigorously support the eradication of the last remnants of colonialism and racial discrimination from the African continent... We subscribe unreservedly to the Charter of the OAU, and it is our resolve to lend effective support to efforts aimed at making it an effective body for the expression and realization of African aspirations (*Republic of Ghana: Our Task*, 1972).

The NRC sought to regain for Ghana the significant role it had played in African affairs. Consequently, Ghana held the 21st Session of the Co-ordinating Committee for African Liberation, at the end of which a 26-point Accra Declaration on African Liberation was issued. At the ideological level, the NRC launched for distribution throughout Africa a book of selected speeches by Nkrumah. It re-established diplomatic relations with Guinea, despite the non-payment by President Sekou Toure of £6 million that Ghana (under Nkrumah) had lent to Guinea. The regime also issued a 19-page Charter of Redemption that denounced all forms of imperialism and exploitation and reaffirmed Ghana's commitment to the African liberation struggle, African unity, and the realization of a continental union government. All Ghana's outstanding contributions to the OAU were paid promptly (Etsiah, 1989/90:46-47). The NRC also successfully negotiated the return from Guinea the remains of ex-President Nkrumah for national burial (Bennett, 1975:310).[1]

Furthermore, the NRC regime decided to advance Nkrumah's dream of African unity: the "United States of Africa". An African High Command had been the brain-child of Nkrumah and his CPP regime, that is, the establishment of a continental fighting force of troops contributed by all African states to support and defend the sovereignty and unity of African states. This idea was in vogue in Ghana in the early 1960s, but never got off the ground, probably because of opposition from or indifference of other OAU member states which feared Ghana's aspiring regional "hegemony" or the loss their sovereignty and independence if the idea became a reality. The NRC, however, revived this idea and established an African Liberation Information Centre. Through this centre, the regime gave publicity to liberation struggles in South Africa, Namibia and Zimbabwe, and whipped up domestic support for its foreign policy objectives. In addition, students from these African countries were invited to Ghana to undertake studies at the expense of the Ghana government. Furthermore, the NRC regime indicated its commitment to a regional economic union, the Economic Community of West African States (ECOWAS), which was only a project at the time.

The NRC also improved relations with Ghana's neighbours and other countries whose citizens had suffered since the implementation of the Aliens Compliance Order in 1972. The improvement in Ghana's relations with these countries was expressed in the visits made by a number of African heads of state to Ghana. Likewise, some Eastern European leaders also visited Ghana (Petchenkine, 1993:68-69). Thus, Ghana re-asserted its leading position in African liberation affairs. Although the regime did not formally cut ties with the West, the re-establishment of links with the Socialist countries in the Cold-War era and the NRC's "anti-foreign capital" policies left no shadow of doubt that "neo-Nkrumahism" had emerged in Ghana's foreign policy.

Economic Policies: The Development of a Commandist Economy

The NRC regime did not face any immediate legitimacy crisis mainly because Busia's policies had antagonized a large section of the politically significant urban groups. Nevertheless, like all military regimes, the NRC was aware that, in the absence of a social base and electoral mandate, sooner or later it had to justify its illegitimate actions. The NRC's way of garnering popular support for itself was to appease all groups that had been angered by the Busia regime. This, in a sense, meant a return to a more or less re-distributive social productive system. One of the weaknesses of the NRC regime, which was to play a major role in its down fall, was that it offered less a rigorous understanding of Ghana's economic crisis. It preferred to simply attribute much of what ailed Ghana to Busia's neoliberal policies and even perhaps to the wickedness of Busia and his colleagues. It this regard, the NRC simply decided to reverse foreign and domestic policies.

The NRC regime had come to power mainly through the exploitation of social discontent generated by previous austerity policies. As such, it had to address the root causes of this discontent. Indeed the economic problems that confronted the NRC were no different from the ones that the Progress Government faced. The difference was only in how these problems should be addressed. The NRC understood the cause of this problem as mainly stemming from the neoliberal economic policies pursued by the Busia regime. Consequently, it assaulted them.

In a provisional-budget introduced for the period from January to June, Col. Acheampong noted that the Progress Government had preferred a doctrinaire attachment to a free market economy which, meant that Ghanaians had to leave their fate in the hands of market forces whose operations were beyond control. His military regime, he maintained, intended to control, rather than be controlled, by the economy. In a speech to the staff of the Ministry of Finance and Economic Planning on 7 February, 1972, Col. Acheampong announced that "the political frame of reference which has guided actions and advice in the past... must be cast into the rubbish heap of history" (see *Speeches and Interviews* by Acheampong, n.d.; 31).

He also noted that his government's policies would mark a departure from *laissez-faire*—the so-called free market economy of the immediate past—and there would be the institution of effective planning in the allocation and utilization of resources (cf. Etsiah, 1989/90:43). Ghana, at this point cannot be said to have evolved a clearly specified development ideology that was acceptable at the national level. The coup manifested this ideological conflict. As Col. Acheampong noted:

> In the fifteen years since Ghana attained independence, we have gone through two separate and *diametrically opposed phases of development*. There was a period when all development was, so to speak, invested upon the state. From 1966, however, an attempt was made to reverse this and in the last three years, we saw a government which believed in total reliance upon the private sector and a free market economy. Since January 13th [1972] the question has been asked often as to what the principle of economic development will be. My answer is simple. I have said that my aim is to use state power to capture the *commanding heights of the economy* and to control it in such a way that its development will not go to benefit only a few handfuls of well-placed Ghanaians (*West Africa*, 23 June 1972:781; emphases added).

What this amounts to in simple terms is that the NRC was seeking to enthrone Nkrumah's economic and political order which emphasized state-propelled development, redistribution politics—what in this study is referred to as Nkrumahism. But the diametrically opposed development strategies that Col. Acheampong spoke of are not without substance. Indeed, Col. Acheampong brings out one key issue raised in the theoretical discourse in Chapter 2: that no nation can effectively sustain growth and development if governments continuously change the institutional framework or the rules of the societal game, for such changes introduce unpredictability and uncertainty in the socioeconomic environment. But Col. Acheampong's solution to Ghana's development problem, that is, to use state power to the gain the "commanding heights" of Ghana's economy, simply entailed dismantling the capitalist economic order that the NLC and the Busia regime sought to construct. The ideological direction of Ghana thus still remained a contested agenda.

The capturing of the "commanding heights" of the economy began with the cancellation of the IMF stabilization programme that the ousted Progress Government was following. The Ghanaian currency, the *cedi*, was revalued in February, 1972, by 42 percent and import licences and foreign exchange regulations were re-instituted. A number of consumer goods were subsidized and price controls inaugurated. The economy had once again come under state regulation. The army officers and most of the civil servants dismissed by the Progress Government were reinstated. The allowances and amenities of the military, police and civil servants were restored. Busia was to later describe the coup as an "officers' amenity coup" arising from their grievances at his efforts to save money, and lamented that the army officers, who—represented the best paid civil servants of the country—could not make a little sacrifice for the development of the country (cf. *New York Times*, 22 January 1972).

Furthermore, the college and university student loan scheme was abolished and student allowances increased. The daily wage of 1.29 cedis was raised to 2 cedis. An NRC decree enacted shortly after the coup annulled the rural development tax. Unemployment insurance was introduced into the social security system, which stipulated that, in the first two months after being let go, the unemployed could receive 50 percent of the previous wages, and, subsequently, could receive 20 percent of the previous wages, per month. This scheme was unprecedented in Ghana. In 1973, the government followed this up with a new social security legislation of sick benefits for all government and formal private sector employees. The price of cocoa was also increased. The social groups that were antagonistic to the Busia regime were appeased. The fragmented worker TUC was legally re-constituted. By these actions, the NRC regime re-imposed Nkrumah-type, re-distributive social productive system. The "ability" of the government to deliver these social benefits further discredited the Busia regime and increased support for the new government. Consequently, the number of strikes declined from 79 in 1972 to 10 in 1973 (cf. Petchenkine, 1993:67).

The NRC also sought greater state control over the leading economic sectors in its bid to gain the "commanding heights" of the economy. The government expanded the state's participation in foreign or jointly-owned companies. Massive nationalization programmes were announced but the government was only able to increase its participation in large extractive companies controlled by foreign capital. Like the Nkrumah government, which was antagonistic to large-scale indigenous private economic activities, the NRC statement on the Investment Policy Decree (NRCD 325) and the Ghanaian Enterprises Decree (NRCD 30) noted that the government detested the concentration of shares in the hands of a few Ghanaian businessmen (cf. Asante, 1993:268).

The debt problem that had been a bane to the civilian government equally confronted the NRC regime. The NRC's solution was to unilaterally repudiate a substantial portion of it, saying that such debts to Western creditor nations were "tainted". Where negotiations were allowed, the government made it clear that the onus of proof that debts were legitimately contracted rested not with Ghana Government, but the creditor nation. Ghana was also to re-examine other credits and loans unilaterally and default on those that the regime judged to be illegal. Furthermore, Ghana was to make payments on debts accruing after 1966 as and when conditions permit. The regime also renounced the decisions reached at previous debt-rescheduling meetings with the Progress Government as invalid. All in all, the government repudiated debts incurred by predecessor governments totalling US$94.4 million (cf. Pellow and Chazan, 1986; Etsiah 1989/90:43).

The abrogation of these debts sent a message to western leaders that Ghana could stand up to her benefactors with all the consequences. But Ghana was not credit-worthy in the Western world, and was not to receive much western assistance, a factor that explains the relatively low foreign debt accumulated by the NRC-SMC regime. The absence of foreign aid could, however, be compensated for by the effective mobilization of social forces to overcome underdevelopment (as the old adage goes: "necessity is the mother of invention"), in that a nation finding itself at loggerheads with the outside world can "look" inward and initiate fundamental changes in productive processes and utilize its own resources towards self-reliance. Indeed, this seems to be what the NRC meant when it declared that it would not pay Ghana's debts. The regime made it clear that Ghana was to go it alone and not depend on western assistance and charity. Consequently, it proclaimed a policy of self-reliance for Ghana. In its *National Redemption Charter*, the NRC stated that its principal aim was "self-reliance" in economic, social, cultural and technological development.

As a demonstration of its intention not to rely on free-market principles, the NRC drew up a Five-Year Development Plan (1975-1980) to centrally plan the economy. The plan emphasized agricultural development to enable Ghana to achieve self-sufficiency in food production and raw materials for its agro-based industries. The regime proclaimed "Operation Feed Yourself" (OFY), a work ethic analogous to the CPP regime's "For Work for Happiness". Like the Nkrumah Government, the NRC set up numerous large-scale farms across the country. Prominent among these were the Ghana National Rcconstruction Corporation (GNRC) and the Regional Development Corporations (RDC) that were to supervize the implementation of the OFY programme. Massive investment in the form of finance capital, agricultural equipment, fertilizer and improved seeds was made in the state's agricultural sector.

This investment led to a dramatic, but brief, recovery in food production. In spite of low foreign investment in the country, the first two years saw significant improvement in the economy, mainly as a result of increased food production and prudent financial management, particularly of the external sector meant to prevent trade deficits as a counter to the repudiation of foreign debts (Frimpong-Ansah, 1991). The NRC regime enjoyed much popularity, at least from 1972 to 1974, mainly because of its bold nationalist stance on debt payment, the abrogation of dialogue with South Africa, and the re-introduction of a re-distributive social production system, but also because of increases in agricultural output. The popularity was, however, a nine-day wonder. After an initial euphoric attempt at self-sufficiency, the regime lost out in continuously implementing its programme. Implementation capacity became weak mainly because the state later came under the crippling effects of personal rule and neopatrimonialism.

From 1975 to 1979, the economy took a dramatic nose-dive and Ghana was to experience a calamitous decline, the scale of which has not yet been repeated anywhere in Africa. Certainly, Ghanaians were not expecting economic miracles from their leaders at this time. After all, the economy had, since the mid-1960s, been undergoing a gradual decline. So, few, if any, expected the self-reinforcing economic débâcle that was to engulf the nation from 1975 onwards, so that, by 1978, a shortage economy had developed whereby virtually every commodity was in short supply. In this period, the Ghanaian economy registered negative growth, huge balance of payment and budgetary deficits, and also an acute shortage of foreign exchange, imported consumer goods and essential agricultural and industrial inputs (machines, spare parts, and raw materials) for local production. Production in all the leading sectors of the economy declined (see tables in the Appendix). A modern economy based on production for exchange is a "mobile" economy. The transport system, once the pride of Ghana, was literally in ruins, limiting economic activity and mobility throughout the country. In this period, it made more sense for people to walk the vast span of Accra city than to wait endlessly for transport. Public workers increasingly refused to report for work, mainly for lack of transport and diminished earnings. Long queues of people waiting for food and transport in the scourging tropical sun were a sight few people in this era could ever erase from their memory.

With such economic paralysis, even the rich had to park their cars and reduce their living standards. The Ghanaian currency, the cedi, was "everywhere" but without much of its value. The purchasing power of a worker's wage fell by one quarter of what it had been in 1972. In many places along Ghana's borders, particularly the cocoa-growing areas, the cedi virtually ceased to be the main medium of transaction as the CFA franc currency of neighbouring countries was increasingly preferred. Hyper-inflation acquired a self-propelling momentum pushing itself into three digits and reducing real incomes to an appalling level. Malnutrition and infant mortality rates rose and the health system collapsed. The education system was also in disarray, evident in the frequent closures of boarding schools for lack of teaching materials, food and other logistics. Most Ghanaian-trained teachers left the country for Nigeria and other neighbouring countries, thus, depriving the schools of personnel (cf. Jeffries, 1989; Frimpong-Ansah, 1991). This situation compelled parents to provide private teachers for their wards and secondary school students to enrol in private informal schools popularly called "Extra Classes" organized by teachers seeking to make ends meet. The economic crisis spawned an extensive informal market economy, trade malpractices and many social evils (sometimes engaged in as a survival strategies), which came to be labelled *kalabule*[2] (see below). On this result of the decline there is little disagreement among scholars. But the key question is, why did the economy decline so suddenly?

In seeking an answer to this question, it is important to remember that Ghana's economic performance cannot be isolated from the dynamics of the world economy. In 1973, the Organization for Petroleum Producing Countries (OPEC) decided to increase prices for crude oil. This action inflicted damage on the economies of non-oil producing developing countries. In 1974, Ghana incurred a large oil import bill, and, hence, balance-of-payment deficits, which undoubtedly negatively affected national economic viability. Yet, in this same period, the international price for cocoa increased by 150 percent, which was an unprecedented record. World prices for other commodities such as gold and timber also increased, but Ghana was unable to offer increased volume of exports for the international market in this period to exploit the favourable opportunities created by the world economy, mainly because exports had shrunk in volume.

Though it is analytically difficult to determine a balance sheet for the oil bill and export receipts, and build a net explanation of this economic decline, there was evidence that internal contradictions caused the sudden decline and that the external environment only aggravated the situation. Some analysts argue that the buoyant prices of Ghana's export products more than offset the effects of the oil prices (see Rothchild, 1991:5; Jeffries, 1989:79). These domestic contradictions seem important because the extent of Ghana's economic crisis was far more than that experienced by Cote d'Ivoire and Togo—the two economically less-endowed neighbouring countries participating equally in the international capitalist system—where Ghanaians had to acquire basic commodities such as soap, toothpaste and sugar. The rapid deterioration of the socioeconomic environment in this period owes much to the patrimonialization of the state. The main preoccupation of patrimonial rulers, as argued in Chapter 2, is the survival of the leader and his regime. This imperative of political self-preservation entailed a developmentally counter-productive logic with its attendant pathologies which undermined the management capacity of the state and, ultimately, the promotion of development.

Since ascending to power, Col. Acheampong had showed little interest in handing over power to civilians. He was thus aiming to rule for life, or at least as long as possible. Like a typical patrimonial leader, Col. Acheampong took actions that would make his rule uncontested. He suspected the military and those around him of unseating his regime. He frequently reshuffled the NRC. For example, all those army officers who had held key positions in the second republic and who could pose a threat to Col. Acheampong's regime were sent abroad as ambassadors. This was another unprecedented development in Ghanaian politics since, ambassadorial posts traditionally go to civil servants of the Ministry of Foreign Affairs. To further consolidate personal power, on 29 October 1975, the NRC passed Decree No. 360, through which all the powers of the NRC were passed on to a new body: the Supreme Military Council (SMC). Acheampong became the chairman of this five-man SMC. Some commissioners were fired and the NRC membership was enlarged to include commissioners of the ministries and the regions, commanders of the 1st and 2nd Infantry Brigades, etc. The establishment of the SMC enabled Col. Acheampong to subordinate the NRC. Some members of the NRC who resigned were given generous retirement grants and loans that enabled them to settle comfortably or to become large-scale commercial farmers.

The patrimonialization of the state introduced patron-client relationships into the ruling process, with pathological effects on the state. This was manifested in the corruption of state administrators and the inability of the state to positively serve broader public interest. Import licensing and foreign exchange control mechanisms introduced to manage the economic crisis at the time became the means through which Col. Acheampong and his colleagues distributed patronage to their relatives, friends and political supporters. Any sort of criticism of the regime's performance or actions was interpreted as subversive, coming from "enemies" of the regime. To pre-empt potential coups, the regime allowed Ghana Army to participate in conflict management around the world, particularly the Middle East. But this did not deter coups.

On 14 July 1972, a plot to overthrow the NRC was uncovered and nine people were arrested. Their trial revealed that some of them were closely associated with the coup that brought the NRC to power but had been left out of the "distribution" of the spoils of office. A decree was hurriedly passed that made coups subversive acts punishable by firing squad or imprisonment from 15 to 30 years. The coup attempt made the regime unsure of its social base, hence society came under greater state control. The military and police mounted barriers on all major roads to locate the "enemies" of the regime. But such roads blocks became avenues for extorting money from mostly drivers and traders. Freedom of the press and association was restricted and private newspapers such as *The Pioneer* and *The Echo* which were critical of government actions, were refused import licences to carry on their activities and were later banned.

The personalization of the state led to increased priority being given to security matters. The police and the military were given preferential treatment and supplied with materials to cow them into submitting to neopatrimonial rule. This was evident in the profligate way in which scarce resources were dissipated on the state's security institutions. The Police Force was given armoured vehicles, Nissan Patrol cars, air-conditioned BMW cars and motor cycles. The government also bought Italian Jet Fighters and German Fokker F-27 transport planes for the Air Force. Between 1971-72 and 1972-74, budgetary allocation for agriculture declined from 6 to 5 percent while that of the defence rose from 8 to 9 (cf. Rothchild, 1980; Gyimah-Boadi, 1989).

One major feature of Acheampong's neopatrimonial rule that was to be a weakness of his regime was that he centralized major decision-making and defied expert advice. The NRC-SMC regime soon became a "secret" organization. It imposed a closure on debates on national issues and its policies simply ignored trends in basic data about the country's socioeconomic performance. The Ghanaian currency, for example, had been traded with other major currencies in the informal market at a rate far lower in value than at what was officially acknowledged. As in 1972, the economic crisis of 1974 called for some domestic policy re-adjustment to increase exports and use state resources prudently.

Even though the positive effects of devaluation on developing countries are still not clear, many analysts agree that devaluation would have been a good policy at this time, but the regime refused to do so. Instead of devaluation, the NRC re-introduced comprehensive import licensing, foreign exchange rationing schemes and controlled prices whose implementation was fraught with favouritism, nepotism and massive corruption. Devaluation was rejected because it was one reason Col. Acheampong gave for overthrowing the Busia regime in 1972. To adopt devaluation would be tantamount to accepting economic failure and pursuing unpopular policies that were likely to generate social discontent against the regime as they had against Busia. The regime was cornered; it refused to reform and held on to its "defective" policies. In order to arrest the shortages in the economy, the regime opened supermarkets through which it sought to sell consumer goods that were now in short supply to the public at government approved prices, but again, these became sources of massive corruption or *kalabule*.

The cocoa industry, the lifeblood of Ghana, was in chaos because the interest and incapacity of cocoa farmers to reproduce themselves under existing domestic marketing conditions were ignored. Between 1973 and 1974, when world cocoa prices went up, the CMB was unable to raise domestic prices to cocoa farmers to reflect the upturn in world market prices that could serve as an incentive to cocoa farmers to continue production or not to sell their produce in neighbouring countries. Cocoa farmers continued to be exploited through the low domestic prices set by the CMB for cocoa. Ghana was, however, not alone in this practice. Indeed Cote d'Ivoire and Togo were also similarly extracting surplus from peasants in this way. But, in these countries, the domestic prices for cocoa rose faster than in Ghana. For example, a 30kg or headload of cocoa in these countries was six times (in nominal terms) and four times (in real terms) higher than in Ghana (Jeffries, 1989:80). For 30 kg of cocoa, a farmer in Ghana received between 900 cedis and 1,200 cedis from smugglers who visited him on his farm, while when he took the same amount of cocoa to the CMB he was paid only 360 cedis. Besides, the CMB was buying the cocoa on credit and issuing chits to farmers redeemable in the future. In neighbouring countries, however, farmers were expeditiously paid and even smugglers could come into Ghana and advance money to farmers pending their harvest (Nugent, 1991). It is little wonder that Ghanaian cocoa farmers, especially those living near the national borders (i.e. in Brong-Ahafo, Western and Volta Region), found it rational to sell their produce in neighbouring markets.

Yet the smugglers did not only come from neighbouring countries. Indeed, the youths in these cocoa growing regions, who had nothing to do and had no skills, and could not even migrate to the cities because the cities had no employment opportunities, took up smuggling as occupation. Most became rich and sometimes disrespectful of traditional authorities. So profitable was the smuggling of cocoa that the police and the border guards were also actively involved. In 1977 alone, it was conservatively estimated that 40,000 tonnes of cocoa were smuggled out of the country, representing well over 12 percent of the total cocoa production (cf. Nugent, 1991). Through smuggling, the peasant cocoa farmer was waging a silent war against the state by refusing to sell to its agent, the CMB. The CMB was supposed to protect the interest of cocoa farmers but its activities rather showed the degree to which they could be exploited.

While accessibility to foreign markets might have kept the cocoa farmer along the national borders in business, this was, however, not the case with those in regions far remote from such markets. In the Ashanti and Eastern regions, where cocoa farmers are far removed from the national borders and thus could not sell in more competitive neighbouring markets, farmers reduced their investment in cocoa economy and channelled their energies into food production. Since food crops are more critical to everyday survival than cocoa, the price of food crops rose faster than that of cocoa, notwithstanding the administrative controls imposed by the state to keep prices low. The returns on investment in food production were far better than those on cocoa. There was, thus, a strong incentive, quite rationally, for cocoa farmers to shift increasingly to food production (see Table 3, in the Appendix for the relative trends in prices of food and cocoa). Although many factors such as overage cocoa trees, lack of agricultural inputs and a poor transportation network, might have affected cocoa production, government policy—or rather non-policy—was mainly responsible for the low cocoa output (see Table 1 in the Appendix, which illustrates the decline of cocoa output over the years).

The NRC-SMC regime was not unaware of this national economic disaster, but its solution was only a moral appeal to farmers and smugglers to refrain from what it termed "unpatriotic" activities. It refused to make new marketing policies for the cocoa industry, or reform existing ones to retain cocoa within Ghana's national borders. The regime diverted attention away from its own policy failures by claiming in the state-controlled media that some Ghanaians were "nation wreckers". Instead of conceiving these smuggling activities as rational within the prevailing socioeconomic context and offering price incentives to cocoa farmers to encourage them to increase production and sell to the CMB, the government defined smuggling as a criminal behaviour, punishable by imprisonment or death and resorted to administrative control of cocoa smuggling.

Army and guard units were deployed along the borders to check these smuggling activities but these units themselves became active participants in smuggling. For example, as this author witnessed on many occasions in the Volta Region, these security officers often charged a standard bribe of CFA 1,000 or 400 cedis for every headload of cocoa that was smuggled out. In an incident in one village, CMB agents themselves re-sold cocoa already purchased dircctly to smugglers from Togo. Although the regime publicly denounced smuggling activities, it was generally believed that some government officials directly benefited from them. There were allegations that some of them sent truck loads of rice and other basic commodities that they bought at controlled prices into neighbouring countries to sell for higher profits. This seems to be one reason for the execution of Major-General Utuka, the former head of the Border Guards and SMC member, in 1979. This behaviour in high places certainly lowered the morale of the border guards who were already poorly paid. Consequently, they actively participated in the activities that they were supposed to check.

Smuggling and bribing became almost standard practice for many state officials. Indeed, to be a border guard (as it was to be a customs officer in Ghana) was one of the most rewarding jobs one could aspire to in this period. But the use of soldiers and border guards to control smuggling imposed a high cost on the state (in terms of salaries, vehicles and other logistics), while the state was getting less and less cocoa to export. Yet, this unproductive administrative control was maintained because it served a patrimonial purpose: it engaged the military and diverted their energies away from engaging in coups against the regime.

Ghana had by now lost its historical position as the leading producer and exporter of cocoa, falling behind other cocoa producers such as Cote d'Ivoire, Brazil and Malaysia. The wisdom of the colonial Governor, Gordon Guggisberg, that the farmer should not be overtaxed to abandon production, seemed to be ignored. By failing to raise the producer price of cocoa to serve as a disincentive to smuggling, it can be argued, the state in this period, played no insignificant a role in "killing the goose that was laying the golden egg" for the national economy. This translated into a loss of foreign exchange for the state, limiting its ability to deliver on its promises, let alone engage in development. In 1977 alone, Ghana lost US$100 million in foreign exchange as a result of cocoa smuggling (cf. Bequele, 1985:234-235).

Yet, it was not only cocoa that left Ghana. Foodstuffs, basic commodities, minerals and timber were also smuggled out in substantial quantities to Cote d'Ivoire, Burkina Faso and Togo because of the artificially low prices the regime imposed on goods and other products in Ghana. In this period, it was reported that Togo was exporting diamonds despite the fact that it had no diamond mines of its own (cf. Nugent 1991). The decline in output was, however, not confined only to cocoa production. Mineral and timber production, as well as manufacturing, also suffered a decline in output (see Tables in Appendix), due mainly to over-valued domestic currency, the frequent breakdown of machines, the lack of spare parts and maintenance, resulting from acute shortage of foreign exchange for imports.

The personalization of state power reached its zenith when Col. Acheampong decided to become (in addition to being the Minister for

Defence) also the head of the nation's strategic institution: the Ministry of Finance and Economic Planning. The re-evaluation—or rather the over-valuation—of the Ghanaian currency enabled the government to import goods cheaply from abroad, but this worked against the interests of export producers. In the wake of falling foreign exchange earnings and foreign loans, the government resorted to deficit financing and outright printing of money.

The failure of the NRC-SMC regime to subject strategic economic policies to bureaucratic expertise along with the reckless dissipation of state resources, led to a disagreement between the director of the Bank of Ghana and Col. Acheampong, with the former losing his office. Col. Acheampong assumed the position of Finance Minister. The re-introduction of free health and education, as well as fringe benefits to civil servants, which the Busia regime sought to limit, bloated the government's expenditure. In this crisis period, Col. Acheampong did not impose an austerity budget, as the Busia regime did, for fear of losing his political followers or becoming unpopular. Increased government consumption in the form of salaries for the bureaucracy, the cost of maintaining the re-distributive system—free education and health services, students allowances and civil service allowances induced budgetary deficits, which were met by the printing of the currency. As the government capacity to print money went unsupervised from 1972 to 1978, the money supply rose from 7.5 percent to 135 percent. This induced runaway inflation. The differentials of the cedi value in the official and unofficial markets widened from approximately 1:2 in 1972 to 1.10 in 1978 (Jeffries, 1989:80). It was estimated that, within this period, Ghana had the most overvalued exchange rate in the so-called developing world. As the value of the cedi shrank, so did confidence in the economy.

The repudiation of foreign debt also had some negative effects on the economy. Ghana's trading partners—mainly western countries—demanded (as they did in the mid-1960s just before the Nkrumah regime fell), cash payment for future imports. Ghana was also taken off the list of the Western aid establishment. The inability to pay in advance as a result of the lack of foreign exchange diminished the state's capacity to import commercial and capital goods. This situation led to chronic shortages of all the basic commodities (wheat, rice, bread, soap, sugar, etc.), spare parts for vehicles and raw materials and other critical components for industries. Industries reduced their activities and many laid off their workers. Hospitals also lacked vital equipment and drugs. This further aggravated the general low level of economic production and the provision of social services in the country. Hunger became a reality in Ghana and the government had to accept food aid from the US and other western countries.

Once again, Ghana was perceived by foreign investors (real and potential) as a hostile and risky environment for investment. The regime's attempt to gain the "commanding heights" of the economy also diminished western interest in Ghana. This was evident in the slow-down of western trade already alluded to. Government policies also created confusion for foreign investors. For example, in 1970 a team of Japanese geologists started prospecting for manganese ore in the Volta River Basin. When the NRC granted prospecting rights to another team from Hungary, the Japanese withdrew, leaving the Hungarians to start from the beginning again. Similarly, Shell, British Petroleum, and Mobil all halted their prospecting for oil and natural gas in protest against the arrival of East European geologists. Some of these western investors did not resume their activities until in April 1975, when the government reversed its pro-East investment policy and granted sole rights to Western capital to prospect for oil (cf. Etsiah, 1989/90:44-45,70). The NRC-SMC's economic and material support for Pan-African ideals also diverted some national resources away from national development.

The patrimonialization of the state not only weakened the state's capability to intervene in the economy positively, it also reduced public accountability within the state itself and the public sector. People who had privileged access to state-allocated import licences and foreign exchange benefited from the artificially low official exchange rate. They imported goods cheaply and made profits sometimes reaching 1,000 percent. Also, few people knew exactly how much Col. Acheampong and his colleagues received as incomes. One could only guess from the reckless dissipation of state's resources that occurred between 1972 and 1979. Even though the economy was by now in dire straits, there seemed to be no financial constraints on the regime, as evidenced by the pomp and elaborate protection that attended every visit the head of state made across the country, the huge entourage and security guards that accompanied him, the fancy cars driven by NRC-SMC members and the lifestyles of their children and families, the substantial sums of money the head of state dispensed on the chiefs and the rural communities he visited and the huge number of houses and mansions built by the regime's henchmen (some of which were to be confiscated by the incoming military regime).

This venality of the NRC/SMC regime impacted directly on the nation's fortunes, because scarce resources were diverted to unproductive activities for securing political allegiance for the regime. Furthermore, there was massive state investment in sports, probably as a means of neutralizing the revolutionary potential of a population that was becoming increasingly destitute. The development of local scientific and technological capacities that the regime announced as the cornerstone of its "self-reliance" policy and was elaborately launched in the OFY programme was not translated into reality. Instead of addressing structural problems that could pave the way for increased productive activities, the government expended its energy and resources on re-distribution issues, particularly the so-called "essential commodities".

Furthermore, massive corruption permeated the NRC-SMC ruling circles. Foreign exchange earned through cocoa exports was diverted into overseas private bank accounts. In one instance, a shipload of cocoa destined for Europe was said to have disappeared on the high seas. It was popular knowledge that top officials ran racketeering operations; lorry loads of rice imported by the state and sold at controlled prices were driven across the nation's borders and re-sold at the instructions of military officers. Officers also issued chits and cars to their favourites, most of whom became road contractors and dubious businessmen and women. It is in this respect that Oquaye (1980) writes that Acheampong's economic policies degenerated in practice into a virtually unbridled orgy of looting of the national treasury and the economy. A highly unproductive kleptocratic regime had emerged, through which officers did not hesitate to privatize office and amass wealth (cf. Chazan, 1983:181; Rimmer, 1992:134-135).

In addition, state administration was reduced to selling and buying. The head of state himself was alleged to have been issuing import licences from his office to his favourites. Cocoa funds were embezzled or used to pay bonuses and allowances to the CMB and government officials while the cocoa industry was collapsing. The CMB scholarship funds that were supposed to go mainly to the children of cocoa farmers (which provides one rationale for why farmers should not get all the world prices for cocoa; the others are the subsidization of farming equipment and chemicals, stabilization of domestic price), went mostly to the children of the elites or those who could easily could afford to pay school fees. Also, a special scholarship fund was established at the CMB, where Col. Acheampong, the Commissioner of Cocoa Affairs, Regional Commissioners and CMB officials could award scholarships directly to their children, relatives and favourites. In one inquiry into the activities of the NRC-SMC regimes at CMB, it was noted that:

> Cmdr. Addo, former chief executive of the Cocoa Marketing Board, told the committee investigating its affairs that the CMB spent nearly 1 million cedis on drinks alone between August 1977 and July 1, 1978. Giving evidence, Cmdr. Addo said during his tenure of office he instituted certain measures to boost the morale of the directors. As part of these measures, he said all the eight or ten

> directors were given a bottle each of whisky, brandy, and gin at the end of every month in addition to receiving a... table allowance [i.e. money] (*West Africa*, 27 November, 1978).

As the state decayed into a patrimonial political machine preoccupied mainly with the survival of the leader, its capacity to manage the public realm and to formulate rational policies that could raise the socioeconomic fortune of the country diminished. In Chapter 2, the argument was made that a neopatrimonial state is anathema to capitalist development. The Ghanaian state at this moment exhibited patrimonial features that undermined the economic progress of the country. When a state that should shape the context for societal development itself turns into a corrupt and ineffective public instrument, citizens are left to fend for themselves outside the state's hegemony and control. In such a context, various economic and social behaviours are likely to sprout in the wastelands left by the failing state.

A patrimonialized economy provided a structure of incentives for all kinds of socially unhealthy behaviours through which everybody seemed to have resorted to fending for themselves. One danger, however, is that once such behaviours—which are in many cases, seen as anti-social or illegal—get rooted in society, they become difficult to eradicate. Chaucer's metaphor quoted above that "if gold rust, what will iron do?" finds an echo here, for, if the state and officials, whom people trust to serve the collective interest and provide leadership by example, could decay to such an extent, what would the ordinary citizens do?

Societal Responses to Economic Collapse: Practical and Strategic Actions

Neopatrimonial governance unleashed by the NRC-SMC regime created a weak state unable to shape its surrounding environment positively and also undermined its capacity to spur socioeconomic development. The weakness of the state was reflected in its loss of control and the destruction of an economy it critically needed for its own reproduction. It could not promote development because it failed to initiate effective policies. Predictably, the formal economy declined, and, as it rescinded, citizens began to disengage from the state and the public sector to provide for themselves in the informal economy, which was mainly based on speculative activities of buying at government approved prices and reselling at sky-rocketed prices. An impressive but perverse transaction economy had developed at the cost of the productive sector. The scarcity of virtually all basic commodities led to the rapid development of this *kalabule economy* as it became known, through which all kinds of malfeasance, hoarding, illegal trading practices, corruption, extortion, and other speculative and moonlighting activities (e.g. currency trafficking) officially labelled as illegal were conducted (Pellow and Chazan, 1986).

At the same time, however, while some people certainly profited from participating in the state or from association with it, some of the activities in which ordinary Ghanaians engaged at this time can scarcely be called "anti-social" at all, as they are rational and practical ways to cope with the crisis inherent in a patrimonial and parasitic economy. Human actors everywhere are rational, but they are rational within the socioeconomic and political context they find themselves. The state, as a rational bureaucratic instrument for managing the economy and society, deteriorated so rapidly under the NRC-SMC regime that the economic and political structure provided an incentive for various actions that did not conduce to the health of society. As the public or national instrument for collective goals collapsed, people were compelled to pursue their selfish interest leading to almost a Hobbesian scenario of a "war of all against all". When cocoa farmers found it necessary to market their produce in neighbouring countries and many times refused transactions in the Ghanaian currency, when market women sold goods that they imported from neighbouring countries with money they bought at the unofficial market rate, and when police officers indiscriminately mount check points and border guards impound commercial goods and extort money from traders or even supervise the smuggling of gold, diamonds and cocoa, they were all taking actions (however injurious they are to health of society) to protect themselves in an environment in which the state had failed to manage the economy for the collective good and set up policies that could elicit productive activities from its citizens.

The failure of the state led many people to take their destinies into their own hands, even if survival meant taking advantage of the state itself (for example, failing to pay taxes, or to outwit it by privatizing office or stealing state property, or by ignoring the state's directives to advance ones personal interests). The informal economy, which got the prices and foreign exchange transactions "right", expanded at the expense of the official artificial economy. But official disdain of it meant that much of the economic activities conducted there was outside the reach of the state, thus, government could not control and tax it for its own revenue. By 1978, about 80 percent of the economic transactions in Ghana took place in this unofficial market and the government's revenue-to-GDP ratio fell from 20.5 percent in 1970/71 to 15.5 percent in 1974/75 and then sharply to 6.5 percent in 1977/78 (cf. Jeffries, 1989:80). In the informal economy, government policies and controlled prices were flouted with impunity, even by its own agents, and it was regulated more by kinship, informal trust, friendship and mutual obligations whose property rights enforcement depended more on private choice and arrangements than rational rules in the public realm.

While the vibrancy of the informal sector or secondary economy emphasizes the degree of state collapse and its development could possibly serve as a fertile ground for citizens to acquire entrepreneurial experience, it is doubtful if such an economy, as it was operated at this time, was conducive to such a positive role. Impressionistic evidence at this time suggests that there were rampant cheating and swindling which justified it being defined as illegal not only by the state but also those who participated in it. For example, it was not unusual for persons to collect money from others, promise to deliver commodities, and then default. A classic example is how land was sold particularly in Accra. Land could be sold by one person to several people, leading to costly litigation or even violent confrontations. These features bear striking resemblance to what is associated with patrimonial economies that undermine economic development. These actions and others of similar nature undermined the trust that should exist in the economic system of the society to make life predictable and future obligations binding. They also eroded the institutional framework that should emerge to support a modernizing economy. Economic activities in this period might have been burdened by the high transaction costs that institutional economists hold as one factor responsible for the inability of some economies to realize their growth potential.

In this period, Ghanaians were not under the Weberian illusion that their state was of use to the nation. This led to a silent detachment of society from the state to survive alongside or apart from formal employment. Urban workers cultivated their own vegetables and produced their own poultry. Women workers traded in all kinds of goods at their workplaces; housewives and children had to do extra trading to supplement their husband's or father's incomes; civil servants and university teachers converted their cars into taxis in the evenings and the police mounted their own barriers and extorted money to supplement their incomes. Most of these actions were, however, mainly urban phenomena and indicate the regress of the modern economy into a subsistence one. In the rural areas, where the farming population had direct access to food, they resorted more and more to subsistence agriculture and dependence on rural resources as a shield from unproductive formal economy. Some urban dwellers sought "self-autarky" and moved from the urban areas to seek sanctuary in their villages (cf. Azarya and Chazan, 1987).

Furthermore, some people sought spiritual escape from alienation, cynicism, emptiness and bleak future engendered by a collapsed economy.

Cultural and psychological adjustment came through gospel revival evident in the spread of Pentecostal and syncretistic churches. Religion became "the opium of the people". Attracting the most vulnerable elements of society—the women and the youth—these churches promised a better life hereafter, but, in some cases these churches were means devised by the "smart" elements of society for making money to survive.

Others, usually the highly educated, chose exit options and migrated from Ghana. Middle-cycle school teachers, doctors and members of faculties of Universities left their posts for other prosperous African countries as well as Europe and North America. The exodus of the better educated elements of society caused a serious brain-drain in the country. Secondary schools lost some 35,000 trained personnel (cf. *West Africa*, 3 January, 1983) and schools had to do with untrained teachers. Even unskilled Ghanaians migrated to Nigeria to become factory hands or to do all types of menial jobs to the extent that Ghanaians developed a "colony" in Lagos. In all, much of the nation's labour force and about one-half or two-thirds of its experienced, top-level professional workforce was lost during this period (Rimmer, 1992:177).

This exodus continued so that, by 1982, out of the 1,500 doctors trained since independence, only 350 were in government service. It was estimated that there were more Ghanaian physicians in West Germany than Ghana, and more Ghanaian lecturers in Nigeria than in Ghana (cf. Azarya and Chazan, 1987:118). Ironically, a country that had attracted economic migrants in the 1950s and early 1960s was now "exporting" its brains. The struggle to escape economic hardship at home was uncontrollable, turning the country's sole passport office in Accra, into a "gold mine" for state officials, who openly demanded bribes of six or ten times the official cost before issuing passports and travel documents. Some Ghanaians who could not afford passport at such prohibitive prices left for neighbouring countries without valid documents which exposed them to harassment from foreign authorities. Husbands often left wives behind and this disrupted many families.

These "exit" actions by teachers, bureaucrats, doctors and the masses, coupled with "de-participation" of cocoa farmers in the formal economy noted earlier, can be construed as a protest against falling living standards in Ghana and the state's inability to serve any common good, other than the interests of those who have patrimonialized it. This belies any argument that the state was acting for the interest of these groups (cf. Chapter 1).

On the other hand, however, Ghanaians did not only resign themselves to their predicament by taking practical actions or exit options to survive. They also sought strategic actions to overthrow the SMC regime. Civil associations consisting of students, lawyers, doctors and nurses, demonstrated to force the military junta out of power. Indirectly, these urban groups were protesting against poor living conditions, thus questioning the "urban coalition thesis", which posits that the state monopolized rural surplus and spent it on itself and urban groups. The thesis, however, sheds light on the NRC-SMC's re-distributive politics of 1972-74, when every group that opposed the Busia regime was placated. But from 1975 onwards, the state pursued or maintained policies that hurt almost everybody in Ghana. These latter state actions can be better understood from the perspective of the political economy of personal rule offered in the theoretical discussion. As the regime's re-distributive resources diminished, the urban groups that had supported the regime now turned against it.

The actions undertaken by civil society and/or the middle class to topple the regime were, however, met with police and army brutality. In this confrontation, the regime defined the "middle class" that opposed it as having been educated at the expense of the taxpayers and a privileged class seeking self-interest. The government-controlled media reminded workers of the Busia regime's allies—the middle class—that attempted to destroy their trade unions. Some element of worker manipulation was evident at this stage. The regime emphasized its record to protect the interest of workers, as evident in frequent increases in wages. But increased money, unaccompanied by increased production, is often a recipe for inflation, which the country was now suffering from as prices continued to rise undercutting living standards.

In an attempt to find an acceptable way to perpetuate his military rule, Col. Acheampong proposed an innovative tripartite political system called the *Union Government* (Unigov), which one American evangelist claimed was ordained by God. Through this system, the army, the police and civilians would share political power. This idea was sold to the nation with massive political campaigns. A referendum manipulated by the regime returned a victory for Unigov proposal but not before the Electoral Commissioner ran for his life when he insisted on proper electoral procedures. In all these attempts to remove the regime, the trade unions and workers remained indifferent, largely because the government had co-opted their leadership through patron-client relationships. And just before the referendum, the regime increased wages by 100 percent. It got their political support, but the increase was swallowed up by runaway inflation.

The destruction of the economy that occurred under the SMC was, however, a source of embarrassment for some of the SMC members. Consequently, some of them decided to do something about it before the social tensions in the country turned into violence or revolution. On 5 July 1978, General Fred Akuffo, a member of the SMC made a palace coup and forced Col. Acheampong to sign a letter of resignation. Col. Acheampong was later put under house arrest. Major Akuffo's main accusation against

Col. Acheampong was that the principle of "collective decision making" which should have guided the NRC-SMC, was violated and that the country was run as a "one-man show". This palace coup opened a new chapter in this military rule: SMC II.

The Supreme Military Council II (1978-1979): The Return to Liberalism

The new regime, SMC II, was headed by General Akuffo. General Akuffo's interlude saw attempts to breathe sanity into the management of public affairs. The new regime promised to hand over power to civilians once the economy had been re-ordered. In order to improve national economic management, policy-making and implementation were returned to where it traditionally belonged: to the experts and the bureaucrats. This led to the creation of a National Economic Advisory Committee, made up of civil servants to advise the government on economic matters. This committee was also mandated to re-establish the nation's credit-worthiness abroad. The domestic currency, which was devalued by the Progress Government and was re-valued by the NRC in 1972 was now devalued by 58.2 percent against the US dollar, from 1.15 cedis to 2.75 cedis (cf. Chazan, 1988). To contain inflationary pressure in the economy and mop out excess liquidity, the currency was demonetized.

This measure was, however, implemented without properly educating the public and without time to enable the rural people, in particular, who had little or no access to banking institutions, to deposit their monies before the deadline for the exercise. All political prisoners of the previous regime were released. Furthermore, a Constituent Assembly was appointed to draft a new constitution and political activities were legalized. Though well intentioned, these measures appeared feeble and belated against the background of the crisis. Ironically, most of these policies were ones the Busia regime had tried to pursue and were rejected. The policies, if they were efficacious at all, were coming too late, as the economic decline had acquired a new fatal dimension and many Ghanaians now appeared to be looking for miracles to end their woes. Nor could the policies satisfy the junior officers and other ranks in the army who were also affected by the general economic crisis of the country, which they felt owed much to the irresponsibility of their superiors.

When General Akuffo took power, he dismissed most of the SMC members who operated such a developmentally counter-productive state. Col. Acheampong was stripped of his military titles and honours and retired to his home-town in the Ashanti region. No accountability was demanded from him or his associates. They were retired and allowed to enjoy their illegally acquired wealth and property. The public thus saw General Akuffo's palace coup as a political ruse to exonerate the SMC members from blame for Ghana's economic decline and shield them against accountability. This was probably so, because during the preparation for up-coming elections, the new SMC members closely associated with the main political parties in an attempt to seek immunity for their actions after leaving the political scene. Such efforts were misplaced, because it was not only civil society that was disappointed by the misconduct of the military elites, but also the junior officers and other ranks of the military.

On 15 May, Flt. Lt. Rawlings led an abortive mutiny against the SMC regime. Interestingly, and as a result of the liberal atmosphere prevailing at this time, the mass-media gave considerable coverage to Rawlings' speech, which emphasized the disgrace brought by the military to the nation, the cancer of corruption and other social evils that had eaten deep into society, the quest for social justice for the ordinary person and the need to go the Ethiopian way (Jeffries, 1989:84). Apparently, this reference was to "socialism" that was current in Ethiopia. This was, however, the first hint that Rawlings shared socialist sentiments, though it is doubtful that he had learnt scientific Marxism.

There was no doubt that many Ghanaians believed in the young officer's simple message that the nation's woes were due to corruption of past rulers and their cronies in civil society. Indeed, his speech struck a strong emotional chord among the urban workers, students and the unemployed. Instantly, Rawlings became a national hero, but he was put in gaol to face further trial. On 4 June 1979, however, the junior officers staged, a successful coup and released Rawlings to become their leader. The SMC II was overthrown and a new military regime of the Armed Forces Revolutionary Council (AFRC) led by Rawlings was inaugurated. Ghana had come under another military rule.

Conclusion

The decade from 1969 to 1979 saw two main regimes imposing two contrasting development ideologies on Ghana. Both regimes failed. The first, the Progress Government, was rejected because its policies that sought economic efficiency directly struck at the interests of the "urban coalition" that wanted to reassert Nkrumah's re-distributive social production system. The other regime, the NRC-SMC, fell because it was more interested in re-distributive politics that could guarantee its survival in a context of declining production and changing fortune. Furthermore, the political leaders of the second regime, patrimonialized the state, undermining its capability and effectiveness in the management and development process.

In Chapter 3, it was suggested that the basic principles that underpin market economy were implicit in precolonial Ghanaian societies. It was also noted that the economic policies of colonialism accelerated Ghana's capitalist evolution. Individualism became accentuated and private wealth accumulation, as well as entrepreneurship, became widely desired and the commodification of relationships approved. In this sense therefore, the Progress Government's neoliberal economic policies, which sought to promote indigenous capitalism, can be said to have come closer to Ghanaian historical reality and social experience. Nkrumah had sought development in a socialist tradition and Busia in a capitalist one. Both were rejected by the military that saw society's progress in terms of its material well-being.

Surely, there can be no indigenous capitalism without a capitalist class. The attempt made by the Progress Government to develop a productive national bourgeoisie and to allow Ghanaians to gain a strong position in their own economy makes some people to see Busia's policies as much more "nationalistic" than those of the CPP which sought socialist patterns of development for Ghana, but did not liquidate foreign capital and was rather antagonistic the development of indigenous private capital. Busia, unlike Nkrumah, was asking Ghanaians to be businesspersons and to control their economy instead of diverting their energies to mainly state sector activities. Since it takes decades and even generations for a society to develop a truly productive bourgeoisie, the potential long-term value of the Progress Government's policies in the development of such a class for future capitalist development (which is currently accepted in Ghana) cannot be discounted.

From the theoretical perspective guiding the study, what seems interesting in Ghana at this juncture is that, a government had recognized that the nation's greatness does not always have to depend on the state monopolizing virtually every facet of the economy, but also on the wealth and advancement of its own citizens. As subsequent chapters make clear, the Busia regime remains, to date, the only one in Ghana's postcolonial history to have behaved like the dynastic rulers of Europe who recognized that their nation's greatness could be served by the activities of their own bourgeoisie. Perhaps the development of a dynamic productive bourgeoisie and capitalism might very well be one way to ward off the predatory international capitalism that has become a vexation to many leaders who seek nationalistic development for Ghana.

Busia's policies were, however, not appreciated in Ghana at this time (and perhaps even after) and the regime was overthrown. The Busia regime's attempt at a "bourgeois democratic revolution" was terminated because it was unable, in the Gramscian terms, to *hegemonize* its bourgeoisie values in a society that had been indoctrinated with socialist egalitarianism and re-distribution undertaken by the state very early in its evolution. As subsequent chapters reveal, it was to take a total collapse of the national economy to recognize the soundness of the Progress Government's home grown "structural adjustment policies".

The 1972 coup manifested the continued contestation in the state over development ideology. That Busia was interested in capitalist development for Ghana, and that the CPP and NRC-SMC were interested in socialism is obvious from the policies each pursued. The coup also indicates the military's antagonism to any policy that reduces its living standards. The coup brought to power a coalition interested in defending their corporate amenities and promoting a re-distributive politics. Through the coup, the military was protecting its living standards.

Once a re-distributive social production system was imposed to appease vociferous urban groups on whom the regime depended for political support, the regime could not reverse such a policy without digging its own grave. The only way to hold on to power was to patrimonialize the state while seeking policies that could continue to deliver goods to a populist constituency. But the patrimonialization of the state eroded public accountability and competence in the public management system, and also undermined the effective utilization of national resource, thus plunging the economy into a crisis that invited another coup. If the failures of the NRC-SMC regime are anything to go by, any attempt in the future to address only re-distributive issues without adequate attention to increasing production would only aggravate existing socioeconomic crisis.

Notes

1 Nkrumah lived in exile in Guinea and "shared" the Guinean presidency with Sekou Toure since the 1966 coup. Nkrumah died in Bucharest, Rumania on 27 April 1972, after an illness. He was buried in Guinea and not Ghana. Following pressure from Ghanaian students, the NRC was forced to bring the remains of Nkrumah home for burial. But this was not before the following conditions from the Guinea Government were met: (i) Nkrumah was fully rehabilitated; (ii) CPP members still under detention in Ghana were released; (iii) an amnesty for those who were with Nkrumah in exile; and (iv) Nkrumah was buried with honours due a Head of State. The intercession of General Gowon of Nigeria in this squabble brought a final settlement and Nkrumah's body was returned to Ghana on 7 July and buried in his birthplace, Nkroful.

2 The term *kalabule* according to Jeffries (1989) probably derived from the Hausa

language *Kere kabure* meaning "keep it quiet". This is probable since in Ghana, the term came to be associated with any activity that promised quick profit and even illicit behaviour particularly in economic affairs.

7 From Populist Revolution to Neoliberalism (1979-1999): An End to Ideology?

> A man will not risk what he has in trade, except for the prospect of very large gains, if he is likely to be robbed by pirates, or to be oppressed by the government if he is successful in business— William Cunningham

Introduction

If the 1966 and 1972 military coups established a tradition that only generals and colonels make coups, the 1979 military revolt led by Flt. Lt. Rawlings suggests that flight lieutenants, sergeants and corporals can also make them. After ruling for a brief period, Rawlings handed power over to civilian politicians in 1979 only to take it back again in 1981. From 1979 to 1983, Ghanaians witnessed a radical and violent attempt at the revolutionary transformation of their society. Since 1983, however, Ghana, under Rawlings, has settled for the most conservative economic policies ever implemented in the country's postcolonial development history.

This chapter focuses on this turbulent period from 1979 to 1999, that is, the turn of the twenty-first century. It discusses how the governments of Rawlings and civilian politicians have understood and grappled with Ghana's development problem and what consequences their policies have had for socioeconomic development. The chapter examines three regimes: the Armed Forces Revolutionary Council (AFRC), the People's National Party (PNP) and the Provisional National Defence Council (PNDC), now called the National Democratic Congress (NDC). It argues that the policies pursued by these governments in this era were incoherent and often destructive of the socioeconomic and political environment for investment and production. This worsened the economic crisis already underway and undercut the state's economic base. In the end, the PNDC decided to embrace the IMF and World Bank's structural adjustment programme (SAP) as a means for economic regeneration.

The chapter sees the last 16 years or so of Rawlings' rule as having achieved some relative political stability, and a predictable macroeconomic environment seemed to be emerging with the consistent implementation of SAP throughout this period. But, it notes that the experience of the recent political reform suggests that democratic rules are only weakly adhered to; neopatrimonialism has re-emerged, possibly creating an environment hostile to indigenous capitalists and thus hindering capitalist development. The account given about the SAP is not exhaustive, as this is a wide subject discussed extensively elsewhere (e.g. Loxley, 1990; Rothchild, 1991). The aim here is to give a panoramic view of its impact and of how the socioeconomic and political environment stands in contemporary times.

The Armed Forces Revolutionary Council: A 112-Day Moral Revolution

The June 5 radio broadcast by Flt. Lt. Rawlings announcing the 4 June 1979 coup, noted that "Countrymen, we the junior officers and other ranks, are very much disturbed about the sunken reputation of the Armed Forces. We have felt that the SMC would do a house-cleaning exercise and put the reputation of the Armed Forces on an even keel before handing over. All attempts to help the SMC do this have failed. In these circumstances, we have no alternative but to take over the administration of the country. In the period at our disposal therefore we have plans for "house-cleaning exercise" and we are going to act on it immediately" (cf. *West Africa*, 25 June, 1979:1013). The announcement also assured Ghanaians that there would be a smooth transition to constitutional rule. The "house-cleaning exercise" was principally directed against senior military officers, particularly those who had supposedly benefited from the NRC-SMC rule.

The AFRC regime did not articulate any policies or development programmes other than: (i) to "eradicate" corruption and bring to justice leaders who have personally appropriated public office or have acquired wealth illegally; and (ii) to revive notions of public accountability and to make Ghanaians conscientious of their rights and duties before restoring democratic rule. The "urban coalition" made of up of mainly workers, the urban unemployed, students and the radical intelligentsia who had once supported the NRC-SMC regime, now endorsed the AFRC regime. They marched in support of the regime with placards reading "Action now", "Death to swindlers!", "Investigate traitors". They also demanded a complete investigation of the "enemies" of the people: the "rich" class of industrialists, businessmen, landowners, etc. (cf. *West Africa*, 25 June, 1979). These popular expressions were, however, confined to the urban areas, particularly Accra, the locus of power. The rural areas, as usual, remained relatively indifferent, as their world is only tenuously touched by politics in the nation's capital. These urban expressions, however, did symbolize support for the regime, which emboldened it to initiate its "revolutionary" measures to deal with the economic crisis.

The AFRC's rule was the shortest so far in Ghana's political history but one of the most politically engaged. Its first major public action was the trial of previous military leaders. On 16 June, Col. Acheampong and Major-General Utukah were found "guilty" of crimes against the people and state and were publicly executed by firing squad in front of jubilant spectators. On 26 June, 1979, another six military officers were executed; these included Generals Fred Akuffo, Akwasi Afrifa, R. Kotei, Air Vice Marshal George Boakye, Rear Admiral Joy Amedume, and the former Commissioner for Foreign Affairs, Colonel R. Felli. The execution of these leaders enjoyed popular support in Accra but such radical politics was unprecedented and appeared alien to Ghana's way of peaceful life. Consequently, the AFRC incurred the displeasure of the professional societies of lawyers and medical doctors as well as religious bodies. They publicly expressed their discomfort with the secrecy and swiftness that surrounded the trials and the failure of the tribunals to follow established judicial process. They even asked the AFRC to hand over power.

The executions might have sent a message to all that moral rectitude and probity are essential qualities required of those who availed themselves of state responsibility. But it takes more than moral appeals and executions to run a national economy for the good of everybody. Young officers were more socially aware of the plight of the ordinary people than officers who could exploit political positions for wealth. Although motivated mainly by the appalling living conditions in society, the AFRC lacked a programme that could arrest Ghana's precipitous socioeconomic decline. Besides blaming past leaders for the crisis, the regime naively attributed the high prices of goods or cost of living to the "greed" and "evils" of traders in the distributive sector, and believed that Ghana's problems could be solved by simply issuing the "right orders".

Consequently, the regime intervened in the market and imposed price controls on "essential commodities". Traders who attempted to evade the government's directives had their goods seized and were severely disciplined. Such traders, mostly women, had their heads shaved and were subjected to public humiliation. They were sometimes stripped, paraded and flogged in public as "economic saboteurs" or "enemies" of the people and the revolution. Others were jailed for making "illegal" profits (see *Daily Graphic*, 5 June, 1980). On 18, August 1979, the largest market in the nation's capital, Makola No. 1, where about 12,000 women traded was blown up by the soldiers because it was defined as the centre for *kalabule*. Markets in Kumasi, Koforidua, and other major towns were dynamited or set ablaze, and goods, spare parts, and property were destroyed.

In addition, the AFRC ordered all private commercial houses such as the UTC, UAC and the state-owned company, the GNTC, to lower the prices of their goods to the so-called controlled prices. A number of Ghanaian, Lebanese, Syrian and Indian businessmen were arrested and accused of various "economic crimes" and their assets and bank accounts were frozen (cf. *West Africa*, 25 June, 1979). The AFRC regime went all out to sell whatever goods it could lay hands on to the general public. The forced sale of goods quickly led to further food shortages in the country because existing stocks were disposed of much quicker than would have otherwise been the case. As women became targets of the AFRC's "anti-commerce" policies, most refused to bring food from the rural to the urban areas, causing further food shortages in the cities (cf. *West Africa*, 26 November, 1979:2182-2186).

The AFRC also extended its "house-cleaning exercise" to civilians who it judged as having "benefited" unjustifiably from the SMC's kleptocratic rule. Civil servants—the senior position holders such as principal secretaries and other officers who acted "corruptibly" in the Acheampong era—had their assets frozen and/or were dismissed. Public and Price Control Tribunals were set up to enforce the new policies. Some "criminals" fearing "revolutionary justice" ran away from the country. It was estimated that some 155 persons were given prison sentences ranging from six months to ninety-five years, and just before the AFRC handed over power to the new civilian regime of Limann, some sixty-eight people who fled the country were tried *in absentia* for corruption and sentenced to various terms of imprisonment (cf. Ray, 1986:102).

The revolt by the junior officers upset the command and hierarchical structure of the army. Having acquired power, some juniors turned to disobey their superiors. It was an achievement of Rawlings and his colleagues and those adversely affected by the politics of this era that the country at this moment did not experience civil war or further bloodshed. There were, however, reports of misconduct among soldiers when they entered the wider society to implement the new policies. They took the law into their own hands and did not hesitate to use force to enforce their will. They commandeered vehicles from civilians and forced market women to reduce prices. They sought preferential treatment from transport owners, extorted money from businessmen, sold weapons in the urban areas, and, in some cases, beat and even killed to enforce the "demands" of the revolution (cf. *West Africa,* 25 June, 1979). The so-called moral revolution had thus acquired a new dimension, threatening life and property.

As a hurriedly-assembled conspiracy group, the AFRC's members did not have much in common except, belonging to lower levels of the military hierarchy. Consequently, their rule experienced strains as a result of personality clashes and accusations of tribalism and favouritism in the dispensation of "revolutionary justice". The absence of a coherent ideology or programme made the AFRC's actions ad hoc, contradictory and conflictive. As much revolutionary excesses were committed, the regime came under increased pressure from civil society to return the country to democratic rule. As a result of these pressures and its own "disunity", the AFRC made good its promise to hand over power and lifted the ban on political activities a bit earlier than scheduled.

The lifting of the ban marked another attempt to re-start democratic politics. A new constitution was drawn which provided for executive (rather than ceremonial) presidency to be elected directly by a majority of the voters. Significantly, too, the constitution provided for a transitional clause which made irreversible the decisions of the AFRC and its revolutionary tribunals. Many political parties, emerged articulating diverse interests. While the initial high number of parties could indicate a revival of interest in democratic politics, only two political parties—the People's National Party (PNP) headed by Dr. Hilla Limann and the Popular Front Party (PFP) headed by Mr Victor Owusu—which constituted new versions of the old CPP (Nkrumah/Left) and UGCC (Danquah-Busia/Right) traditions respectively, carried political weight. For the first time in the nation's political history, no political restrictions were put on anybody. All parties were allowed to contest the elections freely.

Unlike the NLC, the AFRC did not covertly or overtly favour any party. It remained largely neutral, making this democratization period so far one of the rarest in the nation's history, where elections were conducted to the satisfaction of all participants. The PNP won the election, and, on 24 September 1979, the AFRC returned Ghana to democratic rule. On handing over power, the AFRC enjoined the PNP regime to "safeguard the achievements of the AFRC" and also continue its "house-cleaning exercise". Indirectly, the AFRC members were asking the incoming regime to protect them from possible retribution from those who might have been aggrieved by their rule. Perhaps this shows that the most productive politics is one that is conducted in an institutionalized context, whereby leaders follow publicly sanctioned rules. Otherwise, the political game becomes unpredictable, uncertain and unsafe, for even those who win the game do so only temporarily. As will be shown presently, the new civilian regime failed to guarantee such provisions. This was to lead to a rift between the PNP and members of the erstwhile AFRC regime, a factor that was to play no small role in Rawlings' second coming to power.

The Peoples' National Party (1979-81): The Search for a Predictable Socioeconomic and Political Context

One key argument advanced in the theoretical framework in Chapter 2 is that a predictable and calculable socioeconomic and political environment is a basic requirement for sustained development. Although some leaders have privatized their offices and brought the economy to a crisis and, thus, probably deserved punishment, the AFRC's policies of seizing bank accounts, confiscating means of production, destroying supermarkets and property can be seen as introducing risk and unpredictability in the society. By now, it should be obvious that frequent change of government, rules, policies and constitutions, has become inimical to the development of a predictable socioeconomic and political environment critical for soliciting investment capital and eliciting long-term productivity. This reasoning gains credence when the PNP, upon assuming power, outlined two major concerns that needed urgent government attention: the creation of an enabling environment for long-term investment and the restoration of Ghana's image abroad to attract foreign investment.

In his speech to mark the second anniversary celebration of the Third Republic, President Limann deplored the situation of low foreign investment in the country. He singled out two main factors as accounting for this. The first was the lack of dependable legal guarantees to ensure safety of any investment. The investment climate, he said, was fraught with bottlenecks, such as very slow and cumbersome procedures, punctuated by diffused investment laws. This obviously was a reference to the recent violation of property rights institutions by the AFRC, and the frequent changes in policies as well as public rules that attended every change in government.

The second factor was the gross and callous mismanagement of the economy in the past and the spectre of political instability that has haunted the country for decades. This situation, he noted, was evident in the imperceptible economic boycott that culminated in the total blockage of the country by its trading partners, and, more especially, potential investors. As he noted, "foreign investment has been particularly sceptical of the *inconsistent* manner in which [Ghana's] economic policies had been implemented and the impunity with which Government[s] always interfered with even these half-hearted measures to suit [their] political whims and aspirations" (emphasis added). He maintained that both domestic and foreign investors require an atmosphere of economic and political stability (cf. *West Africa*, 21 September, 1981:2183-2190).

In general, democratic regimes, unlike authoritarian ones, are slow in taking actions on public issues given their relative sensitivity to the demands of numerous competing interest groups inherent in democratic polities. Civilian politicians democratically mandated to rule are likely to engage in longer debates in trying to "understand" the root causes of national problems than authoritarian or populist regimes, which can more "ignore" these plurality of interests or which are eager for quick results to justify their rule. A civilian regime such as Limann's that boasted a lot of intellectuals was more likely to systematically analyze the country's economic problems than was the populist regime of the AFRC. As it was, this situation led to some hesitancy in finding solutions to the national crisis and was perceived as indecisiveness of the regime in tackling the socioeconomic crisis. In some analyses of this period, the Limann regime has been accused of indecision, listlessness and inability to solve the country's problems. Some even argue that there is so little to be said about this regime (see for example, Chazan, 1988). These views, however, seem unfair if it is recalled that the Ghanaian economy had by now reached such a deplorable state that no government, however committed, could reverse it easily. It would require time and dedicated government to turn the clock around.

Nevertheless, the regime cannot be totally exonerated. It displayed a lack of purpose to tackle the national crisis. Its claim to the CPP's heritage was almost hollow to those who still believed in Nkrumah's vision because it refused to identify strongly with socialist ideology, preferring to portray its ideology as "to save Ghana". It also lacked the CPP's mobilization spirit to do things. While this probably could be due to lack of resources at this moment in the nation's history, there were, however, scandals of corruption, and embezzlement of state money as revealed after the regime was toppled. Civilian politicians, operating in a politically unstable environment and probably unsure about the future, could hardly settle in office before appropriating the perquisites. The truth and extent of the accusations against the regime cannot, however, be ascertained, since by now it has become customary in Ghana to smear any regime overthrown illegally as corrupt. But the popular belief was that the regime had its hands soiled and was ineffective in addressing the national crisis.

In spite of these limitations and accusations, the PNP regime could boast some modest achievements. President Limann visited Ghana's major trading partners—the Federal Republic of Germany, Britain, Yugoslavia and Rumania—to assure them of his government's intention. The vice-president also visited Brazil. These state visits yielded some positive results for the regime. The Federal Republic of Germany gave Ghana DM40m; out of this, DM7m was an outright grant. Britain also gave an Export Credit of £40m, out of which £25m was allocated for British capital goods and services. The International Development Bank of India also granted Ghana a credit of $6.5m to enable the regime buy Indian TATA buses to improve the public transport system that had virtually collapsed. Furthermore, India granted $8.5m for the development of cottage industries. Other examples of bilateral economic relations were negotiated with other countries (cf. *West Africa*, 21 September, 1981).

The PNP regime also promulgated a new Investment Code (Act 437), which was to encourage both foreign and domestic investment. The Code specified the areas in which the state, Ghanaians and foreigners could participate and where joint activities could be undertaken. The Code assured foreign investors, protection and fair return on their investment. It also provided, among other things, a complete exemption from payment of customs duties on machinery, equipment and guaranteed personal remittances. It also sought clarity for setting up manufacturing industries. In addition, the government established an Investment Centre to coordinate policies and to develop an institutional infrastructure to support investment and productivity. The Code also sought foreign participation in agriculture such as livestock, cereals and tree crop production. An International seminar on Ghana's Mineral Endowment was held to attract investors into the country (cf. *West Africa*, 21 September, 1981:2184). The regime sought a mutually reinforcing tripartite relationship with foreign capital and local investors: perhaps a *dependent development*. From the theoretical perspective guiding this study, the Investment Code, can be seen as an attempt to create a calculable and predictable socioeconomic and political environment for investment and development. This concern was, however, to contrast with the anti-business ideology and destruction of wealth and property that followed the regime's overthrow.

The PNP regime could hardly settle in office and put its programme into action when, on 31 December, 1981, the army, led again by Flt. Lt. Rawlings, seized control of the state's apparatus. The 31 December coup, or "revolution" as its architects called it, brought an end to not only another attempt at democratic rule, but, more significantly, to efforts to develop some stable and calculable socioeconomic and political environment in Ghana. For critical observers of the Ghanaian society, the coup did not come as a surprise; the president himself later said that hardly an evening passed when he did not think that the soldiers would seize power. Indeed, events in the country at this time indicated that a coup might be imminent. It is useful to recount them to better understand the motives behind the coup.

New Political Conflicts and the Problem of Civilian Control of the Military

Military disengagement from politics should be easy when soldiers have ruled or have been perceived to have ruled honestly, when they have not over-stayed their time, have not committed atrocities and when there are enough guarantees that the incoming civilian regimes will not harass them. As discussed in Chapter 5, in the case of the NLC these conditions were generally met. The kleptocratic NRC-SMC regime was, however, not sure of its future; hence its clumsy attempts to hang on to power, for which it paid dearly. In the case of the AFRC, none of these conditions seemed to hold. This is mainly because junior officers and other ranks, who had just killed and shaved the heads of their officers, would certainly have problems adjusting to the rigid life of the military establishment.

Furthermore, as a result of the confiscation of people's life savings and property and the imposition of severe sentences, some of AFRC's "victims" came to feel that they were unjustifiably treated and wanted their cases reviewed. There was, however, a transitional clause in the Constitution that forbade the review or reversal of AFRC actions; a request that appears a little odd in a liberal democracy. For how could a constitution be unamendable? And, more importantly, how could people tried *in absentia* and condemned for forty to ninety-five years imprisonment not have their cases reviewed? These were indeed problems for the civilian regime of the PNP and their resolution was to generate conflict between it and the former AFRC members.

When Rawlings demanded that the PNP regime continued the AFRC's "house-cleaning exercise", he was indirectly asking the democratic civilian regime to engage in radical or populist politics, something that is formally uncharacteristic of democratic regimes. Democratic regimes are, at least in principle, rule-governed and not dominated by personalities. The AFRC rule generated new conflict in society, and a section of Ghanaians—though by no means the majority—had been left embittered. In this democratic context, for the Limann government *to ensure* justice, it must also be perceived by the citizens as *actually* doing justice; that is, subjecting cases to established judicial procedures. The AFRC's combative politics simply had no place in the Third Republic. Given this scenario, the Limann regime sought accommodation with the AFRC members and those aggrieved by their "revolutionary" politics.

It can be recalled that the AFRC's tribunals were very swift in their judgement, their punishments were often severe, their procedures were held in secret, the accused were sometimes absent, and when present, they were sometimes not allowed enough time to prepare proper defence. In some cases, those who administered these courts were the prosecutors and judges at the same time. As a result, some segments of the population, particularly those who lost property, and the relatives of those jailed or killed, were dissatisfied with the AFRC rule and sought redress. The PNP regime came under increased pressure to review legal cases and right the wrongs committed by the previous military regime. The PNP's handling of court cases and the future of the former AFRC members was to precipitate the 31December coup.

When the PNP took power, it was aware of the constitutional crisis that the review of AFRC cases could precipitate as well as the risk to social peace posed by the presence of the ex-AFRC members. Its solution to these potential problems was to persuade AFRC members with financial inducement to go abroad to settle or study. Some accepted this and moved to Britain; Rawlings and some others, however, did not. The regime then discharged Rawlings and the other ex-AFRC members: Brigadier Joseph Nunoo-Mensah, Brigadier Arnold Quainoo (Army Commander) and Mr Charles Lamptey (Inspector General of Police) from the service. The PNP's decision can be interpreted as an attempt by the civilian politicians to deny Rawlings and his sympathizers access to the military institution that they could possibly use to destabilize their rule.

By now, it was obvious that the military was a source of political instability and a menace to civilian politicians who were anxious to see democratic governance take root in Ghana. Like Busia, President Limann was also worried about the danger the Armed Forces posed for civilian rule. This is evident in the President's speech delivered at a graduation ceremony of military cadets. There, he cautiously lambasted those in the army who made life so difficult for ordinary Ghanaians and he admonished the soldiers to play a more positive role in society, rather than disrupting the ruling process. His government also set up a commission to inquire into the conditions of service in the Armed Forces, how the command structure favoured or discouraged stability and discipline, and the effect that military involvement in civil government since 1966 had on morale, discipline and stability within the armed forces (cf. *West Africa*, 3 December, 1979). Obviously, the civilian regime was seeking to re-assert control over the army which had now become to them something like a "Frankenstein monster" gone awry.

The PNP's decision to discharge Rawlings and his colleagues from the army certainly disillusioned them, because it terminated their military careers. Most of them were in their thirties or early forties. It is likely that their ambitions were ended prematurely. Rawlings' refusal to go into exile and his high public visibility were perceived by the PNP as a nuisance and even a threat. Rawlings was alleged to have warned the civilian politicians that they were on probation (cf. *West Africa*, 8 February, 1982:373). There were also allegations that Rawlings was planning to overthrow the civilian regime with the help of Libya. The civilian politicians were thus ruling under the threatening shadow of the military.

To pre-empt a coup, the PNP regime put Rawlings under twenty-four- hour surveillance. But this only made Rawlings more visible and popular in the eyes of a public that continued to admire him for his moral rectitude and his ability to persuade them that Ghana could be a society in which there would no more exploitation of the "suffering" masses. He became a hero and continued to dominate the political scene, holding informal rallies and articulating the interests of the common people. Consequently, in this conflict, the PNP regime was perceived by the public as the villain.

As the PNP regime began reviewing the AFRC cases, Rawlings was supposed to come to court and testify. The transitional clause in the constitution, which disallowed the review of AFRC's decisions, had become a paper tiger, unable to protect previous leaders. New conflicts seemed to have emerged in the society and body politic as a result of the AFRC rule, which the PNP regime needed to address if it was to create the social peace and stability critical for national development. By agreeing to review cases, the PNP regime was not defending the transitional clause and, thus, the constitution.

Meanwhile, the AFRC coalition had broken apart amid allegations of corruption, embezzlement or disappearance of the monies collected (as taxes, penalties, fines, etc.) by the AFRC, favouritism regarding court cases and death penalties. This situation must have been very frustrating for Rawlings and his colleagues. By now, the tension between Rawlings and his sympathizers, on the one hand, and the PNP Government, on the other, had escalated. During this period, Rawlings came into close contact with Marxist intellectuals of the departments of Law, History, and Political Science at the

University of Ghana, where he was exposed to socialist, "dependency" and "neocolonial" ideas that were to shape the PNDC's policies initially, even though it is doubtful that Rawlings actually shared their views. Rawlings was a young officer who was highly disturbed by the developmental failure of the postcolonial state leading to suffering of many Ghanaians. This was what he sought to rectify, and socialism and dependency ideas are not necessarily the only ones that could bring about an effective state apparatus as some of his intellectual sympathizers understood it at the time.

As already noted, a shortage economy had developed in Ghana since the late 1970s. The economy had become increasingly bankrupt. In 1979, the central bank claimed that international financial institutions were refusing further credit to Ghana because it owed over 1,000 million cedis in arrears in short-term loan repayments (cf. *West Africa*, 26 November, 1979). The PNP government could not flood the market with consumer goods to alleviate the shortages. It therefore asked Ghanaians who had foreign exchange abroad to import goods into the country. Goods, however, began appearing on the market, but prices and social discontent remained high. Solving the Ghanaian economic problem would not be an easy task. Being a democratically elected regime, the PNP's actions lacked the "Action Now" and "Good Society Now" politics that defined the AFRC regime.

The Limann regime would have a long way to go in restoring viability to the national economy. But this would require more than a liberal state content with laying a minimalist predictable socioeconomic and political environment. Ghana at this moment probably needed an activating or a "developmental state" with a clearly specified developmental programme, the type that the Limann administration could not construct. But the PNP's concern to establish basic law and order and an enabling environment for investment as a fundamental prerequisite for development was significant given the chaotic and anarchical ruling process so far.

The economic problems notwithstanding, political events in the country during the PNP rule indicated some prospects towards the development of a two-party system: a prelude to political institutionalization. Ghana was evolving towards a two-party political system, reflecting the Left and Right divisions that characterized the independence struggle. As already noted, the PNP was an outgrowth of Nkrumah's CPP, while the PFP was of Danquah-Busia's UGCC-PP. By the latter part of 1981, the PFP was seeking to unite all the opposition parties whether middle or liberal, in preparation to challenge the PNP in future elections. Given this, it can be conjectured that had this political development continued, Ghana would probably have developed and consolidated a two-party democratic political system and thus putting an end to the bitter inter-elite and ideological struggle that had become so inflammatory in the postcolonial period. But the conflict between Rawlings and the PNP did not ease, nor did the economic crisis abate. On 31 December 1981, Rawlings and his Marxist supporters from the University took over the state. This time, they promised a transformation of society in a socialist direction.

Since the last day of 1981, Flt. Lt. Rawlings had been ruling Ghana to the time of this writing. His rule has become, so far, the longest in Ghana's political history. The Provisional National Defence Council (PNDC) that he set up was a military regime, although some of the regime's cabinet positions were held by civilians. But, since 1992, the regime has transformed itself into a "democratic" one. Ghana has achieved political stability in the last twenty years or so mainly as a result of Rawlings' rule. Consequently, a new political tradition based on Rawlings' rule has emerged in Ghanaian politics vying for hegemony with the Nkrumah and J. B. Danquah-Busia traditions. As far as economic policy goes, Rawlings' rule can be divided into two analytically distinct periods: (i) PNDC I (1981-1983), the so-called revolutionary period; and (ii) PNDC II (1983-to the present, i.e. 1999), the neoliberal period. The next section examines these two phases of Rawlings' rule.

The Provisional National Defence Council (1981-1983): Revolution or Destruction?

The 31 December 1981 coup, brought young military officers as well as elements of the radical intelligentsia into politics. Most of these people belonged to the June Fourth Movement that emerged after the June 4 Uprising and the New Democratic Movement, and some of their senior members had been strong opponents of Busia's and Limann's neoliberal policies. Notable among them are the two Tsikata brothers and Dr. Kwesi Botchwey, lecturers at the University of Ghana of Legon. The younger generation included Chris Bukari Atim, ex-president of the National Union of Ghana Students (NUGS), and now vice-Chairman and leader of the June Fourth Movement, and Sergeant Alolga Akata-Pore. Their understanding of the Ghanaian problem shared close affinity with Nkrumah's views that emphasized the dangers of foreign capital domination and "neo-colonialism". They were also profoundly shaped by dependency theory and Marxism which were propagated in numerous evening seminars organized by the Marxist scholars at the University of Ghana at the time.

These radical groups, which formed the core of the PNDC, initiated a new discourse on Ghana's development problems. The government-controlled mass media adopted an anti-western stand and explained the evils of imperialism and why there should be a revolution. Throughout 1982, at every political rally in the cities, workers and students were told that the nation's economic problem was caused by imperialist powers (the US and Britain, in particular) and the activities of the imperialist's agents: western capital, the international financial institutions, such as IMF and World Bank in collaboration with local "comprador" groups in commerce, industry and state bureaucracy. This was to make them see the necessity of a socialist revolution. The national radio blasted out the workers' defiance: "we no go sit down make they cheat us every day" [sic]. This referred to the wealthy people and foreigners supposedly cheating the workers. Rawlings maintained that the revolution was meant to cut off the stronghold that these foreign companies exerted on Ghana (cf. *West Africa*, 28 June, 1982).

Radical scholars and students joined in the "new" discourse construction. For example, Graham (1984:4), at this time, wrote that "the Ghanaian economy exhibits all the classic features of neo-colonialism..." and related the problem to the " ...decisive presence of foreign capital especially monopoly capital in key areas of economic activity such as banking, mining, timber production, manufacturing and large scale commerce...". The National Union of Students (NUGS) also noted that, for Ghana to develop, it was vital that the monopolistic control and domination that transnational corporations had over Ghana's economy be liquidated and foreign companies, such as Barclays Bank, Merchant Bank, Standard Bank, distribution agencies such as UAC, UTC, SCOA, CFAO and the petroleum companies such as BP, Shell and Mobil, be nationalized (cf. *West Africa*, 16 August, 1982:2138).

The PNDC defined Ghana as made up of two classes—the "rich" (citizens) and the "poor" (people)—and maintained that the rich were exploiting the poor. It also claimed that there was no social justice, and without social justice, there would be no peace (cf. *West Africa*, 11 January, 1982:70). The economic crisis was thus politicized and translated into "class conflict" between the "haves" and "have-nots" of society. Consequently, the regime took a number of actions reminiscent of the AFRC rule to initiate a class struggle. They addressed re-distribution rather than productive issues. What was to be done once political power was captured was formulated in the September 1981 issue of the *Workers Banner* (a newly established revolutionary newspaper). There, it was stated that:

> ...whether there will be adequate food for all the people or not cannot be

> determined merely by working harder, or going back to the land as some self-centred politicians tell us. It will be decided principally by who effectively wields state power— the rich men or the poor workers, soldiers, policemen and farmers— the power to determine who should get what share of the national wealth. This is the crux of the matter, the hub of the whole question. Any discussion of the problem of how the people of Ghana are to free themselves from their present dilemma which does not take this as the point of departure is meaningless.

The State and Revolution: Foreign Policy Revision and Dependency-Motivated Actions

Having conceived the economic problem as originating mainly from the external world, the PNDC took some measures, most of which can be said to fit the policy recommendations of dependency theorists and goals of socialism. The PNDC revised foreign policy in favour of revolutionary regimes and socialist countries: Cuba, Libya, Eastern Europe. The new Foreign Minister, Obed Asamoah, noted that, "in the past, we have tended to lean more to the Western countries and to ignore the possibilities of fruitful co-operation with the countries of the East. Now we are going to explore those possibilities" (cf. *West Africa*, 5 July, 1982:1754). A government delegation led by Chris Bukari Atim visited the USSR, and other Eastern European countries. These visits did not yield any significant financial or material benefits, probably because these socialist countries at this time were in crisis. But Chris Bukari Atim told workers that the "socialist countries are true friends of Africa", and also assured them that the PNDC would end the monopoly of multinationals in Ghana (cf. *West Africa*, 12 April, 1982:1032). Rawlings also visited Libya, Nicaragua and Cuba. Ghana signed a 15-year trade agreement with Libya to supply tonnes of crude oil on credit. This was followed up with food aid. Colonel Khadafy also visited Ghana to cement the new relationship between Ghana and Libya. This led to the claim that the coup had a Libyan connection.

The revision of foreign policy was accompanied by anti-imperialist propaganda in the state-controlled media. The regime "uncovered" many attempts carried out by local "counter-revolutionaries" and the West to de-stabilize its revolution. This conflict reached its apogee when the regime deported some US diplomats from the country on charges of "spying" and of behaviour unbecoming of diplomats. The US, in turn, also took similar actions and imposed aid blockade on Ghana.

The early part of 1982 was a period of consolidating the PNDC regime. Parliament was dissolved and all political activities proscribed. Ongoing projects and plans of the previous regime were abandoned. For example, the Hotel Complex being built at Labadi, in Accra, was abandoned. It is only now, some 12 years after, that the government has allowed the reconstruction of the project. In the PNP's programme, however, this project was supposed to have been finished before the silver jubilee celebration of independence. The regime cancelled the PNP's plan for the silver jubilee celebration of independence, claiming to conserve national resources. All the cars and commemorative items imported by the PNP regime to celebrate the silver jubilee were confiscated. These cars were later distributed to state functionaries, revolutionary cadets, and mobilization squads for revolutionary activities. As it was, many of these people only projected themselves as revolutionaries in order to get free cars. It is, however, not clear what the regime did with the other items.

As usual, politicians were arrested and their bank accounts and those of their wives and children were confiscated. Some 200 bank accounts were frozen. The Third Republic Constitution that could provide a basic framework for state actions and behaviour (which Rawlings had endorsed), was now discredited as not expressing the "general will" of the people and containing "abstract paper guarantees". Rawlings promised that nothing would be done from Christianborg, i.e. the Castle (the seat of oppressive colonial rulers and ironically, also the seat of most governments of independent Ghana) without the consent and authority of the people (cf. *West Africa*, 11 January, 1982). The laws of the country were now decrees made by Rawlings with the help of an informal network of Leftist intellectuals and friends around him.

The PNDC reinstated all the soldiers that the previous civilian regime had dismissed and most of them became key members of the PNDC regime. Rawlings assured military officers that, this time, the coup was not against them. The PNDC accused the previous regime of having turned the hospitals and clinics into graveyards and having pursued neoliberal policies that led to the imposition of a neocolonial economy (cf. *West Africa*, 19 April, 1982:1046). It also blamed the PNP for using the judiciary to undermine the constitution and the gains of the AFRC, and for converting parliament into a self-interest club for politicians. It also charged that the PNP regime had not held democratic elections and maintained that, the PNP' Investment Code had "..opened the country to rape"by multinational companies, and the Tono Irrigation Project, for example, in the hands of the British firm, Messrs Tata & Laye, was a "sell out" (cf. *West Africa*, 15 February, 1982:482). It claimed that, "instead of the proper harnessing of the nation's resources, the government was bombarding the people with the claim that Ghana needed to attract foreign investment to ensure development". Thus, according to Rawlings, the PNDC would examine all past foreign agreements to ensure that future investment was closely related to Ghana's needs (cf. *West Africa*, January, 1982:68-73).

The PNDC sought to evolve a new radical state with new macroeconomic management system and institutions to transform society and "de-neocolonize" the economy. Rawlings maintained that his coup was a "people's revolution" for "social justice" and a "holy war" against international imperialism and local exploiters. He also promised popular participation in decision-making forums. The choice of the regime's name, the Provisional National Defence Council, with the pre-fix "provisional", was to suggest that his second coming was not a self-imposition. Rawlings proposed that a popular democracy must be the ultimate goal of the revolution. Power was to be exercised temporarily by the PNDC until a more durable national forum for participation radiating from the local, regional and the national levels, was developed. The PNDC thus conceived itself as a vanguard party leading the peasantry and a proletariat class to a revolution.

In simple terms, a *revolution* can be seen as a radical break with the past and a change in institutional orders, and this calls for new structures and ways in society. In this respect, the PNDC established Peoples' and Workers' Defence Committees (PDCs and WDCs) in local communities and at workplaces. The PDCs were also organized at district and regional levels. An Interim National Coordinating Committee (INCC) was set up to do the work of a National Defence Committee. The main goals of these new structures were to "defend the rights of the ordinary people; expose and deal with corruption and other counter-revolutionary activities... maintain collective national discipline and supervision of national resources ... afford everyone the opportunity to participate in the decision-making process in the country" (cf. *Daily Graphic*, 12 January, 1982). They were to be organs for "participatory democracy". The WDCs were, at the same time, to be organs of industrial democracy by ensuring worker participation in the decision-making process in workplaces in order to bring about increased production and efficiency, and to defend the interests of workers against arbitrary or unfair management practices (cf. *Legon, Observer,* April 1982:6-9). These new institutions also enabled the PNDC to penetrate the wider society and to *hegemonize* its rule.

The structure of the PDCs was to be modelled on Libya's *jamahiriyya* system of government based on Muammar Khadafy's "Third Theory".[1] Consequently, a centre for the study of the *Green Book* was opened at Dzowolu in Accra. Students and revolutionary cadres were sent to Libya and

Cuba to learn revolutionary practice. In this phase of the PNDC rule, state actions and policies remained highly arbitrary, uncoordinated and ad hoc. Yet, some were notable because of their adverse impact on society.

Soon after coming to power, the PNDC withdrew all 50 cedi notes (the highest denomination at the time) from circulation, a measure it claimed was intended to mop up excess liquidity in the economy. But it was also a strategy to reduce the wealth of the middle class, whose activities and life-style the regime had defined as a drain on the economy and one cause of the national malaise. The demonetization measure was implemented without allowing people enough time to deposit their monies in the banks before the deadline. This wiped out much of the wealth of the rural population, which did not have easy access to banking facilities. The government, however, promised paying compensation to those who managed to deposit their 50 cedi notes in the banks, but it defaulted. In 1983, a year later, the government introduced new 50 cedi notes, and lager denominations of 100 and 200, thus making nonsense the previous exercise (cf. Agyeman-Duah, 1987).

The old AFRC thinking that those in commerce were exploitative of the ordinary Ghanaians was revived. Once again, the escalating prices of goods in the country were attributed to the evils and wickedness of the sellers. Task forces were formed to force sellers to reduce prices to the levels set by the regime. Again markets in the cities were burnt by soldiers and revolutionary cadres when traders refused to reduce prices (cf. *West Africa*, 1 February, 8 March, and 26 April, 1982). Traders were beaten and their wares and capital were confiscated. They were also brought before Price Tribunals and were often sentenced to various terms of imprisonment for making "illegal profit" (cf. *West Africa*, 20 March, 1983: 487). Some traders, however, recalled the destruction of markets in Accra, Kumasi and Koforidua and the AFRC's policies and unilaterally reduced prices. The government set up People's Shops to distribute consumer goods at government approved prices and women were banned from selling goods other than foodstuffs. Transport costs were also reduced throughout the country. It was estimated that prices of goods went down by 15-20 percent and transport fares by about 40 percent (cf. *West Africa*, 19 April, 1982:1042). But these actions brought the economy into further crisis. In its 18 January 1982 edition, the *People's Daily Graphic* reported that much of the trading activities in Accra came to a complete halt. Many stores remained closed with the notice "stock-taking". Those that opened displayed only empty shelves.

The PNDC also set up Citizens Vetting Committees (CVCs) to investigate the rich—mainly parliamentarians, managers and their wives and children—as well as those whose life styles were considered "incompatible" with their income (cf. *West Africa*, 11 January, 1982). The criteria for what constituted a "compatible" life style was, however, not objectively specified, except for the idea that one should not have a bank deposit of more than 50,000 cedis (i.e. US$1250) by December 1981 (at which time the unofficial market rate of exchange was about 40 cedis to the US dollar). Consequently, bank deposits exceeding 50,000 cedis were frozen and those who could not account for their monies forfeited them to the state (cf. Agyeman-Duah, 1987). Government's revenue in this period increased mainly as a result of the improved tax collection system and the imposition of taxes on people as well as penalties for alleged tax evasion. Furthermore, "illegally" acquired property (houses, cars, etc.) was also confiscated by the state. Millions of cedis in taxes were collected from those who took "advantage" of the weak state of previous regimes and "refused" to live up to their tax obligation. But this also reflects the decline of the postcolonial state's performance and also its administrative capacity to tax its citizens for its own revenue.

In addition, special courts known as People's Public Tribunals and extra-judicial bodies were established with extensive powers to try those who committed crimes against the people. The traditional modern courts were defined as bourgeois institutions: expensive, cumbersome and lacking in social justice for the masses, and were denied powers to review the cases tried by the Public Tribunals. The Public Tribunals had power over life and were swift in their deliberations, delivering quick and sometimes cruel sentences. There were arbitrary arrests and detention without trial; the accused were sometimes denied the due process to prove their cases (cf. *West Africa*, 15 February, 1982). Decrees were made retroactive and used to undermine the modern courts or oppress political opponents. Laypersons were allowed to present cases in the tribunals as a means of bringing justice to ordinary people and avoiding the confounding technicalities characteristic of the traditional modern courts. The regime also threatened "economic criminals" with death by firing squad. In their first year of their operation, the tribunals handled over one thousand cases and imposed long sentences of imprisonment and executions for crimes ranging from stealing, selling above controlled-prices, smuggling and fraud (cf. Ray, 1986:103).

The PNDC also intervened in the market of rental housing. Rents were lowered below prevailing market prices by the PNDC Rent Control Decrees of 138 and 5. These were followed by PNDC Decree 7, which gave "every Ghanaian who has no where to sleep" the right to occupy any empty house or rental unit if the owner could not be identified (see Ahiakpor, 1985). Property owners who resisted these actions had their houses confiscated by the state. Later, the Accra Landlords and Ladies Associations unilaterally reduced rents by 15 percent (cf. *West Africa*, 5 July, 1982:1787 and 15 April, 1982:483). The rent measure did lower the cost of housing for the poor, but it also generated conflict between tenants and property owners. Rental housing, which was already an acute problem in urban Ghana, was largely in the hands of the private sector. As such, the PNDC's rent policy paradoxically reinforced the practice of advance payment for several years before tenancy could be granted; a practice that normally works to the disadvantage of the poor.

Also set up were Interim Management Committees (IMCs) made up of workers' representatives to manage public corporations and state enterprises. The WDCs were also empowered to supervise the activities of managers of commercial houses, public corporations, and industries (cf. *West Africa*, 15 April, 1983). But these committees and the WDCs became a source of conflict between workers and managers. Having acquired new political power, and having understood their predicament as caused mainly by their managers representing the interest of foreign capital and agents of imperialism, workers did not hesitate to take control of companies and protect their interests. For example, in the Ghana Textile Printing Company at Tema owned by UAC, workers chased out the managers and occupied their offices and bungalows as an expression of the people's power. Some 560 workers on compulsory leave because of lack of raw materials, were reinstated. These workers accused the management of corruption and economic sabotage. A government statement later confiscated the assets and accounts of the company and held in the name of the people (cf. *West Africa*, 15 November, 1982:2991).

Numerous physical attacks against managers and companies were also widely reported in the newspapers. For example, the warehouse of the American company Agripetco, in Tema was burnt down. Also, in Tema, one foreign factory manager was ordered to leave the country within 48 hours. Later, a PNDC statement described the action of workers as a "clear example of the historic vigilance of the working class to protect themselves and national interest", and showed the resolve of working people to liberate themselves from exploiters which must be the primary goal of the period (cf. *West Africa*, 29 November, 1982). The WDC also locked their managers out or paraded them before Peoples' Tribunals, where they received stiff fines and prison sentences. In one other instance, the WDC of University of Ghana, dissatisfied with the University administration, led University workers on strike and demanded the resignation of some of the administrative personnel. When the Accra Regional Secretary appealed to them to go back to work, they refused, indicating the lack of control some leaders had over workers at this period.

The PNDC's revolution also sought to destroy the power that indigenous Ghanaian businessmen and industrialists were said to have in society. Most of them were also said to have evaded taxes and had their assets confiscated by the state. For example, the TATA Brewery Company, in Accra was seized from its Ghanaian owner, Mr. J. K. Siaw and turned over to the state because of tax "evasion" and other business "irregularities". The company was later sold to foreigners as Achimota Brewery Company (ABC). In Kumasi, Dr. Kwame Safo-Adu, who secured a loan from the African Development Bank and the World Bank to build a pharmaceutical factory to employ about 30 workers, had his factory vandalized by revolutionary cadres (cf. Ayittey, 1992:334).

Thus if the Ghanaian business class was actively supported during the Busia period or benignly ignored by other regimes, now it was under direct attack by the PNDC government. A further blow to the middle class life-style came when the government banned importation of luxury cars. The WDC of the Customs Department also impounded some 100 cars, imposed a prohibitive fine of 20,000 cedis, and gave a deadline after which the cars were to become state property. Consumer durables, such as VCRs and video cameras were defined as luxuries and super taxes were imposed on them accordingly.

The regime also threatened the nationalization of foreign-owned companies. VALCO, Agripetco, (an American petroleum-drilling company), the UAC, UTC, CFAO, SCOA and the British mining subsidiary, Lonrho, were threatened, although none was nationalized. But the regime successfully renegotiated existing contracts with VALCO and Agripetco. It made the importation of foreign goods the monopoly of the GNTC and established new agencies such as the State Procurement Agency to import and distribute goods to challenge foreign-owned companies in distributive trade. The Banking Act of 1970 was amended, barring foreign banks from retail banking and redirecting them to other specialist banking. The government also declared its intention of raising its shareholdings in foreign-owned banks and insurance companies from 40 to 80 percent, effective for Barclays and Standard Bank (Ghana) Ltd. The seizure of working capital and means of production, as well as restrictions placed on private economic actors, undermined the trust that could exist between the state and its business community.

Yet not all the PNDC members endorsed these so-called socialism and dependency-motivated actions. Worried about the deteriorating environment and the social anarchy in the country, Brigadier Nunoo-Mensah, Chief of

Defence Staff, and member of the PNDC regime, warned that the politics of the era were not in the interest of the nation; that Ghana could not go on antagonizing all nations; that Ghana was too small and weak to engage in big politics; and that it was wrong to think that Ghana could develop without indigenous and foreign investment. He also warned of the dangers of importing other political systems or ideologies that had nothing to do with Ghanaian realities (cf. *West Africa*, 12 July, 1982). On 22 November, 1982, the Brigadier resigned from the government, claiming that he could not influence politics. Another dissenter was the PNDC's own secretary, B. B. D. Asamoah, who argued that Ghana faced a "human problem and not that of ideology", and whether one was a capitalist or socialist did not matter because the only way to overcome the economic problem was through hard work. The Secretary lost his job after making this statement (cf. Ahiakpor, 1985). In calling for justice and fighting for the poor, the PNDC was easily perceived as a virtuous regime that had overthrown supercilious and non-progressive civilian politicians held as responsible for an unjust order. In this context, such a dissentious view seemed a defence of an exploitative status quo and its holders mischievous and misguided.

The good intentions of the WDCs and PDCs (checking smuggling, cleaning their environment, etc.) did not last because these institutions became avenues for distributing "essential commodities". As such, they attracted people with diverse interests and of dubious character: hooligans, school drop-outs, criminals and the opportunistic rich men. These people took advantage of their recently acquired political power and quasi-judicial and coercive abilities to impose their own rules on society. They thus came into conflict with the police and the army since their roles were parallel. The PDCs also imposed their own rents on property owners. The PDCs are known in the collective memory for mounting their own road blocks, disrupting day-to-day economic activities and travels, seizing goods they designated as meant for *kalabule* activities and redistributing them sometimes among their ranks. In some instances, members became active participants in the very activities they were supposed to check since they often did not have regular sources of income. Indeed, most of their rewards depended on what they were able to seize or confiscate. Some people also tried to use the PDCs to settle land disputes, family feuds and other personal scores.

In other circumstances, they became something like the "anti-chief movements" in the colonial period, attempting to usurp traditional political powers. They thus came into conflict with traditional rulers. In the Ashanti region noted for its long-standing resentment of radical Accra politicians, the

PDCs and WDCs were crushed long before even Rawlings could come to think of abolishing them (cf. Jeffries, 1989). This phase of PNDC's rule challenged all hitherto hierarchies of authority, threw society "upside down", and introduced new societal conflicts.

Given the conflict and social anarchy created by the revolutionary organs, Rawlings was personally disturbed about their negative activities and charged that the middle class, managers, and chiefs should be members of WDCs and PDCs. But these groups had long been alienated from the revolution and remained indifferent, if not openly hostile, to the socioeconomic transformation the PNDC was seeking. In December 1984, the PNDC dissolved the National Defence Committee and replaced the WDCs and PDCs by Committees for the Defence of the Revolution (CDRs). The government also replaced the IMCs of public boards and statutory corporations with Joint Consultative Committees (JCCs). These JCCs were to act only as advisory bodies to managing directors who were now to assume full responsibility for management (cf. *West Africa*, 4 February, 1985:201).

The PNDC's revolutionary politics no doubt hit the middle class hard. Worried about an actual or potential "counter-revolutionary" response, the regime resorted to undemocratic practices and censorship of the media, propaganda, persecution, and sometimes executions of opponents, to tighten its grip on civil society. It maintained vigilance to determine anything fishy within its own circles and society at large. This period saw the deployment of a substantial number of soldiers, police, border guards, PDCs and resources (money, weapons, vehicles, etc.) for state security. Numerous road blocks were mounted across the country by the military, police and PDCs to look for threats to the regime. A dawn-to-dusk curfew was imposed in this period, making it illegal for anybody to travel between six in the evening and six in the morning. Travel in the country was disrupted by these security measures, not to mention the waste of time and the attendant vexations involved in the numerous body checks travellers had to undergo. In this first phase of its rule, the PNDC had to ward-off a number of coup attempts. The regime accused neighbouring countries of harbouring dissidents and coup makers and closed Ghana's borders. Ghana's relations with its neighbours remained strained. This undermined any legitimate commercial intercourse that Ghanaians could have with their relatively better-off neighbours, thereby lessening the economic crisis.

The Socioeconomic Impact of the Revolution and the Reactions of Civil Society

By the end of 1982, the cost of the revolution had come home. The revolution did not reverse Ghana's socioeconomic decline. On the contrary, it exacerbated it because the PNDC's policies dramatically reduced incentives for productive or entrepreneurial activities. Hunger became a reality of everyday life, and daily survival a serious challenge. In this early phase of PNDC rule, about one-half of the urban and two-thirds of the rural population were below the poverty line. The estimated daily intake of calories per capita met only 68 percent of requirements, and previously eradicated diseases such as yaws and yellow fewer had resurfaced. Infant mortality rates, which fell from 132 per 1,000 in 1960s to 86 in the late 1970s, rose to 107-120 per 1,000 during the 1980-1983 (Kraus, 1991). Many boarding schools were closed down for lack of food. The expression "Rawlings' Chain" was used symbolically to describe the gaunt-looking and emaciated people of the era. Between January 1982 and April 1983, prices of locally produced food rose by 600 percent; the Urban Consumer Index (1977=100) averaged 363 in 1980, 800 in 1981, and, by May 1983 was at 2,222.6. The GDP per capita had fallen by 16 percent of the 1972 levels (cf. *West Africa*, 19 March, 1984:618).

In 1981, per capita income was US$410. By 1983, it had dropped to US$345. Bread, sugar and other basic items completely disappeared from the market and, ironically, the government had to receive tonnes of food and medical aid from the imperialist countries: the US, Britain, West Germany, Italy, and the Red Cross as well as the World Food Programme (cf. *Peoples Daily Graphic*, 17 February, 1982; Ahiakpor, 1985) to feed the nation. Foreign exchange transactions in the parallel market were 22 times that of the official rate. Sky-rocketing prices and triple digit inflation undermined the real incomes of formal sector employees. As a result, most urban dwellers had to engage in multiple jobs to cope with life, reducing the hours spent in office and factories. It was estimated that the average public sector worker spent only 15-20 hours a week at his or her job by the early 1980s. The lowest paid public servant's real income was said to be 40 percent of its 1975 level, while the best paid public servant was, astonishingly, 11 percent. The national transport and health system were virtually inoperative. This resulted in the severe curtailment of production, investment and savings, as well as institutional decay.

By this time, Ghana's share of the world cocoa market fell from 35 percent in the mid-1960s to less than 10 percent by 1982, and the country's significance in the international arena declined. Cocoa production, which reached 538,000 tonnes in 1964/65 crop year, had declined to 247,000 in 1980/81 and then to 150,000 tonnes by 1983/84 (see Table 1 in Appendix). Throughout the early 1980s, average capacity utilization in industry was not more than 20 percent; in textiles, it was below 10 percent (cf. Sandbrook and Oelbaum, 1997). This accelerated underemployment and unemployment. The formal economy further contracted, and increasingly became informalized. Such then were the socioeconomic conditions prevailing in this era, prompting the World Bank to declare that, the average Ghanaian was at this time poorer than he or she had been at the time of independence.

To aggravate this precarious situation, in 1983 Ghana experienced a serious drought with bush fires that destroyed crops. This resulted in further food shortages. The Ghanaian borders, however, remained closed so that Ghanaians could not engage in lawful trade with their neighbours to ease the crisis. Still, in 1983, over one million Ghanaians were repatriated from Nigeria. Although the regime made efforts to the transport them back home, the efforts were haphazard as Ghana's borders remained closed intermittently for fear of political de-stabilization. The arrival of this huge number of people put further pressure on a collapsed economy, dilapidated socioeconomic infrastructure, and weak informal social support system. This further reduced Ghanaians to appalling living conditions. This period also saw a significant rise in criminal activities: armed robbery, fraud, moonlighting and cheating increased in society as a whole.

Deepening economic crisis generated disenchantment and even hatred towards the PNDC regime. The urban groups that supported the revolution in its virgin hours fell apart. University students now opposed the regime and openly demonstrated against it. The state-owned media portrayed them as an ungrateful privileged class educated at the expense of taxpayers—in this case, the workers. Consequently, the workers, in turn, sided with the PNDC regime and defined the students as "enemies" of the revolution. Workers from Tema mobilized and openly confronted the students. Later, workers at the University of Ghana disconnected utility services, vandalized them and closed down the university. Other university students joined in these challenges to the PNDC and the government had to close down the universities for one academic year (cf. *West Africa*, 12 July, 1982:1842; 1 November 1982:2859; 3 January, 1983:55) to avert further social anarchy.

The anarchy of the era was also manifested in a series of murders reported across the country, allegedly perpetrated by state security agents, most of whom took the law into their own hands. In one high profile murder case, three judges and a retired army officer were abducted and murdered on 30

June 1982. It was alleged that some key officials in the government were directly involved and that the murders were meant to settle some earlier grievances. The PNDC state can be seen at this moment as failing in the most elementary state function: to protect life and property. Thus, the regime was rapidly losing popularity and even legitimacy, as evident in the calls by civil associations, such as the Ghana Bar Association, the Association of Recognized Professional Bodies and the Catholic Bishops and Protestant Christian Councils for an end to revolution and return to democratic rule.

Although, the PNDC promised "grassroots democracy", it failed to devolve power, at least at this moment. Whenever PNDC members were asked when they would hand over to an elected government, they often retorted, "Hand over to whom?". Later, Mr. Justice D. F. Annan said, during one of the PNDC's political rallies, that the "time-frame [for democracy] will be determined by the speed with which Ghanaians come to realize the meaning of democracy" (cf. *West Africa*, 27 October, 1986:2255). Arguably, these new postcolonial leaders like the colonialist before them, did not judge their people or citizens as competent enough to practise liberal democratic politics. Rawlings also stated that there would be no handover of power (cf. *West Africa*, 1 February, 1982).

The refusal to hand over power further brought the regime into conflict with pro-democracy civil societies thus aggravating an already chaotic economic and political life. As a result, the government developed authoritarian and neopatrimonial structures to deal with the chaos and also counteract pressures for democratic rule. Power was concentrated in Rawlings who also personally dominated the PNDC. Trusted relations and friends were given security and other strategic positions to ensure the regime's survival. Authoritarianism and more importantly neopatrimonialism has thus begun to emerge in Ghanaian politics along with its attendant patron-client relationships that weaken the state and institutions and legal norms and reward unproductive and parasitic economic behaviour. Consequently, the state was unable to transform society in the revolutionary direction it officially proclaimed. Socialist ideology was simply a self-serving rhetoric used to maintain political power and exclude other rival elite groups who were likely to offer alternate development programmes.

The politicization of the economic crisis, the seizure of assets and property, and what could be regarded as "Robin Hood" politics further undermined the productive infrastructure of the economy, and soon there was virtually nothing to re-distribute. The WDCs did not increase productivity as the regime had wanted, because among other things, they spent their working hours attending political gatherings where they were lectured about the injurious effects of imperialism, the benefits of autarchical development, and the workers' revolution. The PNDC's actions that flowed from its ill-digested Marxism or socialist principles and "dependency" understandings hurt the middle class and foreign capital interests, but somehow indirectly they also hurt the poor, because they were destructive of the productive infrastructure of the society. Consequently, the regime was accused of "distributing" poverty.

There is a certain irony here. This first phase of the PNDC's rule sought socialist transformation of Ghana and to lessen its dependence on the external world, but the regime was unable to realize this goal. Rather, it actions further destroyed the ailing economy, an economy that it needed to build in order to ward-off or forestall imperialism and its agents. A common lesson of life is that it is easier to destroy than to build. This revolutionary phase did not last more than a year, but it destroyed the socioeconomic and political environment for production and investment and undoubtedly left a scar on the psychological foundations of society that would come to affect the development of politics and society in the future. A paralyzed economy was unable to ensure the state's reproduction, let alone, permit it to shoulder developmental responsibilities. Ghana's economic decline had finally reached its nadir, with tragic human dimensions and something drastic had to be done to reverse the situation.

In terms of the theoretical framework of this study, these revolutionary actions bear striking similarities to actions that resulted from arbitrary rule in premodern Europe which Weber and other economic historians such as Jones conceive as having held back economic development. In this respect, one can argue that the PNDC's actions, however justified, were also inimical to development because they destroyed the institutional and productive infrastructure of the society. But since it was mainly the activities of foreign capital and local business community that defined governmental actions in this period, it will be useful to examine further, the extent of control multinational corporations exercised over Ghana's economy, their role in the decline of Ghana's economy and the necessity of class struggle at this early stage of the country's capitalist development.

Ghana, Multinational Corporations (MNCs) and Class Struggle

As noted in Chapter 3, as a result of colonialism, a number of western multinational companies have come to dominate the mining, timber and commercial sectors of the economy. When the CPP regime proclaimed socialism for Ghana and Ghana practised economic nationalism, the power of multinational companies did not diminish. But the threat of nationalization during the Nkrumah era did not bring significant direct foreign investment into Ghana. This situation was not rescued by the short-lived liberal regime of Busia that sought greater foreign participation in the Ghanaian economy. This scenario is not difficult to explain because there are many factors that influence the decisions of investors to invest in foreign countries. Apart from such things as infrastructure, banking facilities, raw materials, skilled labour and markets, there are also invisible assets that come into the picture. It takes time for investors to observe a country and calculate that an environment for business is emerging. It takes more time to make decisions to move in, to locate sites, bring in equipment, install it, explore local markets, begin producing, and make profits. All this requires at least several decades of political stability, supported by predictable rational rules and institutional guarantees, which were identified in Chapter 2 as basic for sustained capitalist development. Given the story so far, few of these conditions can be said to have emerged in postcolonial Ghana.

Indeed, the odds against the MNCs were not lowered in the post-Nkrumah period when the local environment became so volatile and almost without predictable and calculable rules and administrative framework. From 1972 to 1979, the Ghanaian state came under the military junta of NRC-SMC, which took measures to increase state's participation in foreign companies. Though these measures were defendable in terms of protecting national interest, the manner in which they were implemented, coupled with the unilateral refusal to pay foreign debts mainly to the West, scared away potential investors. As it was made abundantly clear in the last chapter, when the PNP came to power, it noted that foreign investment had bypassed Ghana and would continue to do so unless policy environment changed. Its investment code was thus meant to reverse the situation, but the regime did not live to see its policy through.

These trends suggest that Ghana's policy environment was becoming risky for foreign investment. In one early study of Ghana's decline, Killick (1978:126) maintains that foreign investors were showing little interest in Ghana. There is also evidence to suggest that, in the later years, Ghana was beginning to experience de-industrialization as far as foreign investment was concerned. Data compiled by Price (1984) indicate that, between 1965 and 1969, direct foreign investment in Ghana stood at US$153.9 million; between 1970-74, this became US$80.8 million; and by 1975-79, it further reduced to US$65.6 million. For this same period, average annual investment was US$30.8 million, US$ 16.2 million and US$13.1 million respectively. Indeed, impressionistic evidence also suggests that some foreign companies and their subsidiaries (e.g. French SCOA) operating in Ghana curtailed their activities in the late 1970s, and by the early 1980s, American Firestone had pulled out of Ghana. This, according to Price (1984) was mainly due to the difficulties these companies were encountering.

When Ghana is seen in a limited comparative perspective, the picture indicates that foreign investment in Ghana during this period is not impressive. It has already been noted that Cote d'Ivoire is one country that is similar to Ghana. Since its independence, Cote d'Ivoire's economy has been closely tied to that of metropolitan France. As such, it was a "neocolonial economy" *par excellence*. Foreign investment has been an ingredient of the Ivorian economy, and, from the dependency perspective, such an economy should decline in the course of time. Yet, this did not happen, or at least the economic decline of Cote d'Ivoire in this period did not occur on the calamitous scale of Ghana's. It is no wonder that Ghanaians went to Cote d'Ivoire and Togo (a relatively obscure and resource-starved neighbouring country) and Nigeria to acquire the basic necessities or settle to make a better living.

Furthermore, when one looks at statistics on direct foreign investment from the OECD countries in Ghana and Cote d'Ivoire, the evidence indicates that Ghana was not benefiting from outside investment. For example, between 1978 and 1980, foreign bank loans to Ghana amounted to US$1.3 million; for Cote d'Ivoire, it was US$141.0 million. The stock of foreign direct investment at the end of 1978 was US$280 million for Ghana; for Cote d'Ivoire it was US$530 million. This heavy foreign investment in Cote d'Ivoire should lead to economic decline, according to dependency theory's prognosis, but it was Ghana that was struggling to survive with low foreign investment. By 1979, Cote d'Ivoire's Gross National Product was US$8,810 million, and per capita income was US$1,070. Ghana in contrast had GNP of US$4,470 and a per capita income of US$400 (Nash and Kantrow, 1987). There are no doubt limitations with the use of per capita statistics as indicators of the level of development. One such limitation is that, as an average statistic, it obscures the actual distribution of income or wealth in society. Yet, it gives a fair indicator of the level of economic development and standard of living in a given country.

While this is not a defence of the activities of multinational corporations in the developing countries, Ghana's economic progress was not helped when such "excessive" understanding of the injuries inflicted by MNCs on local economy defined the scope of governmental actions. There was too much at stake: not only the financial investment of the MNCs but also the economic health of the nation, the activities of indigenous entrepreneurs critical for increasing national wealth and the expectations of better life for citizens. The PNDC's anti-MNC discourse stressed dependence and exploitation and ignored the interdependence that exists among nations. It denied the possibility of a developing country such as Ghana with a small national economic base benefiting from the advanced technology, skills and capital originating from the metropole. It also ignored the fact that once societies begin to modernize, they necessarily become connected to the outside world for good or bad. The issue then is not so much one of "dependence" as of "interdependence". The concept of dependency often prompts the call for "self-sufficiency", which, in the modern context can be a recipe for material poverty. A modernizing process entails dependence, and each country today is locked into a complex international commercial and industrial interdependence, whether "capitalist" or "socialist" or mixed. Indeed under some circumstances the penetration of multinational companies is compatible with economic growth and *development*.

The idea that indigenous businessmen, middle class or bourgeoisie (as these terms have been often used interchangeably) are conniving with foreigners to exploit their country was more often assumed *a priori* than demonstrated empirically. There was, and there is, no reason to assume that Ghanaian businessmen and industrialists would never seek their own advancement and rather be content to work in a subordination position with foreign investors in their own country. Nor for that matter, a government, be it liberal, not protect the interest of the nation, if the 1966 to 1972 pro-indigenous business policies of the NLC and Busia regimes are anything to go by. Ghanaian businessmen have looked up to the state to protect them against more mature foreign capital. Indeed, in one study on local businessmen in the Third World, for example, Kennedy (1977) has questioned what he calls several "myths" about indigenous capitalism in the Third World that derive from dependency theory. Using Ghana as case study, he argues that a local capitalist class was developing and expanding in manufacturing, and the connections this class has with foreign capital (far from limiting the expansion of local capital) has helped it in the acquisition of managerial and technical skills, leading to independence from and competition against foreign capital.

Kennedy's study suggests that a national capitalist class was developing and gradually asserting its independence from foreign capital. The argument that liberal civilian regimes and local businessmen were working with imperialist agents to exploit Ghana needs to be taken with a pinch of salt, if not totally laid to rest. Price contends that what seems striking about the Ghanaian economy in this postcolonial period, is the limited role of foreign firms, the difficulties for operating there, the general lack of interest they showed in Ghana as a site for investment, and, most significantly, the *inverse* relationship between the size of the foreign sector and the performance of the economy (Price, 1984:117). But it must be added here that, some of these problems have equally confronted the domestic entrepreneurial class. The decline of Ghana over the postcolonial years increased absolute poverty. While some people might have been corrupt and greedy, the general trend was such that the so-called rich of the society became too few in number to constitute any significant exploitative class; or for that matter, to be responsible for all the economic ills that afflicted the Ghanaian society. Like the foreign investor, the indigenous business class also had difficulties in undertaking long-term productive activity as a result of politically induced risks in the society.

Furthermore, as the Busia regime's policies show, there was no reason to think that only "radical" regimes can protect the economic kingdom for Ghanaians and liberal regimes are necessarily neocolonialist. A fundamental development challenge that Ghana faces today, is how to raise productive capacity, (i.e. modernize agriculture and industrialize) so as to break "the vicious circle" of poverty in which the majority of the people are trapped. In this quest, the interests of labour cannot be held to be more important than those of capital, since both are necessary to raise the fortune of the nation. Thus labour and capital need not be perceived as always contradictory and antagonistic categories; they can also be complementary.

In addition, the current low level of capitalist development makes the sociological profile of Ghana's social stratification system quite open. The existence of traditional mutual obligation or support systems in the wake of underdeveloped formal welfare system, the partial proletarianization of workers resulting in the straddling of labour between the modern and traditional productive sectors, the informal transfer of wealth from the rich to the poor predicated upon kinship obligations, the existence of a small industrial labour force and a large but politically conservative peasantry, as well as a large number of intermingling actors in the formal and informal sectors (Peil, 1981; Sandbrook, 1982; Ahiakpor, 1991) decompose classes and class consciousness, and question the partition of society into two classes and attributing to them political significance.

But according to the "dependency theory" which guided government actions in this so-called revolutionary era, the more a country depends on foreign capital or fails to lessen its dependence on the international capitalist system, the more it must suffer economic decline. Ghana's decline confirms this prediction, because Ghana has not been able to break or lessen its dependence on the international capitalist system. But just as correlation in statistical analyses does not always mean causation, the plausibility of a theory, and in this case dependency theory, as Price (1984) argues, might not suffice as a test of its validity.

By early 1983, it dawned on the PNDC regime that if Ghana was to survive, there was a need to reverse or improve policy and secure foreign loans. This was, however, not before missions to the Soviet Union, Eastern Europe failed to secure loans. These countries advised the regime to seek help from the IMF and World Bank while trying to hold on to the revolution (cf. *West Africa*, 25 February, 1985). The regime formally renounced its anti-capitalist and imperialist ideology and accepted the interception of the IMF/World Bank structural adjustment programme. The SAP was, however, not publicly debated. Its acceptance led to a split in the regime itself, between those who still favoured the radical Marxist-Leninist approach (the nationalization of all means of production) and those "pragmatists", who like Rawlings now, wanted a practical way now to pull the country out of its economic quagmire.

In 1983, Rawlings called for an end to what he termed "empty theories" expounded by the local Marxists who had hijacked the national discourse and made enemies of everyone who offered any alternate approaches. Indeed Rawlings blamed these Marxists for many of the mistakes of his regime noting that "they failed to offer any solutions, let alone realistic ones, to the problems that confronted the country in 1982.... All they do is talk, talk, talk ..." (Rawlings cited in Ahiakpor, 1991:559). The destruction of the revolution was acknowledged when Rawlings noted that "we can no longer postpone the time for halting the populist nonsense ...and making a noticeable leap forward.... We must not get into the way of thinking that revolutionary activities are a substitute for productive work" (cf. *West Africa*, 12 September, 1983).

The Secretary of Finance and Economic Planning, Dr. Kwesi Botchwey, who was an advocate of Marxist solution and critic of neoliberalism now accepted that the IMF and the World Bank's prescriptions were good for Ghana. His new "realism" was articulated this way:

> the real question is what should a Marxist do faced with the real situation in Ghana... the objective of feeding the people... not from the point of view of any ideal conditions existing in one's head, but the real conditions on the ground... (*West Africa*, 28 January, 1985).

Rawlings and his Finance Minister's approach to the nation's economic problem was rejected by the radical Left (mainly the June Fourth

Movement), which still believed in complete state-ownership of all the means of production (Ahiakpor, 1991:597) and the re-introduction of a re-distributive social production system of universal free education, health care and legal services. When Rawlings dismissed their claims, they staged an abortive coup led by Sergeant Akata-Pore in 1982 to unseat the PNDC regime. Most of the coup makers later escaped to England, where they could enjoy the liberty and political freedom, as well as, high material standard of living associated with late capitalism. From there, they continued to launch criticisms against Rawlings for having abandoned or betrayed the revolution. Their departure paved the way for the implementation of the IMF/World Bank's structured adjustment programme in Ghana.

Structural Adjustment Programmes: Its Impact and Politics

The structural adjustment programme, locally known as the "Economic Recovery Programme" (ERP), was accepted in early 1983. Although the SAP is generally regarded as one policy package, in terms of implementation, a period of economic stabilization entailing the strengthening of balance of payment positions, reduction in domestic financial imbalance, reduced government deficit financing, elimination of price controls and deregulating international trade, normally precedes a more fundamental adaptation and/or planned process of structural change and economic transformation aimed at providing the conditions that sustain long-term economic growth. This other phase involves privatization of state industries, removal of subsidies, the downsizing of public sector employment, improved administrative capacity, institutional and legal reforms, the encouragement of the private sector activity, reduced role of the state in the economy, and subjecting economic processes to the logic of the market. By its thrust, the SAP aims at capitalist development in the long-run.

As a result of the SAP, the economy, in general, was liberated from the rigid political control characteristic of the pre-adjustment era. The cedi, whose value had been administratively regulated, was now allowed to seek its value vis-à-vis other major world currencies on the market. In 1983, the year Rawlings came into power, the US dollar was officially worth 2.75 cedis (Nugent, 1991), but by 1989, it was 650 cedis, and by 1999, 2400 cedis. The drastic depreciation of the cedi has diminished its importance as a store of value and most people resorted to store value in the dollar. This has resulted in the increased "dollarization" of the economy in recent years. But relaxation of foreign exchange and control and the opening of private foreign exchange bureaux enabled Ghanaians abroad to now transfer hard currency home. It has been estimated that they send home some US$350 to US$400 million annually (cf. *Economic Intelligent Unit*, 4th Quarter Report of 1998). These figures amount to approximately the same sums of money the IMF/World Bank sponsored donors' conferences were annually pledging Ghana during the SAP period.

Accurate data collection on these transfers has however been hampered by the prevalent use of informal and non-banking institutions. The banking system of Ghana has undergone successful restructuring and confidence has steadily grown, but many Ghanaians, especially those who have left to live overseas, still remember the seizure of assets by the government in 1979 and 1982, including the freeze on all bank accounts in excess of 50,000 cedis in 1982, during widespread investigations of ill-gotten wealth. The remittances have increased purchasing power in the country generally, and have moderated the negative impact of the SAP, at least for those who have relations abroad. The current economic openness contrasts sharply with the "closedness" of the past, when it was a "criminal" offence to carry on foreign exchange transactions if one was not a government official.

The economy has been stabilized and financial discipline restored, but in recent years, among other things, the NDC government's desire to ensure its rule during the recent democratization period led it to exercise less restraint in the expansion of money. Trade restrictions in agriculture, especially taxes were reduced to promote exports. International trade has been liberalized and many commodities and goods previously lacking on the market have reappeared.

As far as laying the conditions for economic recovery is concerned, the SAP, first and foremost, aims at restructuring incentives in favour of the productive sectors of the economy. The most notable achievement in this regard is in the cocoa economy. Over the last 16 years or so, the PNDC—now the NDC (see below)—has periodically revised cocoa prices so that cocoa farmers now get about 40 percent of the world price, which sharply contrasts with the 15 to 20 percent of former years. As a result, farmers have decreased cocoa sales to neighbouring countries. Cocoa farms were rehabilitated, and, by 1988/89, cocoa production had risen to about 289,000 tonnes representing 88 percent increase over the 1983/84 level which was 150,000 tonnes. This author's own observations and interviews in the cocoa-growing areas in the Volta Region in 1995 indicated that cocoa smuggling has declined. Whether Ghana can retain all the cocoa produce within its borders in the future will depend mainly on the prices paid to farmers in Ghana vis-à-vis those in the neighbouring countries.

The IMF/World Bank's SAP has thus rendered some "social justice" to cocoa farmers who are historically the backbone of the national economy but who have also been exploited by previous regimes. Rawlings himself puts this claim better: "We are acknowledging the historic debt of the whole nation to the farmer and have thus repudiated the monstrous injustices of a past in which we virtually ran the machinery of the state on the tired backs of rural producers and provided little for their basic needs" (cited in Jeffries, 1989:96). But, although agriculture accounts for more than 40 percent of the country's GDP, the government's emphasis in this area was less on positive intervention than on government withdrawal. Subsidies on agricultural machinery and fertilizers have been removed. At the same time, government divested itself from agricultural ventures, including fertilizer production. Production in gold has also increased, competing with cocoa as number one export product. For example, in the first three quarters of 1995, receipts from gold totalled US$467.788m, while cocoa generated US$362.26m for the same period (cf. *West Africa*, 4-10 March, 1996). Production in other sectors have also improved (see *The Economic Intelligence Report*, 1990-91; 1997).

Manufacturing saw early modest recovery, following reduction of import taxes that enabled companies to bring in raw materials and critical equipment for production in existing industries. But more economic "glasnost", and later, exposure of the manufacturing sector to world competition, led to the flooding of Ghana's market with relatively cheap products from the outside world, particularly the East Asian countries resulting in a steady deceleration of industrial growth. For example, employment in manufacturing fell from a peak of 78,700 in 1987 to 28,000 in 1993. Import competition has undermined small scale industries' products for the local markets so that labour intensive exports such as garments, footwear, toys, and other light consumer goods that led to the initial export thrust of the NICs are conspicuous by their absence in Ghana (cf. Lall and Stewart, 1996). The "dumping" of relatively cheap products from the outside world is often cited as one reason for the decline or slow growth of the manufacturing sector in recent years.

Furthermore, public sector employment has been scaled down. The inefficient monopolistic CMB, now the Ghana Cocoa Board (Cocobod), had some 40 percent of its total employees laid-off (Roe and Schneider, 1992). "Ghost" workers have also been removed from the salary list of Cocoa Marketing Board and other public institutions. Public sector employment has been downsized, including reductions of about 60,000 staffs between 1987 and 1992, (Gyimah-Boadi and Van de Walle, 1996) although some analysts argue that this has increased in recent years as a result of neopatrimonialism (see below).

In line with the requirements of the SAP, the state has also lessened its participation in some companies and sectors of the economy. Twenty-two percent of state-owned companies and utilities were privatized by mid-1994 or were in the process of being privatized. The government has sold part of its holdings in the state-owned Social Security and Ghana Commercial Bank in 1995 and 1996 respectively. It also sold 30 percent of the its share in the Ashanti Goldfields Corporation, and this brought the regime some US$316 million. By January 1997, Ghana Telecom was being offered for sale to a Malaysian telecommunication company in partnership with a local consortium. The NICs countries of Korea, Hong Kong, China, Singapore and Brazil, like the mature western industrialized countries, are now not hesitating to enter Ghana's economy with their products and even acquire investments. By the standards of the political rhetoric of the pre-adjustment era, Ghana has now become a "neocolonial economy". There is an element of surprise here in that the earlier Rawlings' regime which portrayed itself as "revolutionary", "anti-imperialist" and "socialist", should be the one to wholeheartedly embrace, the IMF and World Bank orthodoxy. Perhaps ideology has come to an end.

In 1988, in response to criticisms by the UNICEF and other NGOs that the demands of the SAPs (e.g. the removal of subsidies on social services, the contraction of formal employment, etc.) were socially costly, especially for the vulnerable groups of society—women, children, the unemployed—the PNDC and the IMF/World Bank instituted a Programme of Actions to Mitigate the Social Cost of Adjustment (PAMSCAD), which was to provide relief to the poor throughout the country but particularly those who became unemployed as a result of the SAP. PAMSCAD's impact, however, was less than anticipated; its resources were limited and its implementation was weak. Also, some of those adversely affected by the SAP decided to find a niche for themselves in the rural economy and the informal sector, which has witnessed a dramatic expansion in recent years.

According to official statistics and World Bank reports, in the past 16 years or so, Ghana has registered an average GDP growth of 5 percent annually and inflation has also reduced well below the pre-adjustment level (see *Africa Confidential*, 28 April, 1995). To the extent that these statistics are good measures, they indicate economic improvement. To those who are familiar with the magnitude of Ghana's economic collapse, such figures are impressive and the performance of the economy during the period of SAP is a remarkable feat, albeit material poverty is still a defining characteristic of Ghanaian society today. But all these only serve to illustrate the degree of socioeconomic decline in the postcolonial era.

The economy has far passed the stage of crisis and basic commodities have re-appeared on the market. The transport system and communication facilities have improved considerably with the increasing use of cellular (mobile) phones. Eager to portray to the world that their "medicine" is working—or can in Africa—the IMF and the World Bank have touted Ghana's adjustment programme as a success and the country as "the front-runner in adjustment". Today, this status is claimed by Uganda. But behind all this "success", lies Ghana's increased dependence on the "aid industry" in recent years. By 1994, total lending by World Bank alone to Ghana amounted to US$2,860 million (see *Africa Confidential*, 28 April, 1995), which critics as well as the IMF/World Bank now say is unsustainable.

On the socioeconomic impact of the SAP, World Bank analysts maintain that the "benefits of adjustment have been large, visible and widely shared" (Leechor, 1994). But Ghana's performance is, however, not outstanding. In a more comparative and long-term perspective, Ghana's performance is not different from that of other countries in the Sub-Saharan region. For example, in one of the World Bank's reports, it was noted that between 1980 and 1990, real GDP growth in Ghana averaged 3 percent annually, behind nine other Sub-Saharan countries including Botswana, Burkina Faso, Burundi, Congo, Mali and Mauritius and comparable to six others, including Malawi, Senegal, Tanzania and Zimbabwe (see Husain and Faruqee, 1994). The rate of domestic savings rose only slightly, from 5 percent to 8 percent of GDP, compared with a current Sub-Saharan average of 13 percent (cf. Ake, 1996:92-3). The aforementioned study also notes that 77 percent of Ghana's growth was registered in the service sector, particularly commerce, and concludes that after a decade of generally favourable adjustment efforts, private sector investment, normally the driving force for vigorous growth, has not been particularly strong (see below).

The SAP is often said to have performed badly on social criteria and there has been some justifiable concern for the socially weak groups. Unemployment has increased as formal employment has diminished. Health and school fees and university and college student loans have been instituted. The introduction of "user-fees" for social services which were previously subsidized, coupled with the persistent depreciation of the cedi, has increased the cost of living for those in the formal sector depending on fixed incomes. But it is possible to agree with Green (1988) that, given the magnitude of the economic decline before adjustment, there was no evidence or indication that Ghanaians would have been better off without the adjustment. Indeed, for the poor, who have no alternative access to basic services, a reduction in social programmes can have severe consequences. Yet, as Leechor (1994) also suggests, this concern presupposes that in the pre-adjustment period, social services were being delivered effectively, which was not the case in Ghana, given the severity of the decline.

Besides the dependence on aid, the regime's "economic success" with reforms has also been often attributed to political authoritarianism. Some authors emphasize that the IMF conditionality has reinforced authoritarianism, because the SAP hurts many people, particularly urban groups (Herbst, 1993; Franklin, 1997). This view can be overstated for Ghana. A closer examination of the Ghanaian case suggests that the PNDC faced little organized opposition to the SAP as such. One reason for this is that pre-adjustment industrial decay led to a substantial reduction in the number of workers in the mines and factories, and charges of corruption and sycophancy levelled at union leaders in the 1980s, as well as a split in the Trade Union Congress, further weakened the response of organized labour to adjustment policies. Furthermore, urban consumers, who elsewhere (Zambia, for example) had resisted the removal of subsidies and price controls, were not deeply affected by such measures. This is because they had already been paying inflated parallel market prices for "free" public services and "controlled" consumer goods (see also Sandbrook and Oelbaum, 1997). On the other hand, the SAP was greeted with relief by the business class because it was indirectly an attack on the PNDC's class struggle and anti-capitalist or business ideology.

While authoritarianism correlates with the SAP, it would be misleading to draw a causative relationship between the two. This is because the PNDC regime was already authoritarian before the coming into effect of the IMF/World Bank policies. Early opposition to the regime came from civil associations, which perceived the regime as destroying human life, liberty and property. Such opposition were primarily political and not necessarily economic or because the PNDC imposed a socially costly adjustment programme on Ghana. After the radical Left went into exile, Rawlings found it easier to make new policies and undertake austerity measures which have historically offended the "urban-populist coalition".

On the other hand, the SAP was not a popularly "contested" issue because the Ghanaian crisis itself offered one advantage: it made many urban Ghanaians realize that some changes had to be made for the good of society and their welfare. Consequently, as Leechor notes, the reform met little internal resistance. This is not to suggest that there was no resistance at all, but on the whole the reform programme got a "warm" reception, even from groups that were supposed to be hurt by it. Toye (1992) also notes that the depth of economic distress seems to have played a contributory role in the relative ease with which the adjustment programme was accepted in Ghana.

Furthermore, as it has already been implied, the Ghanaian economy had long become informalized as a result of the collapse of the formal economy. Many people tried to find a way to survive outside the formal economy or regarded their participation in it as part-time. The state and society had long been estranged from each other. The government retrenched some 40,000 people but in reality most of them were already outside the formal economy. The PNDC regime also "got away" with the SAP because most of those who were affected by retrenchment were absorbed by the informal welfare system.

In addition, generally, retrenched workers accepted their fate in good spirit as a popular view often expressed at this time was that government employment was no longer profitable, which also shows the degree of economic collapse and the drift of general society from the state and the formal economy. Many government employees had long maintained that they were actually doing nothing in their offices and government employment itself was unprofitable. There was an understanding in society that formal employment had to be trimmed down. Indeed, during the "democratization" process that followed a decade-old SAP, few opposition leaders argued for the necessity to reverse the IMF/World's SAP, although some would have liked to see its modification to enable particularly local industries to be more viable.

Given that Rawlings forcefully usurped power from a democratically mandated regime, his regime was seen as illegitimate, especially by those who were negatively affected by his so-called revolution. They put up a challenge to his power through various civil actions. This made the regime resort to authoritarian tactics to deal with opponents, leading to what Professor Adu Boahen and indeed Ghanaians dubbed "the culture of silence", which refers to the silence of the population over government actions and policies for fear of being detained, liquidated, dragged before CVCs or being molested (cf. Nugent, 1995:163). Furthermore, given the destruction of wealth and property of the middle class and the alleged politically motivated killings that occurred under the regime, it is doubtful that PNDC regime would have been democratic without the IMF/World Bank programme. The SAP aims at capitalist transformation of Ghana. In a sense, this is a revolutionary development in a country where, officially, socialism has enjoyed a pride of place in the market of ideas and where populist elites and groups have held politics hostage throughout much of the postcolonial period for addressing re-distributive rather than productive issues.

Economic Liberalization and Private Sector Investment

One main argument underlying SAPs for rolling back the state's hand in economic processes is to promote the private sector as the main force for economic development. The PNDC has officially accepted this view. As a result, in 1985, it promulgated a new Investment Code to regulate and entice foreign investors into the country. In early 1990, a stock market was established to accelerate the pace of foreign investment and mobilize local savings for capital investment. In September 1995, another investment code was promulgated: Ghana Investment Promotion Centre (GIPC) Act. Unlike the Investment Code of 1985, which was regulatory of foreign investment, this new one was supposed to be promotional. It lists a number of incentives to make Ghana attractive to foreign investors. For example, the Act reduced the minimum foreign capital requirement from US$60,000 fixed by the 1985 code, to US$10,000 in the case of joint venture, while for a wholly foreign-owned enterprise, the minimum level was lowered from US$100,000 to US$50,000. Investors are free to establish their companies without prior approval, except in mining and petroleum industries, which the government must approve. Other incentives include zero-rated taxes (duty free) on machinery and equipment and their parts and tax holidays.

Through this code, the government hopes to encourage foreign investors into both agriculture and industry. The GIPC has organized a forum for investment (see *West Africa*, 17-23 July, 1995:1128). In order to "sell" the country to investors, Rawlings and his economic advisors have also visited the US, Britain, Germany and Japan to meet government leaders and business communities. Ghana has been advertised in the *Financial Times*, *The Economist* and *West Africa* as well as on the Internet as an investment haven. By and large, this Code and advertisements are similar to those initiated by the Limann government, which the PNDC overthrew and accused of permitting the neo-colonization of the economy. Ghana is thus starting anew the construction of a predictable socioeconomic environment for foreign investors.

Paradoxically, however, while the government is anxious to create trust with big foreign investors, this cannot be said as far as its relation with domestic business is concerned. The lack of confidence in the business environment within the private sector continues. Ghanaian businessmen, in particular, have often described the business environment as characterized by mistrust, harassment and the absence of institutional support and, politically, a risky affair. Terms such as "ambivalent attitude" or "issue of governance" are used to characterize the risks that confront businessmen (Leechor, 1994:177). The foreign investment that has come into Ghana in this adjustment period is mainly confined to the traditional foreign investment sectors of mining (gold, diamond, etc.) and timber production. There has been little investment in the other sectors of the economy, for example, heavy and light manufacturing. Indeed, in general, little significant investment has come into the country. Ghana currently seems to be still suffering a backlash of previous mistrust on the part of foreign capital. Local newspaper reports are littered with the frequent disenchantments and frustrations experienced by foreign investors who have shown interest in the country. There continues to exist conflicting rules in the business environment, excessive bureaucraticism and lack of seriousness in many government officials and public institutions to promote investment in the country.

But Ghana may also be suffering from the general lack of interest of foreign investors in Africa. Recurrent political instability, civil wars, low human skills, lack of a reliable physical and institutional infrastructure, and a dependable administrative rule systems in Africa in general means that many African countries are not as attractive to foreign investors as the NICs are at this moment. Consequently, Africa is portrayed in much of western media reports and in some academic analyses (see e.g. Hyden, 1983), as an environment swallowing the monies of Western businessmen with little or no return.

Yet, lack of foreign interest is just the tip of the iceberg of the investment problem. Africans themselves may also be increasingly moving their capital abroad to avoid the unpredictability and arbitrariness of governments. For example, in a recent study devoted to explaining Africa's low economic performance, Collier and Gunning (1997) use capital flight data collected since 1970, computed for East Asia, South Asia, Latin America, the Middle East and Sub-Saharan Africa to conclude that, "despite a lower level of wealth per worker than any other region, African wealth owners have chosen to locate 37 percent of their portfolios outside Africa, a higher extra-regional share than any other region". They go on to predict that "if Africa reduced its capital flight to that of Asia, its capital stock would increase by a half" (Collier and Gunning, 1997:3).

Continued uncertainty about the future, arbitrary rule and lack of the rule of law and independent enforceable laws may help explain the involuntary marginalization of many countries such as Ghana. The IMF/World Bank are said to have also attributed the low of investment in Ghana directly to the actions of Rawlings and his regime (cf. Gibbon, 1992:157-9). Thus, Beckman's observation that the new focus of the World Bank is on restructuring the African state in order to make it supportive of its long-term strategy for the liberation of the market forces and entrepreneurial potentials of African society (Beckman, 1992:83) seems not to be case of Ghana. The regime of Rawlings today is considered formally democratic; in reality, however, it is somewhow patrimonialized (see below). Numerous private newspaper reports and editorials suggest that the regime's actions are undermining the activities of local entrepreneurs, particularly those who do not belong what some analysts describe as the NDC's patrimonial "political machine" that has now emerged in Ghana (cf. Sandbrook and Oelbaum, 1997).

During the 1992 political elections, for example, there were incidents in which the state was used to harass businessmen and the middle class who belonged to opposition parties. President Rawlings was quoted as having asked Ghanaians to boycott the goods produced by indigenous industrialists, such as Mr Kwabena Darkor (a renowned poultry farmer) and Appiah-Menka (a producer of household detergents) because they would get money to finance opposition parties. In 1989, long before this incident, a privately-owned company International Tobacco (ITG) Ghana Ltd. established by a Ghanaian industrialist was also forfeited to the state because of disagreement over company taxes to the state. In the rumblings between the government and the owner, the government converted the company into a joint venture whereby the State Security and National Insurance Trust (SSNIT) now owns 54 percent shares and UK Rothmans International holding the remaining 46 percent. During the election of 1996, which this author observed, the presidential candidate, Mr Kuffour of the opposition party, the Great Alliance, lamented the continued antagonism between the NDC regime and the private businessmen. In one of his campaign speeches, he challenged Rawlings' regime to come out with a programme for domestic investment and not only for foreign investment. This is interesting, because during the early part of his political career (i.e. in 1982) Rawlings asked thus:

> What sense does it make to imagine that foreigners will simply come and develop our resources for us if we ourselves do not put our shoulders to the wheel? The foreigners who have really cared about us will tell you how even aid programmes and relief supplies became caught up in the network of local private profiteers. We believe that, instead of putting our faith in foreign investors, we must rather encourage our own industrialists who are capable of building for us a strong industrial base... It is clear that the failure to encourage our own production in both industry and agriculture is a cause of the immense strains on

> our foreign exchange reserves and hence on our currency at home. With the absence of adequate production, we are thrown into the claws of powerful multinational industrial and trading firms from whom we have to import virtually all we need (Rawlings quoted in *West Africa*, 11 January, 1982:73-4).

When the Economic Recovery Programme was launched in 1983, the Finance Minister, Dr. Kwesi Botchwey, maintained that one main goal was the "re-structuring of [the] neo-colonial economy away from undue export dependence into a more self-reliant economy where there is some integration in the pattern of domestic demand and the use of our own resources" (cf. *West Africa*, 13 January, 1986:68). The regime also noted that "it is essential that we are consistently aware that foreign capital cannot be the main vehicle for our economic independence and development" (cf. *West Africa*, 4 February, 1985:201). With these statements, Rawlings and his finance minister have acknowledged that the progress of a nation depends on the "economic progress" of the national bourgeoisie. But the regime's actions cannot be said to have matched such statements.

Ghana has made some socioeconomic progress in the last 16 years or so, but Ghanaian society is still riddled with much political and social conflict, mainly because socioeconomic life has become so politicized. A new crop of politicians lacking an independent economic base can easily feel insecure about the rise of their own national bourgeoisie. They thus consciously or unconsciously impede the development of their nation by frustrating a significant section of economic actors who could contribute to national development.

Critics of the NDC regime say that a large proportion of available credit continues to go state-owned enterprises and only a minuscule share goes to the private sector. This does not sit comfortably with the government's own rhetoric that the private sector should be the motor of development. Leechor observes in broader Third World comparative terms that, compared with the more dynamic economies—Thailand and Malaysia, in the 1960s, for example—Ghanaian private businesses have one-third or one-fourth as much credit to work with. But, as one leading local industrialist has noted: "successful investment growth ... relies heavily on good linkage between the local businessman and the foreign investor" (cf. *West Africa*, 19-25 July, 1993:1261). Writing on the perpetual tension between independent economic actors and the state, Jonah notes that, under the Fourth Republic verbal attacks on Ghanaian businessmen by government is not an uncommon phenomenon. According to him, the Council of Indigenous Business Associations (see below) set up by the regime is therefore as much meant to garner political support for the regime as to provide a counter weight to the big time Ghanaian businessmen who are anti-Rawlings (cf. *West Africa*, 10-16 January, 1994;18).

Political Liberalization: Recent Socioeconomic and Political Environment

In response to the IMF/World Bank's "political conditionality" and internal pro-democracy pressure, the PNDC regime introduced a multi-party system in 1992. The democratization process in Ghana led to the continuation of the PNDC rule. The regime transformed itself into a political party, the National Democratic Congress (NDC), and set the timing of the election as well as the framework for the conduct of democratic politics. It allowed a limited period of transition between 1991 and 1992, during which it initiated a formal debate on the country's democratic renewal at the various regional capitals. Few individuals opposed to the regime were given opportunity to present seminar papers on the country's political future so that the majority of the participants supported a no-party option. But the demands for democratic rule seemed so overwhelming that the regime had to succumb to the pro-democratic wave.

Following this, the government set up a Consultative Assembly to draft a new constitution, some of whose members were appointed by the PNDC regime. The constitution was thus seen by some as mainly expressing the wishes of the PNDC regime rather than the diverse opinions reflecting the plurality of society. The constitution was approved by a referendum in April 1992, the ban on political activities was lifted in May, and elections scheduled in December. This means that opposition parties had only seven months to establish themselves and organize against a regime that had power for about a decade. By the time the PNDC announced that Ghana would return to a multi-party system, the regime was ready to transform itself into a civilian political party and was poised to successfully contest the 1992 election. The transition period thus began with suspicion by opposition parties about government's sincerity to democratize politics in a fair and free manner.

The 1992 elections were judged by foreign observers to be generally free and fair. But the main opposition party, the National Patriotic Party (NPP), built on the Danquah-Busia political cleavage claimed that the NDC rigged the election and frustrated the activities of the opposition parties. The NPP denounced the outcome of presidential election and boycotted the parliamentary elections.[2] This, however, enabled the NDC to dominate the

National Assembly and further consolidate its claim to power.

That the NDC regime also exploited the advantages of political incumbency has been widely discussed (see, e.g., Nugent, 1995). For example, the regime increased the pay to civil servants by 60 percent (an action the World Bank claimed disturbed macroeconomic stability and undermined the gains of the SAP), in order to win votes. The NDC was said to have financed its election campaigns with more public money than the opposition parties. It also made more use of the media and government vehicles. NDC officials travelled to the rural areas and liberally dispensed patronage to local communities in the form of increased cocoa prices, electricity extension, water supplies, feeder roads and school improvement. These were meant to capture votes for the NDC. In a sense, politics turned "pork-barrel". On the other hand, these projects can be regarded as important to the improvement of living conditions of the rural people, particularly if they were provided systematically and effectively to be sustainable. The main opposition parties were, however, not advantageously placed. Supported mainly by businessmen, their coffers were limited because there was a law that limited individuals' contribution to one million cedis a year (Sandbrook and Oelbaum, 1997).

These practices seemed to be generally repeated during the 1996 elections. The NDC regime was said to have imported "election gifts": T-shirts, watches, bicycles, sewing machines and TV sets, which were distributed in the rural areas in return for votes. There were also reports that alleged that NDC agents purchased registration cards from NPP supporters in order to vote in their stead. Assaults on opposition groups also occurred from time to time and politicians traded insults. The political language did not foster compromise and reconciliation indicating continuous inter-elite rivalry. Rawlings was quoted to have characterized the opposition members as thieves and rogues who were desperate to capture power to enrich themselves. He was alleged to have warned that if the NDC lost the election, perhaps they would have to get power through another June 4 coup, and if "4 June 1979", repeated itself 10 times, he would do what he did 10 times (see *Ghanaian Chronicle*, 15-18 November, 1996). It was also alleged that, during the 1996 elections, the NDC had distributed weapons to its cadres in preparation for war should it lose the election. These statements indicate that the NDC was poised to win political power at all cost and by all means. Arguably, although the macroeconomic management system has been stable for the last decade and half, the political framework that could complement such an economic system to make Ghana's socioeconomic and political environment more predictable and calculable, has not yet effectively emerged.

As a result of Rawlings' being in power for the last 18 years, the state has come under personal rule with its main imperative being the survival of the NDC regime. This means that the NDC's political practices enable it to dominate society. This is manifested in the "culture of silence", and exclusionary politics that seek to neutralize the activities of political opponents and business class who challenge NDC's hegemony. In their assessment of Ghana's recent "democratic" reform, Sandbrook and Oelbaum (1997) write that, in some ways, little has changed from the days of the PNDC. According to them, Rawlings continues to dominate the political scene and rails against the restraints imposed on his power by civil society and institutions that demand public accountability from his government.

Rawlings has also maintained significant control over the army and has shielded the military from financial scrutiny. Although the Auditor-General has been mandated by the constitution to audit the accounts of the military, President Rawlings has blocked the implementation of such a measure and often seeks to weaken the powers of the formal institutions that could demand accountability from the government (see Sandbrook and Oelbaum, 1997). For example, in 1998, parliament passed the Assets Declaration Law, which enjoins senior civil servants to declare their assets and liabilities every four years. But the Rawlings' regime has made an exception to this rule for army officers (cf. *Free Press*, 14-16 April, 1998). Corruption has re-surfaced on an extensive scale within state institutions, prompting Nana Ato Dadzie, the Acting Chief of Staff and Presidential Adviser, to deliver a stern speech in 1998, saying that the level of corruption and waste in the civil service was unacceptable and could endanger Ghana's economic development if not abated. He went on to say that government departments have not bothered keeping proper accounts of their finances and officials have sometimes deliberately diverted resources into other areas. The patrimonialization of the state in recent times has brought in its wake inefficient use of public resources.

The patrimonialization of the state has also led to the development of new forms of patron-client relationships. Many journalistic and impressionistic accounts maintain that government contracts, particularly for road-building and drainage works, continue to go mainly to NDC party supporters, and government officials continue to ask for kickbacks and political favours. The extent of this clientelism has compelled an NDC member to ask in parliament "whether the Government was trying to develop a new class of businessmen where a Political Party card qualifies one for assistance" (cf. Sandbrook and Oelbaum, 1997:627).

New organizations have also been created to serve as a support system for the NDC regime. The most visible is the 31 December Women's Movement (31DWM), formed as a pillar of the PNDC (which came to power on 31December 1981) to advance the interest of women. This organization is incidentally directed by Rawlings' wife, Nana Konadu Agyeman. Originally formed to work under the existing independent National Council of Women and Development (NCWD) established in 1975 to coordinate women's groups and advance women's economic and social interest, 31DWM's activities have practically overshadowed those of the NCWD. The 31DWM has taken up women's issues bringing them to national level for political consideration. It has also undertaken local industry such as soap making, food processing and pottery. But many Ghanaians perceived it as an organization actively supported by the NDC and used to secure women's support for the regime. It has, however, defined itself as Non-Governmental Organization (NGO), which enables it also to qualify for external funding to support its programmes.

Like the Nkrumah regime, the NDC has also created and mobilized civil organizations and made them dependant on the state's patronage system, thus compromising their independence. In return, they have been used to neutralize other groups that did not endorse NDC politics. For example, the government created the Council of Independent Business Associations (CIBA) in 1993 which is chaired by the First Lady. CIBA is supposed to represent the interest of small entrepreneurs before governmental agencies, but as Sandbrook and Oelbaum note in their assessment of Ghana's recent democratization experience, it is one of the legs of the President's clientelist network. The Government was said to have donated 150 million cedis to CIBA (which also poses as an NGO) as a "capacity-building grant", but these funds and those given to the 31DWM led to a well-publicized scandal prior to the 1996 elections, when leaders of NDC and the 31DWM demanded that CIBA account for the billions of cedis it had allegedly received from the government. CIBA is mandated in some areas such as Kumasi to collect taxes from their members on behalf of the Internal Revenue Services and, or a municipal authority in exchange for commission. In August 1966, CIBA imported 330 containers of election gifts—umbrellas, hairpins, plastic chairs and hairdressing equipment—with foreign exchange granted by the Bank of Ghana (see Sandbrook and Oelbaum, 1997).

Another notable civil association that the regime has co-opted in recent years is the Ghana Private Road Transport Union (GPRTU), made up of drivers and their assistants. This Union provides "jobs for the boys" by assuming the responsibility of collecting daily tolls from drivers at the lorry parks. The GPRTU runs its own fleets of buses and it is alleged that the NDC regime gives it money or subsidizes the cost of vehicle, tire, lubricants and spare parts. In return, the GPRTU members gave political support to the regime, using, for example, their buses to campaign for the NDC (cf. Sandbrook and Oelbaum, 1997:622-7). We thus see state generated resources given to political followers, and uncertainty is created, likewise an incentive structure that rewards unproductive political behaviour and waste of scarce public resources. The political environment, as constituted under the NDC regime cannot be said to be very conducive local entrepreneurial activities, at least not to those businessmen who seek to act independently of the NDC regime. Consequently, it might be holding back the development of a vibrant local productive class that could supplement government efforts and act as a positive social force in any possible *dependent development* with external investors.

The resurgence of neopatrimonial practices during this recent phase of Ghana's history reflects the legacies of structural conflicts in the political realm and the desire of a new group of ruling elites bent on keeping power at all cost. Democratization is now accepted as one means to foster institutional renewal, good governance and sound macroeconomic management. But the above evidence suggests that neopatrimonialism is re-emerging, threatening the consolidation of an enabling environment for sustained development. These neopatrimonial rules of the game weaken the bureaucratic evolution of the state necessary for promoting development. Centralization of power, personal loyalties, clientelism, growing corruption, and unofficial presidential control over state institutions undercut the evolution of modern institutional framework that can support long-term development. The danger, as Sandbrook and Oelbaum (1997) note, is that resurgent neopatrimonialism will again impede the capitalist development that Ghana is supposed to now undergo unless it is mitigated by institutional reforms. This political sociology of recent times point to the dominance of personal rule in the political economy which we argued in Chapter 2 as one cause of underdevelopment of Ghana.

This current scenario confirms Ake's (1996) general observation that in many parts of Africa, to become wealthy without the patronage of the state is to invite the unpleasant attention of those who dominate society by virtue of political rather than economic power. For those outside the hegemonic faction of ruling elites, it is generally difficult to harbour any illusions of becoming wealthy through entrepreneurial means. As he noted further, for those who are part of ruling factions, entrepreneurial activity might be unnecessary, for one could appropriate surplus with less risk and less trouble by means of state power. This current paradox is similar to those which occurred under the CPP, and other preceding regimes.

It is interesting to note that the NDC government is pursuing one of the most liberal economic policies in Ghana's postcolonial development history, and its policies should be supportive of the activities of the local business community to fortify an environment for capitalist development. Yet, there is continuous tension between the government and the local business community. Thus, beneath the surface of democratic politics lurks a new form of patrimonial domination. This neopatrimonial political environment creates an unpredictable and incalculable socioeconomic environment and can impede productive activities in the modern sector of the economy. It can also undermine socioeconomic development, unless state reform is undertaken to establish trust and reciprocity in state society relationship.

Yet, in spite of these shortcomings, Ghana has enjoyed political stability in the last 18 years or so under Rawlings, with a stable macroeconomic management system and economic rules, thus an enabling environment for long-term development is incipient. Indeed recent newspaper accounts and institutional reforms, as well as, the gradual acceptance of the NDC regime and opposition parties to abide by liberal democratic rules, suggest that some of these developmentally counter-productive neopatrimonial features outlined above are on the decline. Ghana thus seems to be on the way to developing and probably consolidating democratic rule with a predictable and calculable socioeconomic environment critical for sustainable development. But the future can still be quite unpredictable as this is an economic and political transition period. Whatever political scenario emerges in the future will surely carry implications for Ghana's socioeconomic development.

Conclusion

The period from 1979 to 1983 can be described as one of national self-destruction and not socialist revolution. During much of this period, there was little rationality in the uses to which the Ghanaian state was put. In terms of the objective of this study, the AFRC and PNDC actions against foreign and local capital were very negative and destructive of a stable environment for long-term investment. Apparently, government's actions were well-motivated to create a good and just society, but the methods adopted were arbitrary, anarchical, and, consequently, damaging for the economy and society. Every revolution must have a leader and dedicated followers who understand what goals are at stake and what strategies are realistic for their achievement. Given the magnitude of Ghana's crisis, a true revolution at this time should have drawn all Ghanaians, irrespective of class, occupation and political affiliation, to its programme. The PNDC regime offered a naive and simplistic understanding of the national crisis and thus ignored the root causes of the Ghanaian problem. It incited many people into political action; however, their energies were not directed to productive activities, but to the destruction of the productive infrastructure of society. The regime politicized the economic crisis, and, consequently, induced conflict in society while it failed to improve the living conditions of Ghanaians, which was its ostensible aim for seizing political power. On the contrary, the regime's actions accelerated Ghana's plunge into further material poverty, a situation which also necessitated the SAP, and made Ghana more dependent on "aid industry" in recent years.

On the other hand, the period since 1983 to the present (i.e. 1999) can be described as one of economic regeneration. In the last 16 years or so, Ghana has enjoyed some reasonable political stability owing mainly to the Rawlings regime's ability to effectively ward off numerous coup attempts and deal swiftly with the coup-makers. This period has also been accompanied by attempts to reform the economy and provide the conditions that could sustain economic growth and development. The results so far seem positive; the rules of the game, the macroeconomic management system and policies have been relatively stable. There appears to be some degree of elite agreement as to development ideology. A predictable and calculable socioeconomic and political environment critical for development thus seems incipient. Democratic rules are also emerging, despite some imperfections and the persistence of neopatrimonialism. These recent economic reforms and positive political developments are important achievements, since in the early 1980s physical infrastructure and basic institutions of modern economy ceased to function or were in varying degrees of disarray. In this same period, basic law and order broke down, assets and means of production were confiscated, which aggravated the environment for any productive investment. Recent political stability and the emerging stable macro-economic management system are thus crucial for sustaining long-term socioeconomic development, since endemic political instability, policy reversals, state control and bureaucratic immobilism have been responsible for Ghana's retrogression and image abroad as a risky environment for foreign investment.

Notes

1 In his *Green Book*, Khadafy postulates the "Third Theory" which holds that "a state of the masses" is the only viable alternative to capitalism and communism systems of government. Through popular congresses, the people are supposed to rule themselves with minimal intervention from the political centre and the Head of State merely presides over innumerable revolutionary committees and meetings. These committees, according to Khadafy are more responsive to the needs of the masses than representations in both liberal and communists societies, hence it is unnecessary to have political parties which divide society.

2 For a catalogue of the NPP's accusations against the NDC, see the *Stolen Verdict.* For a more academic assessment, see for example, Mike Oquaye, "The Ghanaian Elections of 1992— A Dissenting View" in *African Affairs,* 1995, 94 (259-275). Compare also the evaluation by Richard Jeffries and Clare Thomas, "The Ghanaian Elections of 1992", in *African Affairs,* 1993, 92 (331-366).

8 Summary, Conclusions and Policy Implications

> The riches, power, and honour of a Monarch arise only from the riches, strength and reputation of his Subjects. For no King can be rich, nor glorious, nor secure; whose Subjects are either poor, contemptible, or too weak through want, or dissention, to maintain against their enemies— Thomas Hobbes

Introduction

If the initial stages of a study narrow down the research problem, the concluding stage is a good time to broaden out again. The purpose of this chapter is threefold: (i) to summarize the colonial and postcolonial development experience; (ii) to draw some theoretical conclusions; and (iii) to suggest some policy implications. The last purpose is justified because development studies has not always been a field of detached academic enterprise. Indeed, most people engaged in it do so with the belief that their knowledge will enlighten the policy-making process and ameliorate living conditions. A study of what is wrong with the status quo must point out how a "better" socioeconomic order can be achieved.

The impetus for the study arose from the intriguing paradox that Ghana's postcolonial development so far presents. Although largely exploitative, colonialism induced some limited economic progress in Ghana such that, in the early 1950s, the economy and society were said to possess characteristics that were advantageous for modern development. Thus, when Ghana got independence in 1957, it was generally assumed that further development would indeed be possible. This optimism, however, did not materialize. By the early 1980s, after twenty-five years of independence, a promising economy had become a basket case, and Ghana faced an uncertain future. Ghana was certainly not alone in this reversal of fortune, but the magnitude of its decline surpassed the experience of other comparable or relatively less endowed countries in the Sub-Saharan African region.

Since the early 1980s, the Provisional National Defence Council (PNDC), and now the National Democratic Congress (NDC) government led by Rawlings, accepted the intervention of IMF and World Bank's stabilization and structural adjustment programmes as a means to rejuvenate the economy and lay the conditions for growth and development. What is sociologically interesting, given the gap between promise and performance of development, is why Ghana did not undergo a socioeconomic transformation as expected. In other words, what failed or what is failing Ghana?

In addressing this research question, the study focused on the interplay between state politics and socioeconomic development. The main assumption of this study is that, to a large extent, the nature of the state and its governing structures, the macroeconomic management system and institutions, determine what type of socioeconomic and political environment will emerge in a society. This socioeconomic and political environment provides the main context within which actors (i.e. peasant farmers, businesspeople, corporations, investors, etc.) would operate or decide to expand economic activities and productive capacities that are crucial to the attainment of socioeconomic progress. In other words, the state, its governing and macroeconomic management systems, as well as the institutions it sets up, are decisive in how societies develop over time.

Chapter 2, offered a theoretical discourse that specified the intimate relationship between the state and socioeconomic transformation, and explored the historical dimensions of this relationship within the context of Europe and other successfully developed or developing countries. Against this background, the nature of the African state and its mode of interaction with the larger society is analytically juxtaposed and some contrasts and parallels are specified that could illuminate some of Ghana's development problems.

The main lesson that came out of this theoretical discourse was that development requires, at a minimum, a socioeconomic and political environment that ensures social peace and facilitates rational calculation as well as reasonable predictability in all fields of human life and activity. This is the barest condition that modern states need to provide if the societies over which they exercise control are to survive and prosper. Beyond this elementary function, however, the state's role in socioeconomic development is debatable. There are those who think the state should not intervene in the economy more than this and there are others who feel it must play a more activating role in development.

The evidence from the experience of all successful development, however, supports the view that the state's role has been prominent in laying the basic rules of the game of society and also directly or indirectly activating the development process. The bureaucratic state has been an important precondition for societal development in terms of setting up a coherent and relatively stable institutional and macroeconomic context, in providing infrastructure and security that facilitates or stimulates the activities of private actors and the state (as in capitalism) and/or that of the state alone (in socialism). Personal rule of the patrimonial type, political instability and arbitrariness, internal war and political anarchy are all inimical to human progress, because, by their very nature, they make social life and systems unpredictable and incalculable, and, consequently, hold development back.

In this study, it was hypothesized that Ghanaian political conditions pose the greatest impediments to the promotion of socioeconomic development. In particular, it was contended that the personal domination of the state and frequent military intervention have undermined the development of the bureaucratic capacity and effectiveness of the state to enable it shape positively socioeconomic processes and also play a critical role in development. How far does the empirical material fit this thesis and the general assumptions formulated in the theoretical framework? In order to specify this linkage between the theory and evidence, it is proper to first recapitulate the empirical material.

Summary of the Postcolonial Development Experience

The empirical part of the study began with a description of the economic and political institutions of precolonial Ghana. It was shown that, in spite of their communal character, the economic systems of precolonial societies were based on market and exchange institutions. Capitalism was thus implicit in the economic activities of precolonial societies. Precolonial societies were also, to a large extent, democratic. Ghana's modern transformation began in earnest with the imposition of colonial rule towards the end of the nineteenth century. The advent of colonialism saw the establishment of a coercive, undemocratic and authoritarian administrative state and the subordination of existing indigenous political institutions. Colonialism was mainly geared towards the exploitation of the colony's resources and markets for the advancement of the metropolitan state and its industries. But a *laissez-faire*, and, later, statist political economy of colonialism indirectly catalyzed Ghana's capitalist development through the introduction of modern money, banking facilities, transport and communication networks, and the incorporation of Ghanaian societies into the international capitalist economy.

This development was manifested in the rapid commercialization of economy and modernization of society with increased structural differentiation and the re-stratification of the society, conflict between the old aristocracy (old elites) and the newly rich common and educated people (the new elites), the proliferation of western education, and a commercial and professional class who, conscious of the injustices of the colonial project, demanded the democratization of colonial state politics and even political independence. The political struggle of this group was animated mainly by their inability to realize their aspirations under the colonial system. Thus, they sought to gain control over the colonial state from alien rulers and use it for the development of their own society.

The tempo of development in the colonial period was largely gradual, as the colonial state did not evolve any radically new economic order that could disrupt the prevailing substructure of the production systems or economic orthodoxy of the indigenous societies. By not consciously or directly intervening in the local economy and society to change the existing rules of the game, the colonial state was able to lure the local population into further capitalist system of production. Peasants cultivated cocoa, people pursued education and then jobs in the colonial public-sector establishment, new professionals emerged, merchants and businessmen sought to enrich themselves individually and their families as well as relatives. Market rationality based on private choice was intensified in society. Thus, it can be argued that the relative success of the colonial economy owed much to previous indigenous institutions and the spirit of capitalism.

In sociological terms, Ghana was becoming a differentiated society, and, more specifically, in Weberian terms, a pre-capitalist society dominated by substantive rationality was gradually yielding to commercial impulse and the demands of formal rationalized capitalist development. At the time of political independence, Ghana was socioeconomically a differentiated society; thus it was a relatively complex one in need of a state that could not only shoulder developmental goals but also mediate or accommodate the interests and conflicts that could arise from such a differentiation.

Yet, there were other intangible aspects of this material transformation. Ghanaian society was "westernized", although this is not to suggest that indigenous traditions were totally swept aside. Indeed, there was a syncretic adaptation of western values to the local milieu, but the resilience of tradition was easily noticeable. English was introduced as the official language and remains so to date; the bureaucracy and the military had British traditions, likewise the judiciary and educational system. It was not only the economy that was "neocolonial", in the sense that British and other western companies came to dominate the modern sector, but also the "national mentality" (as often expressed in the preference for western foreign goods, attitudes, values). Indeed, a large number of the country's leaders, lawyers, civil servants, university lecturers were trained in the United Kingdom and the United States. Such intangible ties are often expressed in the commitment to a western democratic system of government, and values such as liberty and freedom, which found favour with many sections of the educated community. This helps to explain the early peaceful request for independence within the liberal democratic framework set up by the colonialists.

The early nationalists were conservative in orientation, seeking to overthrow the colonial regime to continue the liberal colonial economic and political order. Although colonialism came to an end, Ghanaian society at large retained a considerable residue of British traditions. This was, however, not easily acknowledged or perceived by the radical nationalist leaders who emerged at the eleventh hour of colonialism and led the country to independence. Such material and non-material transformation of Ghanaian society means that colonialism had created socioeconomic groups with vested interests in the maintenance of the colonial order. Indeed, that Ghana led the way to Africa's independence can be attributed to the relatively early maturation of a small "bourgeois" class, powerful enough to challenge oppressive colonial or racist rule. Given this, from the Marxist perspective of class struggle, any attempt to smash the old order and impose a new set of rules was bound to be conflictive and perhaps contested.

As it came to pass and indicated in Chapter 3, the nationalist struggle towards independence led to a split in the anti-colonial front: radical and conservative elites favoured two contrasting models of development:—"structural transformation" and "gradualism", or, in more conventional terms, socialist and capitalist development, respectively. These two groups sought to capture the state and implement their models of development. In the inter-elite contest for independence, the radicals won the day and achieved early independence for Ghana. But their victory did not end the ideological contestation within the state as the conservatives sought to destabilize the radicals and even derail their programme. Ghana's first postcolonial leader, Kwame Nkrumah, had high ambitions and sought a rapid structural transformation of Ghanaian society. A socialist development strategy was preferred, as Nkrumah judged that the capitalist evolution unfolding was too slow for Ghana. Nkrumah and his CPP regime sought a rapid development based on socialist principles. Indirectly, they sought new imprint for society. This search for a new order, however, generated sociopolitical conflict in society because of the resistance posed by other elite groups interested in the continuation of the old order.

The development strategy that was implemented by the CPP regime emphasized capital intensive industrialization, rather than agriculture. But, within agriculture, the stress was placed on large-scale capital intensive schemes aimed at fundamentally changing the technology and organization of the peasant production system. Peasant agriculture had relatively little government attention. Import Substitution Industrialization (ISI) was the cornerstone of the strategy. The outside world was seen as dominated by imperialist machinations and neocolonial exploitation was feared and emphasized; thus, there were vigorous attempts to reduce Ghana's external dependence through a rapid domestic industrialization programme.

There was also the distrust of the role of the market in resource allocation. The state was assigned the role of mobilizer, allocator and manager of resources and opportunities. The state became the main, if not, the sole actor in development providing socioeconomic infrastructure and serving as an important entrepreneur and industrialist in an environment that was defined as lacking a dynamic bourgeoisie that could accomplish its historical mission of transforming the productive forces of society. The state sought uniformity for society by suppressing socioeconomic and political diversity and pluralism. The state also established, at an early stage, a comprehensive free education and health care system and other social services. There was, thus, an intolerance for inequality and, consequently, an orientation towards re-distributive social production system with egalitarian principles.

With initial assets (foreign reserves) and a reliable revenue base ensured by monopolizing the marketing of cocoa through the CMB, the state could tax cocoa farmers to provide extensive economic and social infrastructure. The public sector was rapidly expanded through the establishment of state-owned industries, publicly-owned institutions and agricultural collectivization programmes. The bureaucracy was enlarged to manage these state-sector activities. The regime's immediate desire was to eradicate underdevelopment and lessen Ghana's dependence on the external world within the shortest possible time. An industrialized socialist state was, however, to be the professed ultimate goal.

Although Nkrumah and his CPP regime achieved significant material development and initiated an egalitarian socialist production system, the development programme set in motion became unsustainable, mainly because it was overly ambitious. The time frame for its accomplishment was short, and when seen against the human and material resources available to Ghana at the time, it is doubtful that the development goals would have been easily attained or sustained without modification or reform of the strategy. It was little wonder that the initial assets were quickly expended, and when world market prices for Ghana's major foreign exchange earner—cocoa—turned sour, the regime was deprived of critical foreign exchange and thus came to experience crisis in the management of the economy and development.

Also, such impatient attempts to develop a small country within an incredibly short time led to mistakes and problems. Consequently, development polices and plans were haphazardly implemented. In theory, socialist principles were to determine the character of development, but in reality much of Nkrumah's personal ideology and CPP's political interest infused development strategy. Many of the collectivization projects initiated to increase national output, were also meant to generate employment for the supporters of the regime, and were thus economically unproductive, at least at the time of the regime. There was also the politicization of the development process and a lack of sound feasibility studies, proper planning and administration. The bureaucracy was politicized, and this undermined its ability to formulate and implement policies professionally. These factors and others such as corruption and mismanagement worked in different combinations to make the social production system the CPP regime set up unproductive. While this could be seen as a temporary crisis in the early part of the developmental cycle that perhaps called for fundamental reforms, it was perceived as a failure on the part of the regime and its development strategy, and the military overthrew the regime in 1966.

Yet although Nkrumah and the CPP's development programme was beneficial to many people, it was also plagued by contradictions and paradoxes that triggered further sociopolitical conflict within society. This was compounded by the economic crisis, which accelerated the regime's downfall. When the Nkrumah regime proclaimed socialism for Ghana, it also quite correctly rejected (in terms of principles of scientific socialism) the demands of the local bourgeoisie to gain strategic hold in their own economy. Paradoxically, however, the regime was unable to nationalize foreign companies operating in the country. The CPP regime "tolerated" them and even "supported" some foreign businessmen financially while it deliberately sought to deny Ghanaians the opportunity to undertake large-scale, private economic activities in their own country. There was thus a gap between socialist theory and its practice in Ghana. In one sense, the independent state was like the colonial state, unconsciously or inadvertently "scheming" with foreign capital to inhibit the rise of the local bourgeoisie. It is not difficult to see a conflict here, because the state now in the hands of Ghanaians was antagonizing a significant section of local economic actors: the bourgeoisie.

Also, in order to develop Ghana at the breakneck speed that the regime chose, the state needed massive resources mainly from cocoa farmers. But as world prices for cocoa was declining, the regime was compelled to capture more and more rural surplus to underwrite its development programme, and thus came into conflict with cocoa farmers. By also seeking rapid development within a socialist framework, the CPP's strategy of development entailed some disarticulation with previous capitalist socioeconomic processes and institutional order. Consequently, many of the claims of the independent state were not accepted by the middle class (in the civil and military society), neo-merchants and the traditional aristocracy who highly priced their traditional values of independence, freedom, individual choice and liberty, that existed in the precolonial and capitalist colonial period. Forming the core of the UGCC party from whom Nkrumah broke to form his CPP party, these bourgeois elements fiercely opposed the Nkrumah state. The reaction of the CPP regime was uncompromising, such that an "unbridgeable" divide among political elites was entrenched in the political system. These conflicts and others stemming from local and ethnic parochialism generated new political groupings seeking to carve a hegemonic space for themselves and to restrain the monopolizing postcolonial state.

The Nkrumah state thus came under mounting pressure from different groups seeking to protect their interests, even if it meant tearing the new independent nation-state apart. In order to hold society together and build a new nation-state, the CPP regime tightened its control over society by cancelling constitutional "checks and balances", proscribing political parties and activities, and, in their stead, establishing a one-party state but which later became personalized and patrimonialized. The patrimonialization of the state and the development of its attendant patron-client relationships gradually undermined the government's capacity to positively influence the socioeconomic and political environment. It also reinforced the colonial coercive and authoritarian structures inherited at independence and further inflamed inter-elite conflicts. Consequently, elite groups vehemently opposed each other and pulled apart. Politics became a zero-sum game and incumbent elites sought to keep power by all means and at any cost. In the same vein, the opposition sought to wreck the CPP regime and derail its development programme. Corruption, nepotism and mismanagement ensued in the political system. Consequently, a domestic socioeconomic and political environment inimical to development began to emerge in Ghana.

Yet, it was not only local forces that the CPP regime had to contend with. It was also at loggerheads ideologically with the Western world. Nkrumah's own academic and political writings and the state-controlled media continuously emphasized the dangers of neocolonialism and railed against imperialist domination of Ghana and the Third World. There were often threats of nationalizing foreign capital. This made Ghana unattractive to western investors, if they were interested in Ghana at all. Also, having paved the way for Africa's independence, Ghana wished to play a key role in the liberation of Africa from colonial rule. Resources that could be used for development at home were diverted into external projects, mainly Pan-Africanism, liberation struggles and the OAU. In flexing its muscles on the continent, the CPP regime became, if not a threat, at least a nuisance to the Western interests in Africa. Ghana was caught up in super-power conflict, for which the regime paid through the refusal of western nations to offer temporary debt relief and long-term concessionary loans. In the absence of such long-term loans, the regime was forced to contract short-term suppliers' credits with high interest rates and short maturation period to import capital and consumer goods. This got the regime into debt, which fed the economic crisis, and, consequently, societal resentment that led to the overthrow of the regime.

Successful socialist development requires large and efficient bureaucracy of dedicated and honest individuals, to process the government's information on resource acquisition, allocation, planning, production and distribution. Ghana's bureaucracy was just developing and was thus quite inexperienced to effectively take on an accelerated socialist development programme. Consequently, there were inefficiencies and mismanagement in the public sector, which aggravated or led to the economic crisis. Given this, perhaps a more sustainable nationalistic development strategy would have been a gradualist one that recognized the country's historical realities, socioeconomic diversity, and structural constraints, and sought to attain development within these parameters. Such a strategy would also have sought to develop and modernize society on the basis of accumulated experience and traditional economic practices, using mainly Ghana's own resources and indigenous institutions. This approach could possibly have created economic space for all strategic actors who emerged on the historical scene wanting to fulfil their aspirations for self-rule. This would have been a truly nationalistic development strategy and need not necessarily fail to ensure the welfare and egalitarian goals that the CPP regime sought to achieve under its radical programme that brought it into conflict with vested interests.

While gradualism need not necessarily be interpreted as *laissez faireism*, and would certainly not guarantee immediate developmental success the kind of structural transformation the Nkrumah regime sought, it is, however, likely to minimize the impact of the difficulties and social strains accompanying rapid development. If this was the case, the terms and operation of the incipient state and its ideology would probably have been less contentious. Also, the state's development strategy would probably have been less conflicting and more compatible with Ghana's resources (and thus probably prolonging the longevity of the Nkrumah regime), and possibly yielding a greater measure of long-term development success than was the case.

The 1966 coup may be explained in terms of the grievances of soldiers who had come to equate their self-preservation and well-being with that of society, but it can also be seen as the culmination point of inter-elite rivalry and controversy over which development ideology Ghana should pursue. As the terms and character of the state, as well as its ideological direction, were not agreed upon, the nascent Ghanaian state was opposed from the beginning by farmers, neo-traditional and military elites, professionals, educated and business classes, and forces of "localism" and regionalism who saw it as threatening their interests. The CPP regime "alienated" and "excluded" too many vested interests from the state, even the most fundamental one—the peasant cocoa economy—that it critically needed for its development programme. For these groups, Nkrumah and the CPP regime had gone too far with socialism; thus they "came" together to overthrow him and the regime. In seeking to create a state that would rise above these vested interests, Nkrumah probably failed to appreciate the extent to which socioeconomic and political groups fostered by many centuries of indigenous capitalism and politics could pose a resistance to a radical new order and challenge to his regime.

Thus, the 1966 coup brought into power military elites unsympathetic to Nkrumah's economic and political order. To the coup makers and their supporters, the Nkrumah order was an aberration and interruption in Ghana's long process of capitalist development and the institutionalization of liberal democratic polity. They sought to dismantle the Nkrumah edifice and move Ghana back to the path it had threaded under colonialism. Consequently, a new liberal system of macroeconomic management (i.e. gradualism) was imposed. Ghana courted the West and closed her doors to the socialist countries that had enjoyed favourable status under the Nkrumah period. The civil service was "de-politicized" and gained considerable latitude to shape and implement policies. The state's arm in the economy was radically rolled back, a free market economy was embraced, and a large number of state-owned companies and "socialist" projects were privatized or abandoned. Experts from the socialist countries were thrown out. Ghana was "purged" of Nkrumahism and socialism. New civic educational centres were established to educate a population that was supposed to have been indoctrinated in socialist ideology. The NLC made a new, elaborate constitution to prevent dictatorial rule by one person or party. Institutions that would allow the functioning of liberal democracy were restored.

The rush and anger to "de-socialize" Ghana's "nationalistic" economy also meant another radical departure from the immediate past, and a new kind of discontinuity was introduced into socioeconomic processes. But this only reflects the inter-elite conflict and the lack of "consensual" or "settled" politics critical for development in Ghana. Satisfied that the basic rules of the economic and political game could be started anew, the soldiers returned power to the liberal politicians in 1969 (a group that had been waiting to make its political début) with the hope that their policies would be continued.

The 1969 governmental change brought a coalition of the middle class and neo-traditional elites into power. This group was mainly made up of older and younger generations of opponents to Nkrumah's regime. The state, under Nkrumah, was populist, urban-oriented and authoritarian. Now, under Busia, it was elitist, rural-oriented and relatively democratic. The Busia state acknowledged the plurality of Ghanaian society and remained untotalizing, compromising and conciliatory to various groups. For Busia, the state needed to provide an enabling environment for economic actors in both the traditional and modern sectors (i.e. peasants, the bourgeoisie, etc.) to pursue their aspirations. The regime sought the development of rural society and used state power to protect the indigenous bourgeoisie against what it perceived as unfair competition from foreign capital. Through a variety of fiscal and administrative policies, the regime sought greater economic space for its allies in industry and commerce. It insisted on repatriation of illegal immigrants and the exclusion of non-Ghanaians from certain economic endeavours.

Since the state under Nkrumah looked "favourably" upon the activities of non-Ghanaians, these groups were seen as having advanced their already historically advantageous position over indigenous Ghanaian businessmen. This was to be reversed. In Marxist terms, the Busia state can be conceived as acting on behalf of the bourgeoisie; becoming a committee for managing the affairs of the those who could dominate the society by virtue of economic power. But Busia's achievements were less spectacular than Nkrumah's because they were more piecemeal; contrast, for instance, his rural water supply programme and educational reform with the CPP's highly visible state farms and workers' brigades.

In 1972, however, the Ghanaian economy once again faced an economic crisis, mainly as a result of the fall in the world market price for cocoa. Much of the foreign debt inherited from the Nkrumah era was also due for repayment. To the regime, this necessitated an adjustment in the domestic economy. It thus re-organized domestic priorities, which amounted to further dismantlement of Nkrumah's "re-distribute" politics and system. Imports were restricted, the currency devalued, major subsidies on social service were cancelled, prices were increased and fringe benefits to civil servants (including the military) were taken away. Furthermore, the size of the civil service was trimmed down; a student loan scheme was to replace free college education and a tax was imposed to support a rural development programme. By and large, these austerity measures represented a new realism in national economic management under changed circumstances, and, with the benefit of hindsight, they anticipated many of the elements of today's structural adjustment programme.

These measures were, however, politically unpopular because they hurt many urban social groups whose welfare was tied to the maintenance of the re-distributive social production system. Predictably, these groups withdrew their support for the regime. Army officers also did not take kindly to Busia's policy that took away their corporate amenities that were expanded by the

NLC regime of 1966-1969, so they overthrew the regime. The attempts to promote capitalist development based principally on the activities of indigenous Ghanaian businesspeople was thus abruptly terminated. From the theoretical perspective guiding this study, like the "mercantilist" rulers of premodern Europe who sought greater national wealth by "encouraging" their bourgeoisie, the Busia regime, unlike Nkrumah's, accepted that Ghana's economic progress not only depended on the state, but also on the advancement of the economic activities of its national bourgeoisie and citizens. The potential value of this development strategy enabling Ghanaians to learn and gain control over their own national economy, as the case is in many modern developed states, cannot be over-emphasized here.

While the military justified their 1966 coup, among other things, on the grounds that the CPP regime could not be removed constitutionally from power, this could not be said of the PP regime. Indeed, there were indications that the PP government would seek a new political mandate and would not default on re-election. Furthermore, compared to the Nkrumah regime, the PP government was less neopatrimonial perhaps as a result of the political de-participation of former CPP elements. Yet even though public criticism against the regime remained high in the wake of its policies, the press remained relatively free from government intervention. Busia's main offence was his attempt to further undo Nkrumah's social re-distributive production system, which many people were enjoying but which to his regime appeared unsustainable for the country under prevailing economic circumstances. Indeed given the economic crisis that had emerged in Ghana by mid-1960s, and the belated statements made by the CPP regime to reform policies, one can conjecture that had the CPP remained in office, it would have been compelled to reform, if not totally change its social production system to make it more sustainable.

The PP regime (unlike the CPP because of ideological reasons), wanted Ghanaians to act as individual economic actors and thus control their nation's wealth and contribute through their microeconomic activities to its development. To the PP regime, a nation-state becomes wealthy, economically strong and also developed mainly through the wealth and economic strength of its citizens. Poor citizens can only beget poor nation-states. But its policies were not favourably received, because new groups had emerged under Nkrumah's rule interested in protecting and defending Nkrumah's "re-distributive" socioeconomic order. In some societies, governments take measures that often hurt their supporters, but they are not overthrown in military coups. In Ghana, however, they could be overthrown. In 1972, Busia was ousted from power. The coup confirms a new development in Ghana's postcolonial politics: the militarization of politics and the desire of military leaders and their urban supporters to continue CPP's socioeconomic legacy that dispensed free education, health and fringe benefits.

The new military regime of the NRC-SMC thus set in motion a new phase of policy reversal: quasi-Nkrumahism was enthroned. The regime denounced Busia's free market economic policies and held Busia and his policies responsible for the economic débâcle. A "commandist economy" was proclaimed, increased nationalization of foreign companies announced, and state participation in foreign companies was extended. Western debts were unilaterally abrogated and the domestic currency re-valued. Foreign policy was revised to enable Ghana to foster closer ties with the socialist countries. Controls on import licences, foreign exchange and import restrictions were re-instituted. Subsidies on social services were expanded. Free education, health care and other social services were integral parts of the new order instituted. In essence, re-distributive social production system was re-established. All social groups that the civilian regime had antagonized were appeased. The new military regime was popular, at least up to 1974, mainly because of the initial economic gains, nationalist policies and the "return" to a re-distributive social production system. Nkrumah's ideas that the state should be strong, centralizing, activating and re-distributive were taken on by the NRC-SMC regime but it lacked the capacity and vision of the CPP regime.

The NRC-SMC regime was, however, only able to turn the economy around modestly between 1972 and 1974. This was mainly because of a combination of fortuitous circumstances: a rise in the world market price for cocoa and other mineral resources that Ghana exported and an increase in production of local foodstuffs that the regime took seriously, as well as good climate. But soon a political economy of personal rule re-emerged, and practices that can be characterized in the Weberian term as patrimonial came to predominate. The leader of the NRC-SMC, Col. Acheampong, assumed personal control of strategic decisions in finance and defence. Bureaucrats and technocrats were swept aside as soldiers "colonized" the strategic decision-making centres of the public sector. Most did not hesitate to privatize their newly acquired offices. The state became a major source of upward mobility, status, power and wealth.

The regime sought to retain power as long as it was possible. This led to the frequent re-shuffling of ministerial posts and shifting alliances of patron-client relationships. Security was tightened and state officials were allowed illegitimate access to resources in return for loyalty. A highly kleptocratic state emerged, through which members of the regime and their allies could advance their private interests. Public accountability was eroded as a result of the patrimonialization of the state. Corruption, graft, malfeasance and smuggling permeated high echelons of society and were reproduced at the lower levels, as well. A neologism— *kalabule*— emerged in the popular discourse to describe the new "social cancer" that had eaten into the fabric of society.

Neopatrimonial domination of the state also eroded its rational-legal or bureaucratic capabilities and weakened it as a public-spirited instrument in the service of national development. As the neopatrimonial regime was busy with personal survival and looting of the state treasury, professional management of the economy received secondary attention. Policies were ineffective and there was uncertainty about policy change. The regime held on to old ineffective policies and refused to undertake fundamental reforms in society that could increase productive activities. Public administration was reduced to trading in import licences and commodities. Productivity declined in all the major sectors of the economy. Consequently, the formal economy contracted and this accelerated the informalization of the economy to a point where the state lost virtually control over the economy and could not even generate revenue for its own reproduction. The result was a self-reinforcing socioeconomic decline, which reduced living standards of Ghanaians below that of the immediate postindependence level.

During this period, the state's monopsony, the CMB, also failed to periodically revise domestic cocoa prices, as Ghana's neighbours Cote d'Ivoire and Togo did. But increasing economic decline meant that the government became hungry for resources and resorted to capturing more and more cocoa surplus to support the re-distributive system it enthroned. As domestic terms of trade adversely affected farmers, a "silent war" ensued between them and the state. The farmers retrenched their participation in the formal economy by selling their products in neighbouring markets for higher incomes or abandoning cocoa production altogether. Farmers, who were unorganized and dispersed could not exert pressure on the state to make fundamental changes in their life, and were thus targets of state exploitation. But, in reality, it was the state that lost foreign exchange in the absence of cocoa. Peasants in Ghana, as in many African countries, *do not need* the state, at least as far as their daily survival is concerned, as perhaps the state needs their export crops to generate foreign exchange to run the modern economy since they provide most of their basic food needs. Moreover, much of the rural surplus extracted from cocoa farmers was spent by the neopatrimonial state as salaries and allowances on the bureaucracy and/or consumed privately. With national economic collapse, over one million Ghanaians—mainly high and lower level professionals and technocrats—voted en masse with their suitcases and the left the country, for more prosperous African countries, Europe and North America. This further deprived the country of much of the manpower it had trained over the postcolonial period.

On June 4, 1979, a military uprising by junior officers and other ranks brought the AFRC to power. The old military rulers were overthrown. The coup was motivated mainly by the venality of the NRC-SMC military regime and appalling socioeconomic conditions into which Ghana had sunk. The new leaders undertook a "house-cleaning exercise" to purge Ghanaian society of selfish, corrupt and incompetent political leaders and to initiate moral reforms. The regime also equated high commodity prices with exploitation and imposed revolutionary and illiberal price control policies. It seized substantial amounts of property, nationalized industries, confiscated bank accounts of businessmen, industrialists and traders. But these actions did not solve Ghana's problem arising mainly from inefficient domestic organization and declining productivity. On the contrary, they deprived citizens of working capital to carry on business and long-term investment. The AFRC could not resolve the economic débâcle, mainly because its actions addressed the issue of re-distribution, and not low productivity, the main problem facing Ghana at this time.

After 112 days of "moral revolution", the AFRC handed over power to a democratically elected civilian regime of the PNP, in late 1979. This new government of Limann was mainly concerned with creating a stable socioeconomic context for putting Ghana on a sustainable development course. But it could hardly find its feet before some members of the former AFRC took power again, this time calling for nothing short of a revolution. The PNDC government that emerged from this military intervention now conceived a class struggle, between the proletariat and the bourgeoisie. Local (neo)Marxist intellectuals and dependency theorists gained control over the state and sought to substantiate their ideas in the empirical realm. At the ideological level, western imperialism was identified as the main cause of Ghana's development problem; at the mundane, however, it was foreign capital and local agents in industry, business, trade, etc. supposedly working in collusion to exploit Ghana. The revolution was thus meant to change the status quo and to destroy the economic power of these groups as a prelude to improving socioeconomic conditions for the poor.

Instead of a genuine revolution that could improve productive sectors upon which increased welfare can be premised, the PNDC policies from 1981 to early 1983, led to the wanton destruction of commerce and the productive infrastructure of the society. A new group of radical elites had emerged on the political scene. With "hegemonizing symbols" such as "class struggle", "revolutionary transformation" and "national development", they declared themselves as the only group capable of representing and defending the national interest. They partitioned society into two main classes: "we" and "they"; the "they" being an alleged rich, self-seeking and exploitative elites (i.e. mainly the rich)—said to be responsible for unjust and oppressive socioeconomic order—and the "we",—representing them, the so-called incorrupt, selfless and progressive elements of society who also represent the poor and can create a just order. Consequently, they sought to weaken the economic and the political base of the so-called rich people and their allies—foreign capital. Private banks accounts, houses and cars were confiscated. Shopping centres were blown into pieces. Companies and businesses (both domestic and foreign) were nationalized, foreign managers were chased out by workers, workers took over the ownership and management of firms, new revolutionary institutions were set up in society.

The regime was, however, unable to reverse the economic crisis mainly because its actions addressed re-distributive rather than productive issues. Furthermore, its so-called revolutionary actions undermined the existence or emergence of any calculable and predictable socioeconomic and political environment that could conduce to long-term investment. The PNDC's actions against the so-called rich of society diminished the size and economic power of whatever bourgeoisie (productive and parasitic) existed in Ghana at this time. On the other hand, however, its "revolutionary actions" signalled to real or potential domestic and foreign investors that Ghana was unsafe for investment. The PNDC regime was only nominally socialist; it lacked the political discipline and executive self-abnegation critical to genuine socialist transformation of society.

Consequently, the PNDC rule throughout 1982 heightened the existing economic crisis. The "class struggle", ostensibly meant to halt the decline of the economy, ironically brought more misery to an already impoverished population. By early 1983, the government acknowledged that Ghana could not continue along the same path and that something fundamentally had to be done in the policy arena, to lay the conditions for productive activities, economic growth and development. With IMF/World Bank loans and advice (i.e. structural adjustment programme), Ghana embarked on a capitalist path of development. Seen against socioeconomic conditions prevailing in the pre-adjustment era, recent economic reforms have brought about some improvements in the economy and society. But Ghana has a long way to go to become once again a middle-income country, which helps to illustrate the magnitude of the postcolonial decline.

Final Analysis and Theoretical Reflection on the Development Experience

What is striking about the foregoing summary of postcolonial development experience is the persistent failure of governments over the years to stabilize the ruling process and establish a calculable and predictable socioeconomic environment that can enable the effective organization of the people, their resources and activities to raise productive capacity upon which broader developmental goals rest. There has been a wide divergence in the conceptions of the development problem, development strategies and economic policies. Ghana has lacked a clearly defined development goal and strategy that would be acceptable to all and that could provide a framework for public policy over the postcolonial period. Governing elites have been unable to achieve a consensus on development strategy, let alone ensure a smooth transition from one strategy to another. On the contrary, they have often created political stalemates and economic paralyses and have subjected Ghana to different and often conflicting development ideologies—"socialism", "capitalism", "commandism" and "populist revolution", "neoliberalism", "structural adjustment programmes"—all suggestive of the fact that Ghana, as a nation, has not carried one model of development to its logical conclusion let alone had one long-term development objective.

Since the overthrow of Nkrumah's regime, hyper-instability has come to characterize the governing process and this also impinged negatively on the development process. Surely, the lack of stability in the governing process, per se, should not necessarily pose a problem for promoting development if the basic rules of society, policies and institutions remain relatively stable, or are evolved systematically. Political instability was, however, detrimental to development because every change in government was accompanied by a shift in policy regime, institutions, and by the abandonment or reversal of plans and projects. The pre-independence inter-elite conflict over ideology and development strategy was exacerbated and sustained over the postcolonial period.

Thus, the peculiar formation of the Ghanaian state and its mode of interaction with society since independence has produced an unstable series of political elites, each failing to grapple with and mediate the conflicts arising from a highly mobilized but fractured civil society. Lacking coherence, governing elites have been unable to achieve a unified developmental purpose and map out a nationally relevant strategy to overcome the development problem. They have been too concerned with staying in power or fending off opposition. Consequently, Ghana has been unable able to draw together its social, economic and political diversity to build an all-embracing coherent state and effective governing system that could be beneficial to all.

It would be wrong to attribute this state of affairs to the wickedness of political leaders. After all, leaders have apparently responded to the declining socioeconomic realities they saw before them and had wanted to change things for better but they were unable. Much of the difficulties leaders encountered, no doubt, derived from the colonial legacy of predatory rule and the newness of the state and other institutions. But it seems leaders have also lacked the vision to link political goals and socioeconomic realities of Ghana. They clung to ideas that were "irrelevant" and "ineffective" in guiding action in new and ever-changing development contexts. Perceptions and ideas about development problems remained relatively static in a dynamic development context. Ghana is a relatively young nation in existence for barely 42 years and one that has experienced colonialism. Most of the ideas of colonial exploitation, neocolonialism, and imperialism that were articulated by Nkrumah and the CPP, but also other nationalists in the UGCC party, were mainly intended to bring colonialism to heel. But they have remained acutely relevant in much of the postcolonial period as they were during the struggle for independence. These ideas, which often find expression in the national consciousness or in some intellectual quarters, continue to exert powerful pressure on the definitions of and solutions to the national development problem. Thus, many political leaders continued to see Ghana's independence as continuously vitiated by imperialist machinations and exploitation. Not that externality does not matter, but the excessive emphasis on "externalities" and reactions to it, has ruled out any serious consideration of internal factors that were militating against Ghana's progress.

Nkrumah was overthrown, but in a sense, "Nkrumahism", understood here as a set of Nkrumah's ideas about Ghana's underdevelopment, did not die. Indeed, Nkrumahism continued to be a force to contend with in the debates about Ghana's decline. As the "founding father" of modern Ghana, Nkrumah's ideas are also the founding ideas of Ghana. Nkrumah's ideas about underdevelopment in Ghana are compatible with those often expressed in neo-Marxist development literature: that colonial and neocolonial exploitation facilitated by the activities of comprador local groups are the major causes of economic backwardness of the so-called Third World. These ideas have laid a foundation for the intellectual and political discourse on Ghana's development or the lack of it, and have come, in no insignificant way, to cloud national perception of Ghana's problems and government actions. Indeed, until relatively recent times, a sizeable percentage of Ghanaian intellectuals and political leaders continued to see the nation's problems in mainly neo-Marxist and neocolonialism terms, which Nkrumah also held as constituting Ghana's strategic problem. These ideas were clearly stated during the reign of the NRC-SMC regime, and were inadvertently carried into what can be described as national self-destruction during the AFRC and the PNDC rules in 1979 and from 1982 to 1983.

This historical legacy then helps to explain why government policies over much of the postcolonial period have been generally "anti-capitalist" and "neoNkrumahist", and why liberal thoughts have been marginalized, if not discredited, in the analysis of the national development problem. However, given the historical evidence that capitalism and market institutions were implicit in precolonial societies, the official aversion of some political leaders to entrepreneurial and commercial activities poses an intriguing paradox.

Certainly, it would be an exaggeration to say that every peasant or person in Ghana was a capitalist at heart. But it is doubtful whether all except a small minority of ruling elites were genuinely committed to a socialist transformation of the economy in which private property had no place. Although Ghanaians were communal at heart, they valued their independence and economic actions determined by private choice. The paradox, however, reveals a tension between the association of capitalism with exploitative colonialism, on the one hand, and, on the other, the perennial desire of a new nation to search for something new. Usually, however, this new thing is more of an imitation or reaction to some foreign ideology rather than something based uniquely on Ghanaian socioeconomic realities and experiences. In this respect, government policies and the aspirations of Ghanaians have diverged: that is, the state and society have pulled apart as governments sought to impose foreign ideologies rather than evolving development strategies based on what the people do.

Granted that socialism might be efficacious, the official denunciation of capitalism was, however, not matched by successful socialist practice. Yet, interestingly, socialism had enjoyed a pride of place in the market of ideas in Ghana. Why this was so, could partly be traced to the founding ideas in Ghana and the early postcolonial development strategy adopted. Nkrumah's regime achieved considerable socioeconomic progress for Ghana under its radical development programme. But the regime also instituted a relatively generous "re-distributive" social system that benefited many people, particularly the urban dwellers. With this, the Nkrumah regime also created a constituency in society, mainly urban, whose material well-being was tied to the maintenance of his socioeconomic order. Indeed, any casual acquaintance with Ghanaians—particularly those who belong to the older postindependence generation—quickly reveals a profound nostalgia for the

Nkrumah regime and his achievements. This also helps to explain why "re-distributive" or "consumptionist politics" have proved difficult to dislodge in the post-Nkrumah years.

Although such a re-distributive system can be construed as social progress that benefited many Ghanaians, its maintenance also entailed a large fiscal responsibility for the state. Its viability rested on sustained economic growth, which successive regimes were unable to generate. At the same time, however, its beneficiaries were mainly in the urban areas— the politically powerful coalition of students, workers and the unemployed, whose protests and actions could undermine the power base of leaders. These beneficiaries of the "consumptionist" politics continued to intensify their claim on the state for re-distribution, and leaders learnt through the overthrow of the Busia regime that they could not ignore them without committing political suicide.

Consequently, governments held on to old, defunct and ineffective policies, or were hesitant or even refused to change policies or reform the economy under new circumstances. If they did at all, their efforts were cosmetic and belated. There no doubt were intense political struggles, engagement and energization to change things for the better, but these attempts met with only superficial success because the goals of development were not objectively based upon what the people did and wanted, but were more often tied to the visions of particular leaders and ruling cliques who did not always identify with the people's activities and wishes. And as these leaders sought to retain power at all cost and by all means, they marginalized whatever development goals they professed to have and which were the ostensible reason for the frequent changes of governments.

By seeking not to offend their urban supporters so as to stay in power, political leaders created deep-seated political inertia and addressed mainly re-distributive issues. Consequently, they allowed the Nkrumahist re-distributive social production system to sit deeply at the substructural level. This dragged on for many years, sending the economy into ultimate decline. As a result of their inability to undertake growth-generating policies, governments forwent the laying of the basic conditions that could sustain increased economic production and growth upon which future commitment to broader social goals could rest. In this respect, instead of effectively shaping the socioeconomic environment and activating the developmental process, the state became a force of economic contraction and deterioration.

Although in a sense it can be said that, structurally, little changed after Nkrumah, the general framework for development was discontinuous and disorderly so that leaders were unable to build cumulatively upon the achievements and experiences of their predecessors. In some instances, government actions and policies were blatantly destructive of property institutions and the economic activities as well as entrepreneurial spirit of the people, as the coercive instruments of postcolonial state were violently used to assault society. Through their policies some regimes (although perhaps inadvertently) exhibited little or no sense of entrepreneurship and of value for property and means of production. Thus, they were unable to elicit hard work from the people or channel their energies and resources into long-term productive activities that could increase national wealth.

As political instability became endemic, policies and/or development ideologies traded places and the bourgeoisie whose role is crucial to the development of the modern sector of the economy, were fragmented. Governing and economic management systems, as well as growth-generating institutions that should have evolved over time to support a viable modern economy, have not materialized. The centralization of power in leaders and other state executives, rendered the state apparatus ineffectual, disorderly and kleptocratic, wasteful of scarce resources and undermining administrative capacity and predictability in the socioeconomic and political environment.

Consequently, de-institutionalization has emerged in the social economic and political life of the country. Some indicators of the lack of institutionalization were the uncontested supremacy of personal rule, a lack of transparency in policy circles and "going by the books", arbitrary and frequent demonetization, the pursuance of short-term goals driven mainly by neopatrimonial political logic, frequent abandonment and re-writing of constitutions or their circumventions with impunity by political leaders, disregard for judicial decisions and the unceasing creation of new institutions to replace old ones. De-institutionalization in this respect created uncertainty in Ghana's socioeconomic environment and limited the prospects for long-term investment.

Incessant political instability, ideological contestation, oscillation of development policies and political arbitrariness led directly to cumulative economic decline. The state did not only fail to lay a rational and predictable institutional framework for modern development. The development process itself, like the cyclical political system, was subjected to perpetual renewal and decay. As each new group captured the state, instituted new rules, and preyed on it, they weakened its bureaucratic evolution and integrity and reduced its fortune and capability for the next group. And, as the elite conflict became entrenched, the state became an arena for struggles for power and it could not rise above conflicts or mediate interests. The inability of successive governments to evolve the state into an effective rational public-spirited instrument has one major consequence: state decay or diminished developmental capacity, manifested in the state's inability to manage the national economy without external guidance.

But diminished state capacity has been mainly a function of the patrimonialization of the state over the postcolonial period. This neopatrimonialism has limited the crystallization of the state into a rational bureaucratic organization and has also prevented it from shaping its surrounding environment in a manner that drives development forward. Patrimonialism, Weber argues, often unleashes arbitrary rule, and this has been evident in state-society relationship in Ghana. For example, the personal domination of the state, the fusing of governmental actions with leaders' personal philosophies, the harassment of businessmen and traders, the seizure of means of production, wealth and property, the usurpation and fragmentation of civil societies by state, the "mystery" and uncertainty that surrounds government intentions; the surprise of policies, the interference or domination of the civil service by political leaders, the privatization of the office, corruption and clientelism, have been exhibited by many regimes.

State decay is also manifested in the corruption of state officials, e.g. the police who now mount their own barriers at will and extort money from drivers and traders, the bureaucrats who deliberately create red-tape and demand money for the performance of public office, and other such things. There has also been a lack of bureaucratic discipline, accountability and transparency in the Ghanaian state and public institutions. Political arbitrariness and the weakness of rule of law also led to the deficiency of independently enforceable property rights and security. This has not permitted economic actors to plan long-term investment.

These pathological features are reminiscent of the arbitrariness in early modern Europe, which institutional economists argue, raise "transaction costs" of economic activities in the society. The Ghanaian socioeconomic and political environment thus exhibited features strikingly similar to those of early modern European societies, which were held in the theoretical discussion (see Chapter 2) as having retarded capitalist development. In terms of the objectives of this study, governance, macroeconomic management systems, and the rules of the game were constantly in flux, contested, re-defined, and radically re-evolved; consequently, over much of the postcolonial era, the socioeconomic and political as well as a legal and administrative framework for promoting sustainable development was distortionary, unpredictable and incalculable. From a long-term development perspective, the socioeconomic and political environment has thus lacked the calculability and growth-generating institutions that Weber and economic historians contend as critical for modern capitalist development.

In this politically turbulent postcolonial period, the entrepreneurial class was atomized and fragmented and could make little or no substantial long-term contribution to the economy, except perhaps in speculative trade. Peasants were overexploited and "reduced" their contribution to the formal economy. The violent and arbitrary domestic environment scared away foreign capital. The state itself failed in its developmental mission.

Consequently, a productive vacuum emerged undermining economic growth and ultimately commitment to broader social welfare goals.

These internal contradictions militating against development were periodically aggravated by unfavourable external factors (e.g. the soaring prices of oil in the late 1970s and early 1980s, declining terms of trade for export commodities, rising interest rates on loans, etc.) beyond the control of all leaders and finally Ghana's once promising economy came to its nadir, thus further entrenching Ghana's dependence on the external world with all the debilitating consequences for socioeconomic development. Hobbes' aphorism that no king can be rich, nor glorious, nor secure, when his subjects are poor becomes relevant in understanding Ghana's development predicament. As the economic kingdom was "destroyed", Ghanaians became poor, and their political leaders ultimately became poor as well, to the extent that they could not wage a war against their enemies. The state could not reproduce itself without the intervention of the IMF and World Bank. Consequently, the state has become a quasi-state, supported more by international law and aid than by a self-standing vibrant domestic economy and structures.

In the theoretical discussion, it was noted that military rule can sometimes be beneficial for society. The evidence in this study, however, refutes any claim that the Ghanaian military was a modernizing and developmental agent. Like their civilian counterparts, military regimes equally lacked a coherent development ideology. In 1966, when the soldiers seized power, they pointed out their antipathy to socialism and set out to dismantle the so-called socialist order of Nkrumah. In its stead, they enthroned a capitalist and liberal democratic order. The civilian regime that took over power in 1969 sought to advance this liberal order, but, in 1972, the military again seized power, reversed the liberal order, and imposed a pseudo-socialist order. The coup also indicated the difficulty of dismantling the Nkrumah socioeconomic order. The genie of military politicization was now loose to haunt Ghana.

As military rule became materially profitable to coup makers, the military acquired an irresistible appetite for the fruits of power. With each coup, the threshold of military intervention was lowered and subsequent coups became more likely. The 1972 coup was conducted mainly for the benefit of senior military officers, to the extent that in 1979, junior officers got angry and decided to wage war on the military itself for having destroyed the economy. This hodgepodge coup was meant to eradicate corruption, *kalabule*, and other social evils from society, but, after handing over power to a civilian regime in 1979, the military again seized the power and proclaimed a popular revolution, and then embraced the ultra-liberal economic policies of the IMF and the World Bank. What these swings suggest is that the military did not have any coherent ideological outlook and strategy of development to bring to bear on society.

Military rule has been mainly a means through which factors internal to military institutions, e.g. competing career ambitions and personality clashes, the desire to defend military corporate interests, living standards or privileges, personal insecurity, and military factionalism are achieved or resolved. Military interests and personal insecurities have been protected and resolved through the type of governance military leaders initiate, the type of institutions and decrees they establish, the type of macroeconomic management strategies they pursue, the share of national budget they claim, the shielding of the military establishment and officers from public institutions that demand accountability. The coups thus manifest the military's perennial antagonism to any policy of retrenchment or any that strikes at its standard of living, no matter how justified it is.

Furthermore, as the motives for coups become narrow and personalistic, military rule becomes corrupt and oppressive, and, like typical patrimonial rulers, military leaders sought to keep others from usurping their power so as to rule forever. The more military regimes became concerned with their survival, the more politically nihilist and autocratic they became and the more economically counter-productive their policies became for the society. This resulted in hostility between the civilian politicians and military elites. As the evidence shows, military rule has not been always motivated by national interest. The often articulated reasons for coups, civilian "crimes" such as corruption, malfeasance, profligacy, and economic mismanagement, are in many respects, *ex post facto* justifications that camouflage covert greed and personal ambition and fears of "men on the horse back". Military rule was thus also responsible for the distortion of the ruling process, and deprived society of institutionalization and ultimately was injurious to the evolution of a predictable and calculable socioeconomic and political environment. Given this, if development had occurred at all under military regimes, it was often *in spite of*, not *because of* the military in power.

This political sociology spells out some theoretical implications, at least in the context of Ghana. For example, a state that is habitually contested and susceptible to any group that can mobilize a credible power to capture it can hardly develop to be an instrument of any dominant class, be it the bourgeoisie or labour. The ruling class, however, requires the existence of a state whose principal role is to preserve and reproduce social, economic and political structures of its dominance, which in this study, means mainly the preservation and reproduction of neopatrimonialism. In view of this neopatrimonial politics, the atomization of the local bourgeoisie comes as no surprise because the patrimonial leader, Weber argues (recall the theoretical discussion in Chapter 2), often feels uneasy about or is even hostile to any autonomous action or economic advancement of his subjects because of their economic power to challenge his authority or press for the democratization of his state. A patrimonial ruling elite forming what can be described as "a political aristocratic class" whose reproduction does not depend on the activities of the bourgeoisie, can hardly be expected to engage in or support entrepreneurial or risk activities that could weaken or destroy its dominance.

This helps to explain why, over much of the postcolonial era, the state has not always been used to encourage the national bourgeoisie to gain a strategic position in the structure of ownership of the means of production of the economy. In this regard, the state can hardly be said to be an instrument of indigenous capital. The state is, however, able to act against the will of the bourgeoisie because its revenue base is assured through the exports of cocoa, minerals, timber, etc. So long as elites can portray themselves as ardent economic reformers and secure aid, and so long as the national boundaries are secured in international conventions and codes, and the Ghanaian state does not face any real or potential competing or aggressive state that could threaten or absorb it, state actors have the luxury not to "aid" domestic economic forces or bother much about fundamental economic reform and scientific or technological change at home that could bring about increased productive capacity. In such a context, development can only be incidental and if pursued at all, it would depend primarily on the personal conviction of the leader and the dynamics of the local social situation and forces.

If the bourgeoisie has not yet "captured" the state to serve its interest, then political rather than economic power might be more useful in explaining domination in Ghana. Early postcolonial development discourse took the general view that African societies lacked a productive bourgeoisie to propel economic development, hence, the state was to spearhead development. This is a view that many African leaders endorsed for expanding state-sector activity. Not that this was untrue, but the problem was that, beyond this claim, few intellectuals and political leaders ever argued that African states could also be used to the nurture national bourgeoisie to supplement statist development efforts. Few political leaders were sympathetic to the advancement of their national bourgeoisie or encouraging of the entrepreneurial activities of their citizens in the modern business and industrial sectors as a means of increasing national wealth. The perennial tension between the state and the bourgeoisie in Ghana had a 'Catch 22' quality: on the one hand, the lack or weakness of a productive bourgeoisie made it easier for the state to justify and control the formal economy, on the other, the state control of the economy and patrimonial politics often discouraged the emergence of this bourgeoisie. As the activities of the bourgeoisie are crucial to modern capitalist development, the state indirectly became an accomplice in the underdevelopment of its own national bourgeoisie, and ultimately, an obstacle to the industrialization of the economy. Consequently, Ghanaians were saddled with a singularly unproductive capitalism through which they endured all the social ills associated with capitalism without benefiting from its historic role in developing the productive forces of society.

In explaining what failed Ghana, one can also pinpoint many factors, particularly external ones, which have been unfavourable to Ghana and thus inhibited the promotion of socioeconomic development in this postcolonial era, but the evidence assembled in this study suggests that state politics has been one contributory factor in the decline of Ghana's postcolonial economy. Given the proviso that political leaders have little control over external factors that may affect the economy and society one way or the other, the conclusion in this study is that, in the decline of Ghana's postcolonial economy and society, domestic contradictions and government actions as well as inactions have played no insignificant role, and this illustrates how crucial state, domestic institutions and macroeconomic management systems are in the development process. In this regard, an urgent developmental task now will be how to resolve these internal contradictions to pave way for a more sustainable development to proceed.

What Can Be Done: Towards a Nationally Relevant Development Strategy

The point of departure for this study was the growing dissatisfaction with the conceptualization of development problems and solutions in existing development literature. The approach of the study suggests that a detailed historical and sociological analysis of the dynamic processes operating in specific African countries and the political conditions that shape or give rise to these processes can be useful in generating more realistic solutions to the development problem than was hitherto available. In this final section, an attempt is made to suggest a way out of the dilemma of underdevelopment. Of course, there is no pretence here to offer a definitive blue print as there can of course be other approaches or policies such as agricultural policies and rural development. The approach here derives from the perspective that has guided this study.

Any nationally relevant development strategy in Ghana must take into consideration three issues: (i) where the country stands today in terms of development; (ii) the nature of existing development discourse and strategies; and (iii) what can be practically done. In other words, solutions to the development problems need to be anchored in the concrete socioeconomic realities of the nation at a specific time, and not simply subjected to foreign ideologies, and wishful thinking of some individuals or groups. Solutions to development problems need to be found in the "contextuality" of development as a product of historical processes that are continuously unfolding. The context of development everywhere is, however, constantly changing at various speeds in different times, creating both new opportunities and obstacles to development in different societies. This requires orderly adaptation of efficacious policies, institutional reforms as well as cosmopolitanism (openness) in the policy environment.

On the first issue, Ghana has come a long way in restoring some economic viability since 1983. President Rawlings has achieved political stability for Ghana, at least in the last 17 years, and has also presided over a democratization process. The NDC regime has renewed its democratic mandate and is now ruling for the second time. Although it was noted that the democratization process was fraught with many obstacles, it nevertheless represents an embryonic attempt to establish social peace and stability in the ruling process. Also, the macroeconomic framework has remained basically the same since the acceptance of the IMF/World Bank's structural adjustment programme in 1983. This is indicative of the fact that some stability, predictability and calculability in the rules of game are beginning to emerge. The consolidation of these conditions is, however, important for sustainable development.

On the second issue, at the threshold of twenty-first century, development discourse continues to be defined from Euro-American perspectives and driven by external interest and agents. External aid has been necessary to cope with economic crises, but this cannot provide a sustainable basis for promoting development. Moreover, it entails costs that impose debt burdens on future governments and generations. Notwithstanding the good intentions behind it, aid has not come cheap. Ghanaian leaders have had to pay a high price: the "handing over" of a significant part of national economic management to IMF and World Bank officials in Washington. They have become more accountable to these institutions than to their own citizens, while, at the same time, they are compelled to "squeeze" their nationals to pay high interest rates on IMF World Bank loans.

This exogenization of development practice also has a psychological dimension. Currently, there is a certain lack of confidence in leadership to tackle development on the basis of local understanding, organization and efforts. This lack of confidence is manifested in the leadership's preference for external aides, the submission of national economic management to foreign "experts", the submissive behaviour of government officials to relatively minor officials of foreign governments, and development agencies, the longing of leaders for approval in the advanced countries, the mentality that foreign aid and so-called experts can solve problems of underdevelopment, and so on. Lacking the confidence to pursue development on their own terms, leaders cannot empower their citizens for self-reliance and realization. This condescending behaviour and actions devalue the struggle for independence, to think and do things as a sovereign nation-state. By ceding the control of development to external agents who are sometimes eager to help but unconsciously reproduce their own societies, these leaders have not only forfeited the opportunity to devise a locally relevant and sustainable development strategy, they have also already aborted the development process. When a people must be developed not by themselves but by others, development becomes benevolence that is largely insensitive to social needs and aspirations.

One salient feature of all postcolonial development strategies is that the state always acted upon society in "top-down" fashion with whatever policies leaders judged to be good for or in the interest of the people. There was little of the or no "bottom-up" approach that could enable the people to express their aspirations and popularly support state policies meant to develop them. This has arrested processes of indigenous development and preempted possibilities for alternative development strategies based on the principle of greater "self-reliance" and on the involvement of the people on the ground. The "top-down" development approach has also made the people passive recipients or objects of development, which has heightened tensions, uncertainties and feelings of anomie. If, however, people are the ends of development, then it is logical that they have the prerogative to decide or influence policies that seek their well-being.

For too long, development practice has been appropriated by elites, divorced from the experience and everyday world of the people, and destructive of their activities, aspirations and dreams. Elites have internalized or have been preoccupied with foreign ideologies, seeking to change people into what they are not instead of identifying with them and using politics to promote their activities, be they in agriculture, trade or manufacturing. Generally, then, development has been *against* the people. This has brought about development crises, disenchantment and even cynicism towards government policies.

This leads to the last consideration: what can be realistically done to promote sustainable socioeconomic development? To be able to specify a realistic strategy of development it is important to reiterate what is meant by the concept *development* in this study. Development is more a technical matter involving the mobilization by the state of the people's activities, national resources, skills, innovations, creativity, and enthusiasm to achieve better living standards for all, than simply ideological proclamation or preference. It is also a process involving the setting up of some basic conditions that have been outlined in this study: a relatively stable and predictable socioeconomic and political environment. Furthermore, it involves setting attainable development goals, mapping out relevant strategies and effectively implementing (or correcting) them on continuous basis to attain these goals. Development is also not simply a matter of drawing up an ideal blueprint based on foreign ideologies to be directed from outside. It is essentially, about mobilizing the people to take the destiny of the society into their own hands.

It is easy to analytically categorize developmental paths as capitalist or socialist, and one can say, for instance, that Japan is a capitalist society and China, a communist society. But there is also something uniquely Japanese and Chinese about these two countries' development that is often easily missed. While the development experience of other cultures can be useful for action in other social contexts, there is a danger when development practices are reduced to simply blindly copying foreign ideologies or reacting to them. A more realistic approach to development is to find indigenous solutions to development problems. It is because societies have to cope with fundamentally similar problems of human living that their social structures and institutions evolve over time to look similar and not simply because they have copied foreign ideas. Attempts at self-reliance are thus crucial to successful and sustainable development. Self-reliance in this context does not mean autarky (because small countries cannot go it alone because of narrow scale of their economies, resources and market limitations, etc.), but the *indigenization* of development process to make the people and their activities the ends and means of development. The people and their activities are the wealth of the nation, and the basic objective of development that needs to guide governmental actions is the creation of an enabling environment for the people to engage in productive activities to fulfil their aspirations and have decent standards of living, and enjoy healthy and creative lives.

To an increasing extent, in contemporary times, the social, economic and political lives of all nations are intermeshed. Nations do draw on one another for sources of their material and intellectual growth, but whatever they acquire as a result of cultural diffusion needs to be anchored in or adapted to the peculiar cultural and institutional context of the societies in question, otherwise whatever was borrowed, or learnt cannot work effectively to advance these societies. Because of this, it is difficult to draw conclusions from the experiences of different societies at different historical periods and apply them in other contexts and in other times. Development strategies need to be based on local experience and on the people's knowledge and interests.

Development needs to be based on the support and full participation of the people who must be the means and the beneficiaries of development. Development needs to grow pragmatically from local experience; it needs to be about *what is* and not what *ought to be* in society. It needs take the people as *they are* and not what *they ought to be* according to the vision and dream of one particular leader or group of elites. This means elevating the activities of nationals to prominence and supporting them. In this way, the development process is more likely to become self-sustaining and reinforcing. There is, thus, the need to build a new philosophy of development based on the foregoing principles.

From the study's perspective, the following considerations are worth considering at the policy level:

(i) Given the debilitating effect of political instability on the promotion of development, there is urgent need for the establishment of political stability, social peace and effective institutions for the purpose of promoting development. This calls for the transcendence of the inter-elite conflicts that have bedeviled Ghana's postcolonial system and made politics destructive of society's socioeconomic progress. Unity can be forged by governing elites fostering a consensus on basic rules of society and encouraging debates on national problems and finding compromising solutions them. Institutions of arbitration, reconciliation and compromise, which were and still are, characteristic of traditional societies need to be encouraged. It is crucial to establish a consensual politics and a coherent long-term national strategy for development that would accord with the people's aspirations and economic activities.

Without an elite committed to developing efficient bureaucratic state institutions, without incentives for productive rather speculative or transaction investment, and without a commitment to increasing national wealth and opportunities for individual progress, socioeconomic development will stall. There have been too many "liberators", "redeemers", "revolutionaries", and "reformers" and Ghana is still languishing in material poverty. Obviously, development cannot occur in a context that institutionalizes political anarchy, arbitrariness and violence. In an unpredictable and uncertain environment, businesses and individuals cannot be sure what will be profitable or unprofitable, legal or illegal, in the future. Political stability is one basic prerequisite for attainment of a predictable and calculable environment. A precondition for development is a bureaucratic state, which is capable of fostering political order, an adequate infrastructure, a calculable law and administration and development facilitating policies.

(ii) There is the need to improve the management and use of public resources, finances, and aid. One way to attain this is for incumbent regimes to promote public debates on strategic national issues, and build consensus over broad policy matters and on what direction development should follow.

The poverty of ideas and serious debates over the national crisis at the policy level hitherto means that even the population is not informed about government intentions. There is the need for a more sophisticated debate among elites about socioeconomic dilemmas. Governments also need to inform the general public of the dilemmas they face and what necessary sacrifices society at large has to make from time to time. In this way, governments are likely to create legitimacy for themselves and even pursue necessary, but unpopular policies that can support economic viability. Also policies and project implementation as well as evaluation capacities need to be made more effective.

Through this strategy, they can also save themselves from "unnecessary criticisms" and possibly pre-empt coups, which have often been staged in the face of economic crisis. As more elites are permitted to partake in debates on crucial national issues, they are directly or indirectly shaping or participating in prevailing policies. They are, thus, less likely to unduly reverse or abandoned policies in the future, if they are in power and thus build upon the achievements, experiences and successes, or correct the failures of predecessor regimes. This will enable development goals and policies to be pursued on a continuous and cumulative basis to attain effective results. But all these also require the reform of the state to make it more rational and effective instrument of public management and development. A reformed state is not only necessary for societal development, but also, in the longer-term, necessary for consolidating state power and the power base of rulers. Only governments and the people can reform the state to make its mode of operation beneficial for society at large.

(iii) Given the deliberateness with which development is pursued in many national contexts in modern times, there is also the need for creating, through public policy, an environment of security of property and predictability for wealth creation. In a sense, development is about people creating wealth. Without the public guarantees that enable wealth creation there can be no socioeconomic development. Social peace, trust and reciprocity or co-operation in state-society relationship are key features of a predictable environment. They are basic to encourage productive investment (both foreign and domestic), which provides a basis for wealth generation. Also vital is the protection of rights of ownership through the existence of the rule of law, the enforcement of contracts and laws independent of arbitrary political power. Without a mutually beneficial relationship, whereby the state trades the provision of security of property and an economically-supportive institutional framework for productive activity, the sustainability of socioeconomic development will be arduous. There is a need for systematic efforts at the policy level favouring investment and production because, at this early stage of the nation's developmental cycle, the creation of wealth might be more important than sharing it. An institutional framework that supports entrepreneurial development and increased political accountability is likely to strengthen the political kingdom and the effectiveness of the state in the long run, which state elites need for pursuing development. Without this, the growth potential of Ghana cannot be realized, and, thus development will be unduly delayed. At the moment, these conditions just seem to be emerging and thus need to be nurtured.

With these three conditions in place, gradually with time, a less conflicting and more predictable socioeconomic and political environment (institutional framework) which is a prerequisite for sustaining long-term development and supporting proper functioning of a modern economy and society, would emerge.

(iv) Yet effective implementation capacity of the state is also crucial. Given the "top-down" approach to development, which seems to have alienated the people, there is also the need for economic and political decentralization, which can ensure the efficient utilization of resources, as well as accountability in management systems at all local levels of society. This will also bring development closer to the people who can put forward their needs and be supportive of the programmes that are to improve their living conditions. Decentralization—both in the political and economic realm—is important for restoring people's confidence in their own capacity to mobilize resources and build society. In this way, a governing structure responsive to the needs of the people would be constructed and make people supportive of governmental policies and actions. Many of these changes will depend on the deliberate actions of political agency and the strengthening of local forces, civil society, organizations and communities that can sustain whatever reform is made in the state and reinvigorate the institutions that will make the state an effective agency in development.

(v) There is also the need for a more activating role of the state through the provision of services that facilitate production. This can take two forms: indirect and direct. The indirect refers to the setting up of an enabling socioeconomic and political environment, and the direct to the provision of physical infrastructure and human skills through technical and vocational training, the development of local scientific capacity to generate new technology and adapt foreign technologies to serve local needs. Ghana, like many African countries, continues to believe in the myth of "appropriate technology", but the reality of technological advancement is that few nations, if any, have developed only through the transfer of "appropriate technology" from one culture to another. That a sturdy domestic technological capacity is basic for supporting manufacturing and agricultural development cannot be overemphasized.

One problem that seems to face Ghana, and, indeed, many African countries is the lagging response to technological (soft and hard) advances or the inability to develop or domesticate foreign technology. Ghana's productive means and systems have benefited very little from the rapidly expanding technologies originating from the outside world. Productivity in agriculture, industry and services continues to be low. Consequently, the economy is not being synchronized with economic and technological developments in the industrialized world. One area where this is manifested is in the lack of dynamism in the industrialization front reverberating in Ghana's continuous dependence on agricultural commodities, particularly cocoa, which was developed to meet the demands of the early phase of the industrial revolution in Europe. The needs of contemporary industries in the North have, however, changed over the last century, such that cocoa is no longer a critical raw material to the advanced world.

Moreover, the innovative capacity of modern capitalism seems boundless, in that it keeps generating synthetic products and raw materials to keep costs of production as low as possible. This has reduced the demands for cocoa over the postwar years. The result is the perpetual deterioration of the international terms of trade for cocoa and decreasing foreign exchange earnings. Ghana thus needs to develop new economic activities to become competitive in the world economy. This will mean, for example, the development of manufacturing industries that a government-assisted business class in partnership with the state and foreign capital can bring about. Government use of public institutions to encourage modernizing elites in the private sectors to learn to exploit opportunities provided by trade, foreign investment and technology to attain a rapid rate of economic and industrial growth is desirable. These foregoing issues seem to be some of the challenges that lie ahead of governments of Ghana. These are some of the features that could define a nationally relevant development strategy.

If capitalism is now being accepted in Ghana as a viable development alternative, then for it to be progressive there is a need for the establishment of an enabling socioeconomic and political environment. Just as there can be no true socialism without complete state control over the means of production, there can be no true capitalism without the existence or creation of a domestic capitalist class. Without this class, capitalism becomes merely a means for the government to use its power to channel resources into the pockets of its own members and their associates acting in private capital. Again, without this class, any claim by the state to promote capitalism connotes little more than an opportunistic exploitation by parasitic, unproductive political and bureaucratic elites of "insider" privileges, which is not likely to lead to the development of the classic risk-taking entrepreneurial behaviour.

Clearly, it is impossible to doubt the crucial role of the state in "late" capitalist development. But it is important to make a crucial distinction between state-led (or induced) capitalist industrialization (which is likely to include a certain amount of public ownership of key activities, at least in the early stages)—but where a vigourous private sector and market competition are also promoted—and state-propelled industrialization, where the state owns almost all the means of production and the private sector is negatively viewed and thus confined to certain limited and inconsequent areas. Given the early stage of Ghana's development, the impetus for creating a capitalist class and development needs to come from the state. Development is also about wealth creation and there is no reason why Ghanaians should be denied opportunities to create or own wealth in their country, more so when foreign investors are being allowed. This goes against the logic of governments existing principally to protect and advance the interest and welfare of citizens.

Governments also need to provide indigenous capital with a favourable domestic environment and financial support, especially since the product of the indigenous private sector is normally strongly oriented towards domestic consumption. Government collaboration with or encouragement of domestic capitalists, is in the long run interest of the state, since it needs the critical resources the capitalists have in the form of taxes. These measures are compatible with government's desire to broaden its revenue base and increase domestic productive capacity. The bourgeoisie and its activities are thus important for creating a sturdy financial base for the state.

Moreover, a vibrant local bourgeoisie is one important indicator to foreign investors that an environment conducive to long-term investment has emerged or is emerging. Also, an expanding bourgeoisie is important for enhancing public accountability and consolidating democratic governance. A nation's development cannot proceed until public policies express the wishes and economic activities of the people who are both ends and means of development. Thus, a nationally relevant capitalist developmental strategy essentially needs to reflect the desire of the domestic bourgeoisie to reduce and eliminate their junior status in relation to foreign capital. This will require considerable statecraft. While encouraging direct foreign investment, so that novel capital, technology and managerial knowledge are transmitted to the country, the state is aiming at using its leverage to secure a larger share of economic surplus for the national productive bourgeoisie.

The indigenization of the ownership, employment and control of sectors of the economy—particularly those that do not require huge capital and sophisticated technology—could be vital for the construction of a nationally relevant development strategy. One other measure would be to restrict foreign ownership and management in key areas of the economy, especially in areas where Ghanaians can confidently participate and also to encourage partnership between Ghanaian businesspeople and foreigners in other areas. These efforts are complementary to that of a state seeking to raise national fortune.

There has been an influential political claim used to justify an expanding state sector: that there is no productive bourgeoisie to undertake economic transformation, but a parasitic business class interested in conspicuous consumption. This claim is only partially true, if not totally unfair, because it ignores the growth paths of a productive bourgeoisie over time, especially their collective ability to learn technical and managerial as well as political skills. This claim also understates the accomplishments of this class and implicitly sets impossibly high standards for them, relative to the length of time it actually took the capitalist classes of today's industrialized countries to become effective social forces in economic transformation. It also fails to recognize that governments do aid the development of the business and industrial class to become a positive force in economic development. It can also, however, be argued that the present economic precariousness of Ghana has resulted from the weakness and inability of a solid productive bourgeoisie to emerge and supplement the development efforts of governments and also serve as an effective link with foreign capital in any kind of positive *dependent development* project.

The formation of a politically influential and productive capitalist class with a solid and mature bourgeois culture to support and focus its economic and political projects takes generations. Even the relatively industrialized countries of the former Soviet Union and Eastern Europe are now discovering this, to their chagrin. They have just acquired some parasitic *nouveaux riches,* realizing that it will take time to acquire serious creators of wealth. Yet as already noted, there is no reason to assume that an indigenous business class will always be willing to collude in a subordinate capacity with foreign investors, as dependency theory sometimes seems to suggest. In reality, the trend in many developing nations has been the invasion of more and more economic sectors (initially the preserve of foreign capital) by local businessmen and state corporations. Equally, one can speculate that, so long as the basic rules of society are not rational and predictable enough to elicit long-term productive investment from the people, the business class will be interested in speculative and short-term gains and will not focus its attention on long-term entrepreneurial activities.

Though there exists goodwill from external donor countries and development agencies, ultimately only Ghanaians, their resources and politics can build a viable modern society for the benefit of all. Thus a new form of politics needs to evolve that builds effective state apparatus, draws rulers and the ruled closely together, and sees the activities of citizens as complementary to those of the state. In this endeavour, state actors are in a vantage position to develop governing structures, macroeconomic management systems and institutions that can create the conditions for long-term development. They are also in a position to reform the state to become more developmental, for by so doing they strengthen their own kingdom.

External factors matter in development. But it is important to note that there are few countries that do not experience shifting fortunes in their external accounts. This is true of rich and poor, industrialized as well as agricultural, countries. As a country passes from a subsistence economy to modern one, it begins to experience changes in fortunes arising from the dynamics of the international environment. These changes require continued effective organization at home to compensate for vulnerability in the ever-competitive global economy. They also necessitate adaptations in economic policy, including sometimes temporary reductions in desired rates of economic expansion and even consumption. Ghana cannot be an exception to this rule, and it does not help Ghana's socioeconomic progress when coups become answers to temporary crises, because they introduce instability in the ruling process and weaken the state's effectiveness to evolve as a rational instrument for managing societal problems.

It cannot be gainsaid that the immense gap between the developed and the less developed economies, along with global market conditions, has made it more difficult to escape dependence in the late twentieth century than it was for developing economies of the nineteenth century. Nevertheless, the phenomenal economic growth of some Asian countries suggests that socioeconomic development is still possible in the periphery and much depends on the arrangement of domestic conditions. Although reform in the international economic environment would be beneficial to Ghana's progress, in all practicality external dependence remains an objective reality that Ghana has to confront. One possible way to do this, is to improve domestic capacities and raise the efficiency of productive forces. At least, on this issue, governments have some level of control. An enabling socioeconomic and political environment for productive activity and an effective "developmental state" seems desirable to facilitate and catalyze the development process. A "developmental state" is a state that establishes its legitimacy mainly on its ability to promote and sustain development, understanding by development, the combination of steady high rates of economic growth and structural change in the productive system both domestically and in its relation to the international economy. And development cannot be promoted and sustained regardless of the global economic factors without domestic peace, predictable environment and state support for productive investment. Without these fundamental conditions, Ghana is likely to have problems promoting and sustaining development.

Appendix

Table 1: Cocoa Production to the Nearest Thousand Metric Tonnes (1947-1984)

Year	Production	Year	Production
1947-48	208	1965-66	401
1948-49	278	1966-67	368
1949-50	248	1967-68	415
1950-51	262	1968-69	323
1951-52	211	1969-70	403
1952-53	247	1971-72	454
1953-54	211	1972-73	407
1954-55	220	1973-74	340
1955-56	229	1974-75	397
1956-57	264	1975-76	327
1957-58	207	1976-77	277
1958-59	226	1977-78	268
1959-60	317	1978-79	281
1960-61	430	1979-80	277
1961-62	409	1980-81	245
1962-63	413	1981-82	247
1963-64	428	1982-83	180
1964-65	538	1983-84	150

Source: Adapted from: (a) Milburn, J. F. (1977), *British Business and Ghanaian Independence*, Table 8, p. 107; (b) 'International Cocoa Organization Estimates', in *Africa Economic Digest*, IV, 50 (Dec., 1983); (c) *Quarterly Digest of Statistics*, Central Bureau of Statistics, now Statistical Service of Ghana (June, 1982); and (d) *Economic Intelligent Unit Report* (1990).

Table 2: Producer Price of Cocoa for Ghana, Cote d'Ivoire and Nigeria (1957-1976)

Producer Price (£) and index				Price Differences as Percentage of Ghana Price	
Year	Ghana	Cote d'Ivoire	Nigeria	Cote d'Ivoire and Ghana	Nigeria and Ghana
1957	138.40	136.22	147.63	- 1.6	+ 6.7
1958	132.28	158.88	147.63	+ 20.1	+ 6.7
1959	117.38	130.66	150.91	+ 11.3	+ 28.6
1960	110.23	137.97	157.47	+ 25.2	+ 42.9
1961	110.23	138.20	108.96	+ 25.3	- 1.2
1962	110.23	101.74	100.06	- 7.7	- 9.2
1963	108.39	102.01	104.98	- 5.9	- 3.1
1964	99.21	102.29	111.54	+ 3.1	+ 12.4
1965	90.94	102.15	104.57	+ 12.3	+ 15.0
1966	75.78	85.97	72.17	+ 13.4	- 4.8
1967	85.02	103.59	90.22	+ 21.8	+ 6.1
1968	101.23	118.09	100.60	+ 16.7	- 6.2
1969	113.74	117.47	143.53	+ 3.2	+ 26.2
1970	119.97	123.80	174.15	+ 3.2	+ 45.2
1971	116.92	126.23	177.93	+ 8.0	+ 52.2
1972	102.62	134.71	177.98	+ 31.3	+ 73.4
1973	135.65	170.30	201.19	+ 25.5	+ 48.3
1974	181.00	223.99	313.35	+ 23.8	+ 23.8
1975	226.29	367.51	443.39	+ 62.4	+ 95.9
1976	327.13	410.06	547.30	+ 25.4	+ 67.3

Source: Bequele, B. (1985), 'Stagnation and Inequality in Ghana', in Dharam Ghai and Samir Radwan (eds), *Agrarian Policies and Rural Poverty in Africa*, ILO, Geneva.

Table 3 (a): Relative Prices for Food and Cocoa Agriculture (1963-1973) with 1963 as Base Year: 100

Year	Prices of Local Food	Price of Cocoa	Difference
1963	100	100	100
1964	126	90	71
1965	173	83	48
1966	200	69	35
1967	170	77	45
1968	184	92	50
1969	200	104	52
1970	210	109	52
1971	236	109	46
1972	259	94	36
1973	313	123	39
1974	363	164	45
1975	474	206	43
1976	770	297	39

Table 3 (b): Growth Rates for Food and Cocoa Agriculture (1963-76)

Growth Rates	Food	Cocoa
1963-70	9.3	2.10
1970-76	22.27	18.86
1963-76	12.56	7.99

Source: Bequele, A. (1985), 'Stagnation and Inequality in Ghana', in Dharam Ghai and Samir Radwan (eds), *Agrarian Policies and Rural Poverty in Africa*, ILO, Geneva.

Table 4: Export Production Volumes in Thousand Units (1975-1981)

Item/Year	1975	1976	1977	1978	1979	1980	1981
Cocoa (thousand tonnes)	397.3	326.7	277.4	268.2	280.8	277.2	246.5
Timber (cubic meters)	623	565	586	591	285	185	222
Gold (kg)	16.3	16.5	15.0	12.5	11.1	11.0	10.6
Diamonds (carats)	233.3	228.3	194.7	142.3	122.6	114.9	836.4
Manganese (tonnes)	415.3	361.9	292.4	316.7	253.8	249.9	223.1
Bauxite (tonnes)	325.2	271.6	279.2	327.9	235.3	225.1	181.3

Source: Hutchful, E. (1985), 'IMF Adjustment Policies in Ghana Since 1966' in *Africa Development*, Vol. X, No. 1/2. See also *Quarterly Digest of Statistics*, Statistical Service of Ghana (June, 1984).

Table 5: Index of Mining Production (1977-1981) with 1977 as Base Year: 100

Item/Year:	1977	1978	1979	1980	1981
Gold	100	83.6	74.2	73.4	70.8
Diamonds	100	73.1	63.0	59.0	43.0
Manganese	100	108.3	86.8	85.7	76.5
Bauxite	100	117.4	84.3	80.9	65.2
Overall Index	100	87.1	74.9	73.8	68.2

Source: Hutchful, E. (1985), 'IMF Adjustment Policies in Ghana Since 1966', in *Africa Development*, Vol. X, No. 1/2. See also *Quarterly Digest of Statistics*, Statistical Service of Ghana (June, 1984).

Table 6: Export Volume Indices for Ghana and Cote d'Ivoire (1967-1976)

Exported Product	1967		1974		1976	
	Ghana	Cote d'Ivoire	Ghana	Cote d'Ivoire	Ghana	Cote d'Ivoire
All Exports	80.2	63.3	77.6	124.6	51.1	125.9
Cocoa	89.3	73.4	83.5	143.4	48.8	136.3
Timber logs	46.5	62.1	50.8	86.3	36.7	93.3

Source: Bequele, A. (1985), 'Staganation and Inequality in Ghana', in Dharam Ghai and Samir Radwan (eds), *Agrarian Policies and Rural Poverty in Africa*, ILO, Geneva.

Bibliography

(a) Primary Sources

(i) Government Official Documents:

Quarterly Digest of Statistics, Central Bureau of Statistics (Accra).
National Assembly Debates, 11 March 1964 (Accra).
Republic of Ghana: Our Task, (1972), Government Printer (Accra).
Speeches and Interviews by Col. I. K. Acheampong, vol. 1, n.d. (Accra).
The 1966 Budget (Accra).
The Gold Coast Annual Report of 1947 (Accra).
The Gold Coast Gazette, 17 March 1923 (Accra).
The Ten Great Years 1951-1960, Ghana Information Service (Accra).

(ii) Newspapers and Magazines:

Africa Confidential (London).
Daily Graphic (Accra).
Ghanaian Times (Accra).
Legon Observer (Legon, Accra).
Peoples Daily Graphic (Accra).
The Economic Intelligent Unit Report (London).
The Free Press (Accra).
The Ghanaian Chronicle (Accra).
New York Times (New York).
West Africa (London).
Workers Banner (Accra).

(b) Secondary Sources

Achebe, C. (1967), *A Man of the People*, Heinemann, London.
Afrifa, A. A. (1966), *The Coup: 24th February 1966*, Frank Cass, London.
Agyeman-Duah, B. (1987), 'Ghana, 1982-6: The Politics of the PNDC', *The Journal of Modern African Studies*, vol. 25, no. 4, pp. 613-642.
Ahene, R. A. (1992), 'Private Investment Incentives in Sub-Saharan in Africa', in R. A. Ahene and S. B. Katz (eds), *Privatization and Investment in Sub-Saharan Africa*, Praeger, New York.

Ahiakpor, J. C. W. (1985), 'The Success and Failure of Dependency Theory: The Experience of Ghana', *International Organization*, vol. 39, no. 3, pp. 535-552.

Ahiakpor, J. C. W. (1991), 'Rawlings, Economic Policy Reform and the Poor: Consistency or Betrayal', *The Journal of Modern African Studies*, vol. 29, no. 4, pp. 583-600.

Ake, C. (1985), 'Why is Africa not Developing?' *West Africa*, 17 June, pp. 1212-1214.

Ake, C. (1996), *Democracy and Development in Africa*, The Brookings Institution, Washington D.C.

Almond, G. A. and Coleman, J. S. (eds) (1960), *The Politics of Developing Areas*, Princeton University Press, Princeton.

Almond, G. A and Powell, G. B. Jr. (1966), *Comparative Politics: A Developmental Approach*, Little Brown and Company, Boston.

Amin, S. (1972), 'Underdevelopment and Dependency in Black Africa:Origins and Contemporary Forms', *Journal of Modern African Studies*, vol. 10, no. 4.

Amin, S. (1973), *Neo-Colonialism in West Africa*, Penguin Books, Harmondsworth.

Amin, S. (1974), *Accumulation on a World Scale: A Critique of the Theory of Underdevelopment*, 2 vols. Monthly Review Press, New York.

Amin, S. (1976), *Unequal Development*, Monthly Review Press, New York.

Amin, S. (1987), 'The State and the Question of Development', in P. Anyang' Nyong'o (ed.), *Popular Struggles for Democracy in Africa*, Zed Books, London.

Amonoo, B. (1981), *Ghana 1957-1966: Politics of Institutional Dualism*, George Allen & Unwin, London.

Amsden, A. (1985), 'The State and Taiwan's Economic Development', in P. B. Evans, D. Rueschemeyer and T. Skocpol (eds), *Bringing the State Back In*, Cambridge University Press, Cambridge.

Amsden, A. (1988), 'Taiwan's Economic History: A Case of *Etatisme* and a Challenge to Dependency Theory', in R. Bates (ed.), *Toward a Political Economy of Development: A Rational Choice Perspective*, University of California Press, Berkeley.

Amsden, A. (1991), *Asia's Next Giant, South Korea and Late Industrialization*, Oxford University Press, Oxford.

Antonio, R. J. and Glassman, R. M. (eds) (1985), *Capitalism and Socialism: Weber's Dialogue with Marx*, University Press of Kansas, Kansas.

Apter, D. (1965), *The Politics of Modernization*, University of Chicago Press, Chicago.

Apter, D. ([1955]/1972), *Ghana in Transition*, Princeton University Press, Princeton N.J.

Apter, D. and Rosberg, C. (eds) (1994), *Political Development and the New Realism in Sub-Saharan Africa*, University Press of Virginia, Charlottesville.

Armah, A. K. (1968), *The Beautyful Ones Are Not Yet Born*, [sic], Heinemann, Oxford.

Asante, K. B. (1993), 'Nkrumah and State Enterprises', in K. Arhin (ed.) *The Life and Work of Kwame Nkrumah*, Africa World Press Inc., Toronto.

Austin, D. (1964), *Politics in Ghana 1946-1960*, Oxford University Press, London.

Austin, D. and Luckham, R. (eds) (1975), *Politicians and Soldiers in Ghana 1966-1972*, Frank Cass, London.

Ayittey, G. (1991), *Indigenous African Institutions*, Transnational Publishers Inc., New York.

Ayittey, G. (1992), *Africa Betrayed*, St. Martin's Press, London.

Azarya, V. and Chazan, N. (1987), 'Disengagement from the State in Africa: Reflections on the Experience of Ghana and Guinea', *Comparative Studies in Society and History*, vol. 29, no. 1, pp. 106-131.

Baran, P. ([1957]/1964), *The Political Economy of Growth*, Monthly Review Press, New York.

Bates, R. (1981), *Markets and States in Tropical Africa: The PoliticalBasis of Agricultural Policies*, University of California Press, Berkeley.

Bates, R. (1987), 'The Politics of Agricultural Pricing in Sub-Saharan Africa', in Z. Ergas (ed.), *The African State in Transition*, Macmillan, London.

Bauer, P. T. (1954), *West African Trade: A Study of Competition, Oligopoly and Monopoly in a Changing Economy*, Cambridge University Press, Cambridge.

Baynham, S. J. (1984/1985), 'Civil-Military Relationships in Ghana's Second Republic', *Journal of Contemporary African Studies*, vol. 4, no. 1/2, pp. 71-88.

Beckman, B. (1976), *Organizing the Farmers: Cocoa Politics and National Development in Ghana*, The Scandinavian Institute of African Studies, Uppsala.

Beckman, B. (1992), 'Empowerment or Repression? The World Bank and Politics of Africa Adjustment', in P. Gibbon, Y. Bangura and A. Ofstad (eds), *Authoritarianism, Democracy And Adjustment: The Politics of Reform in Africa*, The Scandinavian Institute for African Studies, Uppsala.

Bennett, V. P. (1975), 'Epilogue: Malcontents in Uniform', in D. Austin and R. Luckman (eds), *Politicians and Soldiers in Ghana 1966-1972*, Frank Cass, London.

Berg, E. (1971), 'Structural Transformation Versus Gradualism: Recent Economic Development in Ghana and the Ivory Coast', in P. Foster and A. R. Zolberg (eds), *Ghana and the Ivory Coast: Perspectives on Modernization*, The University of Chicago Press, Chicago.

Berger, P. and Luckmann, T. ([1966]/1991), *The Social Construction of Reality: A Treaties in the Sociology of Knowledge*, Penguin Books, London.

Berman, J. B. and Leys, C. (1994), *African Capitalists in African Development*, Lynne Rienner Publishers, Boulder.

Bequele, A. (1985), 'Stagnation and Inequality in Ghana', in D. Ghai and S. Radwan (eds), *Agrarian Policies and Rural Poverty in Africa*, International Labour Organization, Geneva.

Bienefeld, M. A. (1982), 'The International Context for National Development Strategies: Constraints and Opportunities in a Changing World', in M. Bienefeld and M. Godfrey (eds), *The Struggle ForDevelopment: National Strategies in an International Context*, John Wiley & Sons Ltd., Chichester.

Boahen, A. (1975), *Ghana: Evolution and Change in the Nineteenth and Twentieth Centuries*, Longman, London.

Boateng, E. A. (1994), 'Ghana: Physical and Social Geography', *Africa South of the Sahara: Regional Surveys of the World*, Europa Publications Ltd., London, pp. 401.

Booth, D. (1985), 'Marxism and Development Sociology: Interpreting the Impasse', in M. Shaw (ed.), *Marxist Sociology Revisited: Critical Assessments*, Macmillan, Houndmills.

Bottomore, T. B. (1985), *Elites and Society*, Penguin Books, Harmondsworth.

Bourret, F. M. (1960), *Ghana: The Road to Independence 1919-1957*, Oxford University Press, London.

Bratton, M. and Rothchild, D. (1992), 'The Institutional Bases of Governance in Africa', in G. Hyden and M. Bratton (eds), *Governance and Politics in Africa*, Lynne Rienner Publishers, Boulder.

Brohman, J. (1995), 'Universalism, Eurocentrism, and Ideological Bias in Development Studies: From Modernization to Neoliberalism', *Third World Quarterly*, vol. 16. no. 1, pp. 121-146.

Brown, D. (1982), 'Who are the Tribalists?: Social Pluralism and Political Ideology', *African Affairs*, vol. 81, no. 322, pp. 37-69.

Burns, R. T. (1996), 'The Crooked Ways of Socio-economic Development: What Argentina Can Teach Us About Sustainable Dynamic Systems of

Social Production', paper presented at the Uppsala Theory Circle Seminar, Uppsala University, Sweden.

Burns, R. T. and Flam, H. (1987), *The Shaping of Social Organization: Social Rule System Theory with Applications*, Sage Publications, London.

Callaghy, T. M. (1987), 'The State as Lame Leviathan: The Patrimonial Administrative State in Africa', in Z. Ergas (ed.), *The African State in Transition,* Macmillan, London.

Callaghy, T. M. (1988), 'The State and the Development of Capitalism in Africa: Theoretical, Historical and Contemporary Reflections', in D. Rothchild and N. Chazan (eds) *The Precarious Balance: State and Society in Africa*, Westview Press, Boulder.

Cardoso, F. H. (1972), 'Dependency and Development in Latin America', *New Left Review*, vol. 74, July/August 1972.

Cardoso, F. H. and Falleto, E. (1979), *Dependency and Development in Latin America*, University of California Press, Berkeley.

Carter, A. (1995), 'The Nation-State and Underdevelopment', *Third World Quarterly*, vol. 16; no. 4, pp. 595-618.

Chabal, P. (1992), *Power in Africa: An Essay in Political Interpretation*, Macmillan, Houndmills.

Chamlee, E. L. (1993), *The Cultural Foundations of Economic Development: A Study of Urban Female Entrepreneurs in Ghana, West Africa*, Unpublished Doctoral Thesis, Beloit College.

Chazan, N. (1982), 'Ethnicity and Politics in Ghana', *Political Science Quarterly*, vol. 97, no. 3, pp. 461-485.

Chazan, N., (1983), *An Anatomy of Ghanaian Politics: Managing Political Recession 1969-1982*, Westview, Boulder.

Chazan, N. (1988), 'Ghana: Problems of Governance and the Emergence of Civil Society', in L. Diamond *et al.*, *Volume Two: Democracy in Developing Countries: Africa*, Lynne Rienner Publishers, Boulder.

Chazan, N. Mortimer, R. Ravenhill, J. and Rothchild, D. (1992), *Politics and Society in Contemporary Africa*, Lynne Rienner Publishers, Boulder.

Clapham, C. (1985), *Third World Politics: An Introduction*, Routledge, London.

Clapham, C. (1991), 'The African State' in Douglas Rimmer *et al.*, *Africa 30 Years On: The Record and the Outlook after Thirty Years of Independence Examined for the Royal African Society*, Heinemann Educational Books, London.

Cohen, I. (1981), 'Introduction' in Max Weber, *General Economic History*, Transaction Books, New Brunswick.

Coleman, S. (1990), *Foundations of Social Theory*, Harvard University Press, Cambridge, Massachusetts.

Collier, P. and Gunning, J. W. (1997), *Explaining Economic Performance*, Working Paper Series, Centre for the Study of African Economies, Oxford University.

Collins, R. (1980), 'Weber's Last Theory of Capitalism: A Systematization', *American Sociological Review,* vol. 45, December, pp. 925-942.

Cotterrell, R. (1992), *The Sociology of Law*, Butterworths, London.

Crook, R. (1990), 'State, Society and Political Institutions in Cote d'Ivoire and Ghana' *IDS Bulletin*, vol. 21. no. 4, pp. 24-34.

Daaku, K. Y. (1970), *Trade and Politics on the Gold Coast 1600-1720*, Clarendon Press, Oxford.

Dadson, J. A. (1993), 'Agricultural Development Strategy Under Nkrumah', in K. Arhin (ed.), *The Life and Work of Kwame Nkrumah*, Africa World Press Inc., Toronto.

Danquah, J. B. (1997), *The Ghanaian Establishment: Its Constitution, its Detentions, its Traditions, its Justice and Statecraft, and its Heritage of Ghanainaism,* Ghana Universities Press, Accra.

Davidson, B. (1992), *The Black Man's Burden: Africa and the Curse of the Nation-State,* James Curry, London.

Decalo, S. (1973), 'Military Coups and Military Regimes in Africa', *The Journal of Modern African Studies*, vol. 11, no. 1, pp. 105-127.

Decalo, S. (1990), *Coups and Army Rule in Africa: Motivations and Constraints*, Yale University Press, New Haven.

Diamond, L. (1988), 'Introduction: Roots of Failure, Seeds of Hope', in L. Diamond J. Linz and S. M. Lipset (eds), *Volume Two: Democracy in Developing Countries: Africa*, Lynne Rienner Publishers, Boulder.

Diamond, L. (1994), 'Promoting Democracy in the 1990s: Actors, Instruments and Issues', Unpublished paper presented at the Nobel Symposium on Democracy's Victory and Crisis, 27-30 August, Uppsala University.

Doornbos, M. (1990), 'The African State in Academic Debate: Retrospect and Prospect', *The Journal of Modern African Studies*. vol. 28, no. 2, pp. 179-198.

dos Santos, T. (1970), 'The Structure of Dependence', in *American Economic Review,* vol. 60, no. 2. pp. 231-6.

dos Santos, T. (1973), 'The Crisis of Development Theory and the Problems of Dependence in Latin America', in H. Berstein (ed.), *Underdevelopment and Development*, Penguin, Harmondsworth.

Dowse, R. (1969), *Modernization in Ghana and the USSR: A Comparative Study*, Routledge & Kegan Paul, London.

Dowse, R. (1975), 'Military and Police Rule', in D. Austin and R. Luckman (eds), *Politicians and Soldiers in Ghana 1966-1972*, Frank Cass, London.

Dumont, R. ([1966]/1988) *False Start in Africa*, Earthscan Publishers, Ltd. London.

Durkheim, E. ([1895]/1982), *Durkheim: The Rules of Sociological Method*, (translated by Halls, W. D. and edited by S. Lukes) Macmillan, London.

ECA, (1989a), *Africa's Alternative Framework to Structural Adjustment Programmes for Socio-Economic Recovery and Transformation*, Addis Ababa.

ECA, (1989b), *Statistics and Policies: ECA Preliminary Observations on the World Bank Report: Africa's Adjustment and Growth in the 1980s*, Addis Ababa.

Eggertsson, T. (1990), *Economic Behavior and Institutions*, Cambridge University Press, Cambridge.

Eisenstadt, S. N. (1966), *Modernization, Protest and Change*, Prentice-Hall, Englewoods Cliffs, N.J.

Emmanuel, A. (1972), *Unequal Exchange: A Study in the Imperialism of Trade*, Monthly Review Press, New York.

Ergas, Z. (1986), 'In Search of Development: Some Directions for the Future Investigation', *The Journal Of Modern African Studies*, vol. 24, no. 2, pp. 303-333.

Ergas, Z. (1987), 'In Search of Development in Africa', in Z. Ergas (ed.) *The African State in Transition*, Macmillan, London.

Eriksen, T. L. (1990), *Afrikas Krise. Finnes det alternativer til Verdensbankens Diagnose og Medisin*?, NORAD and NUPI, Oslo.

Esseks, J. D. (1971), 'Government and Indigenous Private Enterprise in Ghana', *The Journal of Modern African Studies*, vol. 9, no. 1, pp. 11-29.

Esseks, J. D. (1975), 'Economic Policies' in D. Austin and R. Luckman (eds), *Politicians and Soldiers in Ghana 1966-1972*, Frank Cass, London.

Etsiah, K. A. (1989/90), 'Military Leadership and Foreign Policy Behaviour in Ghana 1972-1982', *Journal of Contemporary African Studies*, vol. 8/9, no. 1/2, pp. 27-54.

Evans, P. (1992), 'The State as Problem and Solution: Predation, Embedded Autonomy, and Structural Change', in S. Haggard and R. Kaufman (eds), *The Politics of Adjustment: International Constraints, Distributive Conflicts and the State*, Princeton University Press, Princeton N.J.

Evans, P. B., Rueschemeyer, D. and Skocpol, T. (eds) (1985), *Bringing the State Back In*, Cambridge University Press, Cambridge.

Evans, P. (1979), *Dependent Development: The Alliance of Multinational, State and Local Capital in Brazil*, Princeton University Press, New Jersey.

Fatton, R. Jr. (1992), *Predatory Rule: State and Civil Society in Africa*, Lynne Rienner Publishers, Boulder.

Finer, S. (1962), *The Man on Horseback: The Role of the Military in Politics*, Pall Mall, London.

Foster, P. and Zolberg, A. R. (eds) (1971), *Ghana and the Ivory Coast: Perspectives on Modernization*, Chicago University Press, Chicago.

Frank, G. (1967), *Capitalism and Underdevelopment in Latin America: Historical Studies in Chile and Brazil*, Monthly Review Press, New York.

Frank, G. (1969), *Latin America: Underdevelopment or Revolution*, Monthly Review Press, New York.

Frank, G. (1991), 'No Escape from the Laws of World Economics', *Review of African Political Economy*, no. 50, pp. 21-32.

Franklin, J. (1997), 'IMF Conditionality, Threat Perception, and Political Repression: A Cross-National Analysis', *Comparative Political Studies*, vol. 30, no. 5, October, pp. 576-606.

Frimpong-Ansah, J. H. (1991), *The Vampire State in Africa: The Political Economy of Decline in Ghana*, James Curry, London.

Fukui, H. (1992), 'The Japanese State and Economic Development: A Profile of a Nationalist-Paternalist Capitalist State', in R. P. Appelbaum and J. Henderson (eds), *States and Development in the Asian Pacific*, Sage, Newbury Park, CA.

Fukuyama, F. (1992), *The End of History and the Last Man*, Hamish, Hamilton, London.

Furtado, C. (1964), *Development and Underdevelopment: A Structural View of the Problems of Developed and Underdeveloped Countries*, University of California Press, Berkeley.

Galtung, J. (1971), 'A Structural Theory of Imperialism', *Journal of Peace Research*, vol. 8, pp. 81-117.

Garlic, P. (1971), *African Traders and Economic Development in Ghana*, Claredon Press, Oxford.

Gerf, H. H. and Mills, C. W. ([1948]/1977), *From Weber: Essays in Sociology*, Routledge & Kegan Paul, London.

Gerschenkron, A. (1962), *Economic Backwardness in Historic Perspective*, Harvard University Press, Cambridge, Massachusetts.

Gibbon, P. (1992), 'Structural Adjustment and Multipartyism', in P. Gibbon, Y. Bangura and A. Ofstad (eds), *Authoritarianism, Democracy And Adjustment: The Politics of Reform in Africa*, The Scandinavian Institute for African Studies, Uppsala.

Gibbon, P. and Bangura, Y. (1992), 'Adjustment, Authoritarianism and Democracy: An Introduction to Some Conceptual and Empirical Issues', in P. Gibbon, Y. Bangura and A. Ofstad (eds), *Authoritarianism,*

Democracy And Adjustment: The Politics of Reform in Africa, The Scandinavian Institute for African Studies, Uppsala.

Giddens, A. (1971), *Capitalism and Modern Social Theory*, Cambridge University Press, Cambridge.

Gilpin, R. (1987), *The Political Economy of International Relations*, Princeton University Press, Princeton.

Girling, J. (1981), *Thailand, Society and Politics*, Cornell University Press, Itchaca, N.Y.

Goodell, G. (1987), 'The Importance of Political Participation for Sustained Capitalist Development', in P. Berger (ed.), *Modern Capitalism, Vol. 2: Capitalism and Equality in the Third World*, Hamilton Press, Lanham.

Goulbourne, H. (1987), 'The State, Development and the Need for Participatory Democracy in Africa', in P. Anyang' Nyong'o, (ed.), *Popular Struggles for Democracy in Africa*, Zed Books, London.

Graham, Y. (1984), *Ghana,the Politics of Crisis: Class Struggle and Organization 1976-1983*, Unpublished paper presented at RAPE Conference, 29-30 September on The World Recession and the Crisis in Africa at the University of Keele, Britain.

Gramsci, A. (1971), *Selections from 'The Prison Writings'*, Lawrence & Wishart, London.

Green, R. H. (1971), 'Reflections on Economic Strategy, Structure, Implementation and Necessity: Ghana and Ivory Coast 1975-67', in P. Foster and A. R. Zolberg (eds), *Ghana and the Ivory Coast: Perspectives on Modernization*, The University of Chicago Press, Chicago.

Green, R. H. (1988), 'Ghana: Progress, Problematics and Limitations of the Success Story', *IDS Bulletin*, vol. 19, no. 1.

Gregor, A. J. (1979), *Italian Facism and Developmental Dictatorship*, Princeton University Press, Princeton.

Griffiths, R. and Griffiths, B. (1987), 'Variables of Success and Failure in Third World Development', in P. Berger (ed.), *Modern Capitalism: Capitalism and Equality in the Third, World Volume 2*, Hamilton Press, Lanham.

Gyimah-Boadi, E. (1989), 'Policies and Politics of Export Agriculture', in E. Hansen and K. Ninsin (eds), *The State Development and Politics in Ghana*, Codesria Books, London.

Gyimah-Boadi, E. and van de Walle, N. (1996), 'The Politics of Economic Renewal in Africa', in B. Ndulu and N. van de Walle, (eds), *Agenda for Africa's Economic Renewal: US-Third World Policy Perspectives*, no. 21, Transaction Publishers, New Brunswick.

Habermas, J. (1987), *The Theory of Communicative Action: (Volume Two) The Critique of Functionalist Reason*, Polity Press, Cambridge.

Habermas, J. (1975), *Legitimation Crisis*, Beacon Press, Boston.
Harris, N. (1987), *The End of the Third World: Newly Industrializing Countries and the Decline of an Ideology*, Penguin Books, Harmondsworth.
Harrison, D. (1988), *The Sociology of Modernization and Development*, Unwin Hyman, London.
Hart, K. (1982), *The Political Economy of West African Agriculture*, Cambridge University Press, Cambridge.
Healey, J. and Robinson, M. (1992), *Democracy, Governance and Economic Policy*, Overseas Development Institute, London.
Herbst, J. (1993), *The Politics of Reform in Ghana 1982-1991*, University of California Press, Los Angeles.
Hermassi, E. (1980), *The Third World Reassessed*, University of California Press, Berkeley.
Herskovits, M. J. and Harvitz, M. (1964), *Economic Transition in Africa*, Routledge and Kegan Paul, London.
Hettne, B. (1988), 'The Role of the State in Development—Problem or Solution?: With a Note on the Potential of Ethnodevelopment', Unpublished Paper for the seminar, Swedish Development Cooperation with Sub-Saharan Africa: State and Nation Under Stress, 6-8 September, Saltsjöbaden, Sweden.
Hettne, B. (1990), *Development Theory and the Three Worlds*, Longman House, London.
Hill, P. (1957), *The Gold Coast Cocoa Farmer*, Oxford University Press, London.
Hill, P. (1963), *The Migrant Cocoa Farmers of Southern Ghana: A Study in Rural Capitalism*, Cambridge University Press, Cambridge.
Hill, P. (1970), *Studies in Rural Capitalism in West Africa*, Cambridge University Press, Cambridge.
Hirschman, O. A. (1977), *The Passions and the Interests: Political Arguments For Capitalism Before Its Triumph*, Princeton University Press, Princeton.
Hobson, J. A. ([1902]/1988), *Imperialism: A Study*, Unwin Hyman, London.
Hodder-Williams, R. (1984), *An Introduction to the Politics of Tropical Africa*, Unwin Hyman, London.
Hoogvelt, A. M. (1976), *The Sociology of Developing Societies*, Macmillan, London.
Hopkins, A. G. (1988), *An Economic History of West Africa*, Longman, New York.
Hoselizt, B. F. (1960), *Sociological Aspect of Growth*, Free Press, Glencoe, Ill.

Hout, W. (1993), *Capitalism and the Third World*, Edward Elgar, Hants.

Howard, R. (1978), *Colonialism and Underdevelopment in Ghana*, Croom Helm, London.

Hume, D. and Turner, M. (1990), *Sociology and Development: Theories, Policies and Practices*, Harvester Wheatsheaf, New York.

Huntington, S. P. (1968), *Political Order in Changing Societies*, Yale University Press, New Haven.

Husain, I. and Faruquee, R. (eds) (1994), *Adjustment in Africa: Lessons from Case Studies*, World Bank, Washington D.C.

Hutchful, E. (1979), 'A Tale of Two Regimes: Imperialism, the Military and Class in Ghana', in *Review of African Political Economy*, no. 14, January-April, pp. 36-55.

Hutchful, E. (1985), 'IMF Adjustment Policies in Ghana Since 1966', *Africa Development*, vol. x, no. 1/2, Codesria, pp. 122-136.

Hutchful, E. (1987), *The IMF and Ghana: The Confidential Record*, Zed Books, London.

Huq, M. M. (1989), *The Economy of Ghana: The First 25 Years Since Independence*, Macmillan, London.

Hyden, G. (1980), *Beyond Ujamaa in Tanzania: Underdevelopment and an Uncaptured Peasantry*, Heinemann, London.

Hyden, G. (1983), *No Short Cuts To Progress*, University of California Press, Berkeley.

Hyden, G. (1996), 'The Challenges of Analyzing and Building Civil Society', *Africa Insight*, vol. 26, no. 2, pp. 92-106.

Iliffe, J. (1983), *The Emergence of African Capitalism*, Macmillan, London.

Ingham, B. (1993), 'The Meaning of Development: Interactions Between New and Old Ideas', *World Development*, vol. 21, no. 11, pp. 1803-1821.

Jackson, R. H. and Rosberg, C. G. (1982), *Personal Rule in Black Africa: Prince, Autocrat, Prophet, Tyrant*, University of California Press, Berkeley.

Jackson, R. H. and Rosberg, C. G. (1988), 'Why Africa's Weak States Persist: The Empirical and Juridical in Statehood', *World Politics*, vol. 35, October 1-24.

Jackson, R. H. and Rosberg, C. G. (1994), 'The Political Economy of Personal Rule', in D. E. Apter and C. G. Rosberg (eds), *Political Development and the New Realism in Sub-Saharan Africa*, University Press of Virginia, Charlottesville.

Jeffries, R. (1989), 'Ghana: The Political Economy of Personal Rule', in D. B. Cruise O'Brien, J. Dunn and R. Rathbone (eds), *Contemporary West African States*, Cambridge University Press, Cambridge.

Jeffries, R. and Thomas, C. (1993), 'The Ghanaian Elections of 1992', *African Affairs*, vol. 92, pp. 331-316.

Jepperson, R. L. (1991), 'Institutions, Institutional Effect, and Institutionalism', in W. W. Powell and P. J. DiMaggio (eds), *The New Institutionalism in Organizational Analysis*, The University of Chicago Press, Chicago.

Jones, E. L. (1987), *The European Miracle: Environments, Economies and Geopolitics in the History of Europe and Asia*, Cambridge University Press, Cambridge.

Kasfir, N. (1987), 'Class, Political Domination and the African State', in Zaki Ergas (ed.), *The African State in Transition*, Macmillan Press, London.

Kay, C. (1993), 'For a Renewal of Development Studies: Latin American Theories and Neoliberalism in the Era of Structural Adjustment', *Third World Quarterly*, vol. 14, no. 4, pp. 691-702.

Kay, G. B. (1972), *The Political Economy of Colonialism in Ghana: A Collection of Documents and Statistics 1900-1969*, Cambridge University Press, Cambridge.

Kennedy, P., (1977), 'Indigenous Capitalism in Ghana', in *Review of African Political Economy*, January/April, no. 8.

Kennedy, P., (1990), *African Capitalism: The Struggle for Ascendancy*, Cambridge University Press, Cambridge.

Killick, T. (1978), *Development Economics in Action: A Study of Economic Policies in Ghana*, Heinemann, London.

Killick, T. (1989), *A Reaction Too Far. Economic Theory and the Role of the State in Developing Countries*, Overseas Development Institute, London.

Killick, T. (1995), 'Economic Flexibility in Africa: Evidence and Causes', in T. Killick (ed.), *The Flexible Economy: Causes and Consequences of the Adaptability of National Economies*, Routledge, London.

Kimble, D. (1963), *A Political History of Ghana 1850-1928*, Clarendon Press, Oxford.

Kirk-Greene, A. M. H. (1986), 'West Africa: Nigeria and Ghana', in P. Duignan and R. Jackson (eds), *Politics and Government in African States, 1960-85*, Croom Helm, London.

Kohli, A. (1993), 'Democracy Amid Economic Orthodoxy: Trends in Developing Countries', *Third World Quarterly*, vol. 14, No. 4, pp. 671-689.

Kraus, J. (1970), 'Ghana Without Nkrumah: The Winter Discontent', in I. G. Markovitz (ed.), *African Politics and Society*, The Free Press, New York.

Kraus, J. (1971), 'Political Change, Conflict and Development in Ghana', in P. Foster and A. R. Zolberg (eds), *Ghana and the Ivory Coast: Perspectives on Modernization*, The University of Chicago Press, Chicago.

Kraus, J. (1991), 'The Political Economy of Stabilization and Structural Adjustment in Ghana', in D. Rothchild (ed.), *Ghana: The Political Economy of Recovery*, Lynne Rienner Publishers, Boulder.

Lal, D. (1983), *The Poverty of Development Economics*, Institute of Economic Affairs, London.

Lall, S. (1975), 'Is Dependence a Useful Concept in Analyzing Underdevelopment?', *World Development*, November-December, pp. 799-810.

Lall, S. and Stewart, F. (1996), 'Trade and Industrial Policy in Africa', in Benno Ndulu and Nicolas van de Walle (eds), *Agenda for Africa's Economic Renewal: US-Third World Policy Perspectives*, no. 21, Transaction Publishers, New Brunswick.

Lancaster, C. (1988), 'Political Economy and Policy Reform in Sub-Saharan Africa' in S. K. Commins (ed.), *Africa's Development Challenges and the World Bank: Hard Questions, Costly Choices*, Lynne Rienner Publishers, Boulder.

Lancaster, C. (1993), 'Governance and Development: The Views from Washington', *IDS Bulletin*, vol. 24, no. 1.

Leechor, C. (1994), 'Ghana Frontrunner in Adjustment' in I. Husain and R. Faruque (eds), *Adjustment in Africa: Lessons from Country Case Studies*, World Bank, Washington D.C.

Leftwich, A. (1983), *Redefining Politics: People, Resources and Power*, Methuen, London.

Leftwich, A. (1993), 'Governance, Democracy and Development in the Third World', *Third World Quarterly*, vol. 14, no. 3, pp. 605-624.

Leftwich, A. (1994), 'Governance, State and the Politics of Development', *Development and Change*, vol. 25, no. 2; pp. 363-386.

Lerner, D. (1964), *The Passing of Traditional Society*, The Free Press, New York.

Le Vine, V. T. (1975), *Political Corruption: The Ghana Case*, Hoover Institution Press, Stanford.

Leys, C. (1976), 'The Overdeveloped Post-Colonial State: A Re-evaluation', *Review of African Political Economy*, January-April, pp. 39-48.

Leys, C. (1996), *The Rise and Fall of Development Theory*, James Currey, Oxford.

Leys, C. and Berman, C. (1994), 'Introduction', in B. J. Berman and C. Leys (eds), *African Capitalist in African Development*, Lynne Riener Publishers, Boulder.

List, F. ([1841/1966), *The National System of Political Economy*, A. M. Kelley, New York.

Loxley, J. (1990), 'Structural Adjustment in Africa: Reflections on Ghana and Zambia', *Review of African Political Economy*, no. 47, pp. 8-27.

Lubeck, P. M. (1987), (ed.), *The African Bourgeoisie: Capitalist Development in Nigeria, Kenya and the Ivory Coast*, Lynne Rienner, Boulder.

Luggard Lord, ([1922]/1965), *The Dual Mandate in British Tropical Africa*, Frank Cass, London.

Maclean, S. (1997), 'Managing Development in Subsaharan Africa in the 1990s: States, Markets and Civil Societies in Alternative Paradigms', in R. A. Siddiqui, (ed.), *Subsaharan Africa in the 1990s: Challenges to Democracy and Development*, Praeger Publishers, Westport.

Marcussen, H. S. and Torp, J. E. (1982), *The Internationalization of Capital. The Prospects for the Third World: A Re-examination of Dependency Theory*, London, Zed Press, London.

Maren, M. (1997), *The Road to Hell: The Ravaging Effects of Foreign Aid and International Charity*, The Free Press, New York.

Marenin, O. (1987), 'The Managerial State in Africa: A Conflict Coalition Perspective', in Z. Ergas (ed.), *The African State in Transition*, Macmillan Press, London.

Markovitz, I. L. (1977), *Power and Class in Africa: An Introduction to Change and Conflict in African Politics*, Prentice-Hall Inc., Englewood Cliff, New Jersey.

Marx, K. (1973), 'The Eighteenth Brumaire of Louis Bonaparte', in K. Marx and F. Engels, *Selected Works In Three Volumes*, vol. 1, pp. 394-487, Progress Publishers, Moscow.

Marx, K. and Engels, F. (1973), 'The Communist Manifesto', in K. Marx and F. Engels, *Selected Works in Three Volumes*, vol. 1, pp. 98-137, Progress Publishers, Moscow.

Mazrui, A. A. and Tidy, M. (1984), *Nationalism and New States in Africa*, Heinemann Educational Books, Nairobi.

McCarthy, S. (1994), *Africa: The Challenge of Transformation*, I. B. Tauris & Co Publishers, London.

McClelland, D. (1961), *The Achieving Society*, Van Nostrand, New York.

Mikell, G. (1989), *Cocoa and Chaos in Ghana*, Howard, University Press, Washington D.C.

Milburn, J. F. (1977), *British Business and Ghanaian Independence*, C. Hurst & Co., London.

Miliband, R. (1974), *The State in Capitalist Society*, Quartet, London.
Moore, B. (1968), *Social Origins of Dictatorship and Democracy: Lord and Peasant in the Making of the Modern World*, Penguin, Harmondsworth.
Morawetz, D. (1977), *Twenty-Five Years of Economic Development: 1950-1975*, World Bank, Washington D.C.
Mosley, P. and Toye, J. (1988), 'The Design of Structural Adjustment Programmes', *Development Policy Review*, vol. 6, pp. 395-413.
Mosley, P. and Weeks, J. (1993), 'Has Recovery Begun? Africa's Adjustment in the 1980s Revisited', *World Development*, vol. 21 no. 10, pp. 1583-1606.
Nash, L. and Kantrow, A. (1987), 'Multinational Corporations and Economic Development', in P. Beger (ed.), *Modern Capitalism: Capitalism and Equality in the Third, World Volume 2*, Hamilton Press, Lanham.
Nelson, J. (1990), 'The Politics of Economic Adjustment in Developing Nations', in J. Nelson (ed.), *Economic Crisis and Policy Choice*, Princeton University Press, Princeton N.J.
Neuber, A. (1995), 'Adapting The Economies of Eastern Europe: Behavioural and Institutional Aspects of Flexibility', in T. Killick (ed.), *The Flexible Economy: Causes and Consequences of the Adaptability of National Economies*, Routledge, London.
Ninsin, K. A. (1989), 'Thirty Years of Development Experience', in K. A. Ninsin and E. Hansen (eds), *The State Development and Politics in Ghana*, Codesria Books Series, London.
Nkrumah, K. (1957), *Ghana, An Autobiography*, Nelson and Sons Ltd., London.
Nkrumah, K. (1961), *I Speak of Freedom*, Heinemann, London.
Nkrumah, K. (1963), *Africa Must Unite*, International Publishers, New York.
Nkrumah, K. (1964), *Consciencism: Philosophy and Ideology for Decolonization and Development, with Particular Reference to the African Revolution*, Heinemann, London.
Nkrumah, K. (1965), *Neo-Colonialism: The Last Stage of Imperialism*, Nelson and Sons Ltd., London.
Nkrumah, K. (1968), *Dark Days in Ghana*, Panaf Publications, London.
North, D. (1981), *Structure and Change in Economic History*, W. W. & Company, New York.
North, D. (1990), *Institutions, Institutional Change and Economic Performance*, Cambridge University Press, Cambridge.
Nugent, P. (1991), 'Educating Rawlings: The Evolution of Government Strategy Toward Smuggling', in D. Rothchild (ed.), *Ghana: The Political Economy of Recovery*, Lynne Rienner Publishers, Boulder & London.

Nugent, P. (1995), *Big Men, Small Boys and Politics in Ghana: Power, Ideology and the Burden of History 1982-1994*, Pinter, London.

Nukunya, G. K. (1992), *Tradition and Change: An Introduction to Sociology*, Universities Press, Accra.

Obeng, S. (1997), *Selected Speeches of Nkrumah*, Afram Publications, Accra.

O'Brien, D. C. (1979), 'Modernization, Order and Erosion of a Democratic Ideal', in D. Lehmann (ed.), *Development Theory: Four Critical Studies*, Frank Cass, London.

Ocran, A. K. (1968), *A Myth Is Broken*, Longmans, London.

O'Donnell, G. A. (1979), *Modernization and Bureaucratic Authoritarianism: Studies in South American Politics*, University of California Press, Berkeley.

Offe, C. and Rogne, C. (1988), 'Thesis on the Theory of the State', in A. Giddens and D. Held (eds), *Classes, Power and Conflict: Classical and Contemporary Debates*, Macmillan Press, London.

Opoku-Agyeman, (1988), 'Setbacks to Political Institutionalization by Praetorianism in Africa', *The Journal of Modern African Studies*, vol. 23, no. 3, pp. 403-435.

Oquaye, M. (1980), *Politics in Ghana 1972-1979*, Tornado Publications, Accra.

Oquaye, M. (1995), 'The Ghanaian Elections of 1992—A Dissenting View', *African Affairs*, vol. 94, no. 375, pp. 259-280.

Owusu, M. (1970), *Uses and Abuses of Political Power: A Case Study of Continuity and Change in the Politics of Ghana*, Chicago University Press, Chicago.

Palma, G. (1978), 'Dependency: A Formal Theory of Underdevelopment or a Methodology for the Analysis of Concrete Situations of Underdevelopment?', *World Development*, vol. 6. pp. 881-924.

Palma, G. (1981), 'Dependency and Development: A Critical Overview', in D. Seers (ed.), *Dependency Theory: A Critical Reassessment*, Francis Pinter, London.

Partiff, T. (1990), 'Lies, Damned Lies and Statistics: The World Bank/ECA Structural Adjustment Controversy', *Review of African Political Economy*, no. 47, pp. 128-141.

Parsons, T. (1951), *The Social System*, Routledge, London.

Peil, M. (1981), *Cities and Suburbs: Urban Life in West Africa*, Holmes and Meier, London.

Pellow, D. and Chazan, N. (1986), *Ghana: Coping with Uncertainty*, Westview Press, Boulder.

Perlmutter, A. (1969), 'The Praetorian State and Praetorian Army: Toward a Taxonomy of Civil-Military in Developing Countries', *Comparative Politics*, April 1969.

Petchenkine, Y. (1993), *Ghana in Search of Stability 1957-1992*, Praeger, Westport.

Philips, E. H. (1979), 'Ethnicity in Ghanaian Politics', *South African Journal of African Affairs*, vol. 9. no. 1.

Pinkney, R. (1972), *Ghana Under Military Rule 1966-1969*, Methuen & Co., London.

Poggi, G. (1978), *The Development of the Modern State: A Sociological Introduction*, Stanford University Press, Stanford.

Post, K. (1991), 'The State, Civil Society, and Democracy in Africa: Some Theoretical Issues', in R. Cohen and H. Goulbourne (eds), *Democracy and Socialism in Africa*, Westview Press, Boulder.

Poulantzas, N. (1973), *Political Power and Social Classes*, New Left Books, London.

Poulantzas, N. (1975), *Classes and Contemporary Capitalism*, New Left Books, London.

Price, M. R. (1984), 'Neo-Colonialism and Ghana's Economic Decline: A Critical Assessment', in *Canadian Journal of African Studies*, vol. 18, no. 1, pp. 163-193.

Ragin, C. C. (1987), *The Comparative Method, Moving Beyond Qualitative and Quantitative Strategies*, University of California Press, Berkeley.

Rawls, J. (1972), *A Theory of Justice*, Oxford University Press, Oxford.

Ray, D. I. (1986), *Ghana: Politics, Economics and Society*, Frances Pinter Publishers, London.

Riggs, F. (1964), *Administration in Developing Countries: The Theory of Prismatic Society*, Houghton Mifflin, Boston.

Rimmer, D. (1991), 'Introduction: The Africas of 1961 & 1991', in D. Rimmer *et al.*, *Africa 30 Years On: The Record and the Outlook after Thirty Years of Independence Examined for the Royal African Society*, Heinemann, London.

Rimmer, D. (1992), *Staying Poor: Ghana's Political Economy 1950-1990*, Pergamon Press, Oxford.

Robertson, H. M. (1959), 'A Criticism of Weber and His School', in W. R. Green (ed.), *Protestantism and Capitalism: The Weber Thesis and Its Critics*, D. C. Heath and Company, Boston.

Rodney, W. (1972), *How Europe Underdeveloped Africa*, Bogle-L'Ouverture, London.

Roe, A. and Schneider, H. (1992), *Adjustment and Equity in Ghana*, OECD, Paris.

Roemer, M. (1981), 'Dependence and Industrialization Strategies', in *World Development*, 9 May, pp. 429-34.

Rostow, W. W. (1960), *The Stages of Economic Growth. A Non-Communist Manifesto*, Cambridge University Press, London.

Rostow, W. W. (1971), *Politics and the Stages of Growth*, Cambridge University Press, London.

Rothchild, D. (1980), 'Military Regime Performance: An Appraisal of the Ghanaian Experience: 1972-1978', in *Comparative Politics*, vol. 12. no. 4, pp. 459-479.

Rothchild, D., (1987), 'Hegemony and State Softness: Some Variations in Elite Response', in Z. Ergas (ed.), *The African State in Transition*, Macmillan, London.

Rothchild, D. (ed.) (1991), *Ghana: The Political Economy of Recovery*, Lynne Rienner Publishers, Boulder.

Rueschemeyer, D. and Evans, P. B. (1985), 'The State and Economic Transformation: Toward an Analysis of the Conditions Underlying Effective Intervention', in P. B. Evans, D. Rueschemeyer and T. Skocpol (eds), *Bringing the State Back In*, Cambridge University Press, Cambridge.

Rueschemeyer, D., Stephens, E. H. and Stephens, J. D. (1992), *Capitalist Development and Democracy*, Polity Press, Cambridge.

Sandbrook, R. (1982), *The Politics of Basic Needs: Urban Aspects of Assaulting Poverty in Africa*, Heinemann Educational Books Ltd., London.

Sandbrook, R. (1985), *The Politics of Africa's Economic Stagnation*, Cambridge University Press, Cambridge.

Sandbrook, R. (1986), 'The State and Economic Stagnation in Tropical Africa', *World Development*, vol. 14, no. 3, pp. 319-332.

Sandbrook, R. (1993), *The Politics of Africa's Economic Recovery*, Cambridge University Press, Cambridge.

Sandbrook, R. and Arn, J. (1977), *The Labouring Poor and Urban Class Formation: The Case of Greater Accra*, McGill University Press, Montreal.

Sandbrook, R. and Oelbaum, J. (1997), 'Reforming Dysfunctional Institutions Through Democratization? Reflections on Ghana', *The Journal of Modern African Studies,* vol. 35, no. 4. pp. 603-646.

Scruton, R. (1996), *A Dictionary of Political Thought*, Macmillan, London.

Seers, D. (1979), 'The Meaning of Development, With A Postcript', in D. Lehmann, (ed.), *Development Theory, Four Critical Studies*, Frank Cass, London.

Seidman, A. W. (1978), *Ghana's Development Experience*, East African Publishing House, Lusaka.

Shaw, M. (1985), 'Marxism, the State and Politics', in M. Shaw (ed.), *Marxist Sociology Revisited: Critical Assessment*, Macmillan Press, London.

Shaw, T. M. (1993), *Reformism and Revisionism in Africa's Political Economy in the 1990s*, Macmillan Press, London.

Shillington, K. (1992), *Ghana and the Rawlings Factor*, Macmillan Press, London.

Sklar, R. L. (1994), 'Social Class and Political Action in Africa: The Bourgeoisie and the Proletariat', in D. E. Apter and C. G. Rosberg (eds), *Political Development and the New Realism in Sub-Saharan Africa*, University Press of Virginia, Charlottesville.

Skocpol, T. (1985), 'Bringing the State Back In: Strategies of Analysis in Current Research', in P. B. Evans, D. Rueschemeyer and T. Skocpol (eds), *Bringing the State Back In*, Cambridge University Press, Cambridge.

Smith, S. (1980), 'The Ideas of Samir Amin: Theory or Tautology?', in *The Journal of Development Studies*, vol. 17; no. 1. pp. 5-21.

Szereszewski, R. (1966), 'The Performance of the Economy', in W. Birmingham, I. Neustadt and E. N. Omaboe (eds), *A Study of Contemporary Ghana: Volume 1, The Economy of Ghana*, Allen and Unwin, London.

Tawney, R. H. (1959), 'Religion and the Rise of Capitalism', in W. R.Green (ed.), *Protestantism and Capitalism: The Weber Thesis and Its Critics*, D.C. Heath and Company, Boston.

Timberlake, L. (1988), *Africa In Crisis: The Causes, the Cures of Environmental Bankruptcy*, Earthscan Publications, London.

Toye, J. (1987), *Dilemmas of Development: Reflections on the Counter-Revolution in Development Theory and Policy*, Blackwell, Oxford.

Toye, J. (1992), 'Interest Group Politics and the Implementation of Adjustment Policies in Sub-Saharan Africa', in P. Gibbon, Y. Bangura and A. Ofstad (eds), *Authoritarianism, Democracy And Adjustment: The Politics of Reform in Africa*, The Scandinavian Institute for African Studies, Uppsala.

Twumasi, Y. (1975), 'The 1969 Election', in D. Austin and R. Luckman (eds), *Politicians and Soldiers in Ghana 1966-1972*, Frank Cass, London.

UN (1988), *World Commission on Environment and Development: The Brundtland Commission Report*, Oxford University Press, Oxford.

UNICEF (1987), *Adjustment With A Human Face*, New York.

Wade, R. (1990), *Governing the Market: Economic Theory and the Role of Government in East Asian Industrialization*, Princeton University Press, Princeton.

Wallerstein, I. (1974), *The Modern World System, Vols. 1 and 2*, Academic Press, New York.

Wallerstein, I. (1979), *The Capitalist World Economy*, Cambridge University Press, Cambridge.

Warren, B. (1980), *Imperialism: Pioneer of Capitalism*, Verso, London.

Weber, M. (1930), *The Protestant Ethic and the Spirit of Capitalism*, Unwin University Books, London.

Weber, M. ([1927]/1981), *General Economic History, with Introduction* by Ira Cohen, Transaction Books, New Brunswick, N.J.

Weber, M. ([1968]/1978), *Economy and Society: Volume One*, edited by Guenther Roth and Claus Wittich, University of California Press, Los Angeles.

Weber, M. ([1968]/1978), *Economy and Society: Volume Two* edited by Guenther Roth and Claus Wittich, University of California Press, Los Angeles.

White, G. and Wade, R. (eds) (1985), *Developmental States in East Asia*. DS Research Report 16, Brighton.

Wilks, I. (1975), *Asante in the Nineteenth Century: the Structure and Evolution of a Political Order*, Cambridge University Press, Cambridge.

Williams, E. (1964), *Capitalism and Slavery*, Deutsch, London.

World Bank (1981), *Accelerated Development in Sub-Saharan Africa: An Agenda for Action*, Washington D.C.

World Bank (1989a), *Sub-Saharan Africa. From Crisis to Sustainable Growth. A Long-Term Perspective Study*, Washington D.C.

World Bank (1992), *Governance and Development*, Washington, D.C.

World Bank (1997), *World Development Report 1997: The State in a Changing World*, Oxford University Press, Oxford.

World Bank and UNDP (1989b), *Africa's Adjustment and Growth in the 1980s*, Washington D.C.

Wunsch, J. and Olowu, D. (1990), 'The Failure of the Centralized African State', in J. S. Wunsch and D. Olowu (eds), *The Failure of the Centralized State: Institutions and Self-Governance*, Westview Press, Boulder.

Young, C. (1982), *Ideology and Development in Africa*, Yale University Press, New Haven.

Young, C. (1988), 'The African State and Its Political Legacy', in D. Rothchild and N. Chazan (eds), *The Precarious Balance: State and Society in Africa*, Westview, Boulder.

Index